RNA
Nanotechnology

RNA Nanotechnology

edited by

Bin Wang

PAN STANFORD PUBLISHING

Published by

Pan Stanford Publishing Pte. Ltd.
Penthouse Level, Suntec Tower 3
8 Temasek Boulevard
Singapore 038988

Email: editorial@panstanford.com
Web: www.panstanford.com

British Library Cataloguing-in-Publication Data
A catalogue record for this book is available from the British Library.

RNA Nanotechnology

ISBN 978-981-4411-64-6 (Hardcover)
ISBN 978-981-4411-65-3 (eBook)

Printed in the USA

Contents

Preface

In the past few decades there has been incredible growth in "bionano"-related research, accompanied by numerous publications in this field. Although various compilations address topics related to deoxyribonucleic acid (DNA) and protein, there are few books focused on determining the structure of ribonucleic acid (RNA) and using RNA as a building block to construct nanoarchitectures for biomedical and health care applications.

RNA encodes genetic information; it can also act as a biocatalyst (enzyme) and perform critical regulatory/control functions. RNA molecules are extremely versatile and are poised to play a central role in nanobiotechnology. *RNA Nanotechnology* is a comprehensive volume that details both the traditional approaches and the latest developments in the field of RNA-related technology. This book targets a wide audience; a broad introduction provides a solid academic background for students, researchers, and scientists who are unfamiliar with the subject, while the in-depth descriptions and discussions are useful for advanced professionals.

RNA Nanotechnology opens with reviews on the basic aspects of RNA biology, computational approaches for predicting RNA structures, and traditional and emerging experimental approaches for probing RNA structures. This section is followed by explorations of the latest research and discoveries in RNA nanotechnology, including the design and construction of RNA-based nanostructures. The final segment of the book includes descriptions and discussions of the potential biological and therapeutic applications of small RNA molecules such as small/short interfering RNAs (siRNAs), microRNAs (miRNAs), RNA aptamers, and ribozymes.

Thirty-eight scientists from four continents have contributed to this book. As Editor, I would like to thank all of them for the high quality of their work. I would also like to thank the editorial staff at Pan Stanford Publishing for their essential contributions toward the publication of this book. Finally, in representing all the contributors, I would like to dedicate this book to our family members, colleagues, and friends for their support.

Bin Wang

Chapter 1

Introduction

Bin Wang
Department of Chemistry, Marshall University, One John Marshall Drive, Huntington, WV 25755, USA
wangb@marshall.edu

Ribonucleic acid's (RNA) probable role in the origin of life is reflected in its multiple functions. RNA encodes genetic information; it can also act as a biocatalyst (enzyme) and perform critical regulatory/control functions [1–3]. The structure of RNA governs its function inside the cell. Therefore, understanding the biological functions of RNA requires an accurate knowledge of its secondary and tertiary structural features.

RNA nanotechnology is an emerging research field that has only existed for about 15 years [4–10]. Unlike double-stranded deoxyribonucleic acid (DNA) molecules, where self-assembled nanostructures mainly rely on base pairing between complementary strands to form particular secondary structures, single-stranded RNA molecules are more mobile and have the ability to fold into complex tertiary structures [11–16]. RNA nanostructures are easy to generate experimentally. Moreover, RNA nanoparticles are more thermodynamically stable compared to their DNA counterparts. Therefore, the use of RNA as building blocks to construct static

RNA Nanotechnology
Edited by Bin Wang

ISBN 978-981-4411-64-6 (Hardcover), 978-981-4411-65-3 (eBook)
www.panstanford.com

nanoarchitectures and/or mobile nanomachines has great potential, especially in the growing field of nanomedicine.

RNA is an excellent carrier for gene/drug delivery because it can penetrate cell membranes [17–20]. Once inside the cell, unlike DNA molecules, RNA nanoparticles can survive in the acidic environment of the endosome and become distributed throughout the cell. On the other hand, the chemical instability of RNA and its tendency to break down in the presence of enzymes have slowed progress in this field. However, it is practical to produce RNA-based nanoparticles that are resistant to ribonucleases (RNases), enzymes that degrade RNA, and are highly stable for applications in nanotechnology [21, 22].

In this book, Chapter 2 introduces the fundamental aspects of RNA biology. Chapters 3 and 4 describe computational approaches for predicting RNA structures, including both its secondary and tertiary configurations. Chapters 5–10 explore traditional and emerging experimental approaches for probing RNA structures, including X-ray crystallography, nuclear magnetic resonance (NMR) spectroscopy, enzymatic cleavage, chemical probing, nucleotide analog interference mapping (NAIM), in-line probing, thermal denaturation, and single-molecule approaches.

Chapters 11 and 12 detail the design and construction of RNA-based nanostructures. The latest developments in RNA nanotechnology are discussed in these chapters, including the assembly of one-dimensional, two-dimensional, and three-dimensional RNA nanoparticles, along with their potential applications.

Chapters 13–18 illustrate the potential biological and therapeutic applications of small RNA molecules such as small/short interfering RNAs (siRNAs), microRNAs (miRNAs), RNA aptamers, and ribozymes. Several other topics are also addressed, including alternative splicing and RNA motifs for sensing metal ions.

Overall, *RNA Nanotechnology* is a comprehensive volume that reviews both the traditional approaches and the latest developments in the field of RNA-related nanobiotechnology. Topics covered in this book lie at the interface of molecular biology, biochemistry, nanomedicine, biophysics, biomedical engineering, bioinformatics, and computational biology, making this volume a valuable addition to the existing body of knowledge.

References

1. Cech, T.R. (1989). RNA as an enzyme. *Biochem. Int.*, **18**, 7–14.
2. Cech, T.R. (2009). Evolution of biological catalysis: ribozyme to RNP enzyme. *Cold Spring Harb. Symp. Quant. Biol.*, **74**, 11–16.
3. Cech, T.R. (2009). Crawling out of the RNA world. *Cell*, **136**, 599–602.
4. Guo, P., Zhang, C., Chen, C., Garver, K., and Trottier, M. (1998). Inter-RNA interaction of phage phi29 pRNA to form a hexameric complex for viral DNA transportation. *Mol. Cell*, **2**, 149–155.
5. Jaeger, L., Westhof, E., and Leontis, N.B. (2001). TectoRNA: modular assembly units for the construction of RNA nano-objects. *Nucleic Acids Res.*, **29**, 455–463.
6. Hansma, H.G., Oroudjev, E., Baudrey, S., and Jaeger, L. (2003). TectoRNA and "kissing-loop" RNA: atomic force microscopy of self-assembling RNA structures. *J. Microsc.*, **212**, 273–279.
7. Chworos, A., Severcan, I., Koyfman, A.Y., Weinkam, P., Oroudjev, E., Hansma, H.G., and Jaeger, L. (2004). Building programmable jigsaw puzzles with RNA. *Science*, **306**, 2068–2072.
8. Shu, D., Moll, W.D., Deng, Z., Mao, C., and Guo, P. (2004). Bottom-up assembly of RNA arrays and superstructures as potential parts in nanotechnology. *Nano Lett.*, **4**, 1717–1723.
9. Guo, P. (2005). RNA nanotechnology: engineering, assembly and applications in detection, gene delivery and therapy. *J. Nanosci. Nanotechnol.*, **5**, 1964–1982.
10. Guo, P. (2010). The emerging field of RNA nanotechnology. *Nat. Nanotechnol.*, **5**, 833–842.
11. Bindewald, E., Grunewald, C., Boyle, B., O'Connor, M., and Shapiro, B.A. (2008). Computational strategies for the automated design of RNA nanoscale structures from building blocks using NanoTiler. *J. Mol. Graph Model*, **27**, 299–308.
12. Martinez, H.M., Maizel, J.V., Jr., and Shapiro, B.A. (2008). RNA2D3D: a program for generating, viewing, and comparing 3-dimensional models of RNA. *J. Biomol. Struct. Dyn.*, **25**, 669–683.
13. Afonin, K.A., Bindewald, E., Yaghoubian, A.J., Voss, N., Jacovetty, E., Shapiro, B.A., and Jaeger, L. (2010). In vitro assembly of cubic RNA-based scaffolds designed in silico. *Nat. Nanotechnol.*, **5**, 676–682.
14. Kasprzak, W., Bindewald, E., Kim, T.J., Jaeger, L., and Shapiro, B.A. (2010). Use of RNA structure flexibility data in nanostructure modeling. *Methods*, **54**, 239–250.

15. Severcan, I., Geary, C., Chworos, A., Voss, N., Jacovetty, E., and Jaeger, L. (2010). A polyhedron made of tRNAs. *Nat. Chem.*, **2**, 772–779.
16. Grabow, W.W., Zakrevsky, P., Afonin, K.A., Chworos, A., Shapiro, B.A., and Jaeger, L. (2011). Self-assembling RNA nanorings based on RNAI/II inverse kissing complexes. *Nano Lett.*, **11**, 878–887.
17. Guo, S., Tschammer, N., Mohammed, S., and Guo, P. (2005). Specific delivery of therapeutic RNAs to cancer cells via the dimerization mechanism of phi29 motor pRNA. *Hum. Gene Ther.*, **16**, 1097–1109.
18. Khaled, A., Guo, S., Li, F., and Guo, P. (2005). Controllable self-assembly of nanoparticles for specific delivery of multiple therapeutic molecules to cancer cells using RNA nanotechnology. *Nano Lett.*, **5**, 1797–1808.
19. Shu, D., Shu, Y., Haque, F., Abdelmawla, S., and Guo, P. (2011). Thermodynamically stable RNA three-way junction for constructing multifunctional nanoparticles for delivery of therapeutics. *Nat. Nanotechnol.*, **6**, 658–667.
20. Afonin, K.A., Grabow, W.W., Walker, F.M., Bindewald, E., Dobrovolskaia, M.A., Shapiro, B.A., and Jaeger, L. (2011). Design and self-assembly of siRNA-functionalized RNA nanoparticles for use in automated nanomedicine. *Nat. Prot.*, **6**, 2022–2034.
21. Liu, J., Guo, S., Cinier, M., Shlyakhtenko, L.S., Shu, Y., Chen, C., Shen, G., and Guo, P. (2010). Fabrication of stable and RNase-resistant RNA nanoparticles active in gearing the nanomotors for viral DNA packaging. *ACS Nano*, **5**, 237–246.
22. Guo, P. (2011). RNA Nanotechnology: methods for synthesis, conjugation, assembly and application of RNA nanoparticles. *Methods*, **54**, 201–203.

Chapter 2

Fundamental Aspects of RNA

Christian Zwieb
Department of Biochemistry, University of Texas, Health Science Center, San Antonio, Allied Health Bldg., 7703 Floyd Curl Drive, San Antonio, Texas 78229, USA
zwieb@uthscsa.edu

In its capacity to process information and catalyze biochemical reactions within each cell, ribonucleic acid (RNA) permeates many areas of the biological sciences. For example, RNA's exceptional versatility has sparked ideas about how life might have originated on planet Earth. Taking hints from nature, the potential use of RNA molecules in bio- and nanotechnology as a sensor, drug, and molecular machine has now been realized and is at the forefront of the applied biosciences. This chapter provides some guidance for understanding the important principles that govern the structure and function of RNA molecules.

2.1 Introduction

RNA stores information in its sequence and can also act as a catalyst. This duality has fostered the idea of an early "RNA world," where RNA perhaps was central to primitive forms of life [68]. The enzymatic property of RNA molecules is linked to their capacity to

RNA Nanotechnology
Edited by Bin Wang

ISBN 978-981-4411-64-6 (Hardcover), 978-981-4411-65-3 (eBook)
www.panstanford.com

function as natural nanoscale molecular machines that are often enhanced in their effectiveness by RNA-binding proteins in the formation of ribonucleoprotein (RNP) complexes. Prototypical RNPs with RNA as their functional cores are the protein-synthesizing ribosome [4, 25]; the signal recognition particle (SRP), which binds to ribosomes and directs nascent polypeptides to the cell membrane [63]; the transfer messenger RNP (tmRNP), which rescues immobilized bacterial ribosomes [32]; ribonuclease P (RNase P), which processes transfer RNA (tRNA) precursors [33]; and the spliceosome, which removes introns [43]. Many more RNA types have been identified, the majority of which are listed in Rfam [19] and fRNAdb [47].

Considering that only about 1.5% of the human genome codes for proteins [38], vast regions of a genome, previously referred to as "junk," are candidates for being transcribed into numerous noncoding (nc) RNAs with yet-to-be-elucidated functions. Efforts are underway to develop the computational tools for identifying the ncRNAs in the large genome data sets and provide comprehensive collections of RNA sequences to the scientific community [5, 30, 46, 48]. Advances in the areas of RNA identification and classification will undoubtedly lead to new discoveries of cellular functions and provide a rich source for the discovery and exploitation of the multifaceted roles of RNA.

This chapter describes the molecular composition of RNA and the naming conventions used for its constituents, and it samples the principal features that govern the RNA structure. The overall aim is to encourage the advancement of RNA research and assist the investigator and the student in a better understanding of the intricate and dynamic biology of these versatile molecules.

2.2 The RNA Chain

RNA is an unbranched polymer of covalently joined ribonucleoside units. A ribonucleoside is composed of a ribose sugar that is covalently bound to one of four different nitrogenous bases. The bases adenine (A) and guanine (G) are derivatives of a heterocyclic aromatic purine (abbreviated as R). Cytosine (C) and uridine (U)

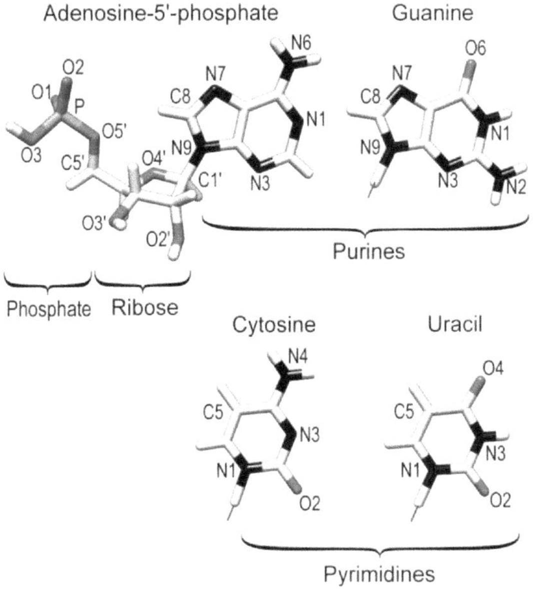

Figure 2.1 The building blocks of RNA.

are pyrimidine bases (abbreviated as Y). The stereochemically important glycosidic bond connects the carbon (C1′) of the ribose with the N9 of a purine base or with the N1 of a pyrimidine base (Fig. 2.1).

A ribonucleotide forms by covalent attachment of one, two, or three phosphates to the ribose 5′ carbon of a ribonucleoside. The 5′ nucleotide triphosphosphates (NTPs, where N stands for any of the four bases) are used within the cell to synthesize RNA in the 5′ to 3′ direction. Each addition of an NTP releases one inorganic pyrophosphate molecule with one phosphate left to join the 3′ and 5′ positions of two neighboring ribose rings.

By convention, the sequence of an RNA molecule is written using single letters and is numbered left to right from 5′ to 3′ in the molecule's synthesis direction. The letters may be grouped as triplets to specify the codons for the amino acid–encoding regions of the messenger RNAs (mRNAs) or in groups of 10 characters for easier readability by humans. The sequence of the bases is referred to as the primary structure of an RNA molecule.

The numbers of ribonucleosides in a natural RNA chain can range from 22 in the micro RNAs (miRNAs) [45] to approximately 17,000 in the Xist RNA [9]. The prototypical tRNAs are composed of 73 to 93 ribonucleosides [12], whereas the 16S ribosomal RNA (rRNA) of the bacterium *Escherichia coli* consists of 1,542 ribonucleotides that are arranged in sequence unique to the species [8].

With few exceptions, most RNA molecules are synthesized in nature by transcription from their encoding deoxyribonucleic acids (DNAs). After the RNA is transcribed, modifications may be introduced at particular sites. Ribonucleoside modifications include methylations and the generation of less common bases, such as inosine (abbreviated as I), dihydrouridine (D), or pseudouracil (ψ). RNA splicing [58] and RNA editing [24] are processes that specifically remove or insert ribonucleosides post-transcriptionally. These mechanisms are the reason for potentially significant deviations of the DNA gene sequence from the primary structure of the processed RNA molecule.

2.3 Into the Fold

2.3.1 *Folding Principles*

The bases of single-stranded RNA tend to interact with each other in two principal ways: either through stacking, where the bases are arranged perpendicular to their planes; or through hydrogen bonding within the base planes. The planar purine-pyrimidine base pairs, A-T and G-C, were proposed originally by Watson and Crick in their elucidation of the structure of the DNA double helix [64]. The equivalent base pairs in RNA are A-U and G-C, where A-U is held together by two and G-C by three hydrogen bonds. The non-Watson–Crick G-U wobble pair is frequently observed in most RNA molecules and is about as stable as the A-U Watson–Crick pair (Fig. 2.2) [21].

The two ribose groups attach to the same side of a base pair and thus define two indentations or grooves. The major groove is delineated by N7 of the purine and the C6 of the pyrimidine, whereas purine N3 and the pyrimidine O2 are exposed in the minor groove (Fig. 2.2). Because the overall dimensions of the three

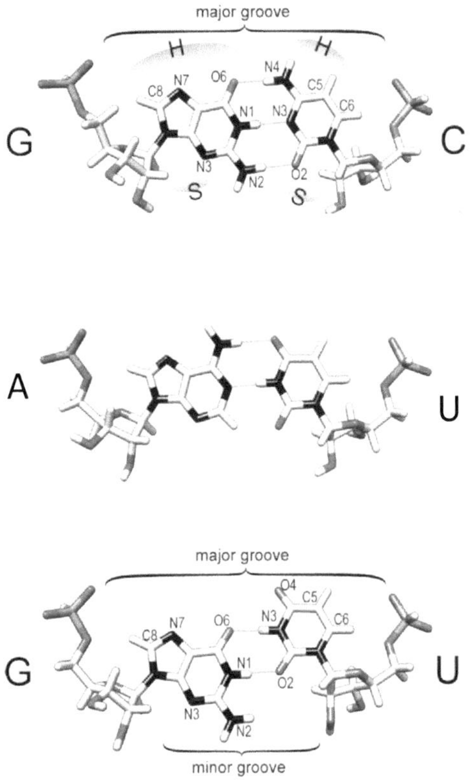

Figure 2.2 Watson–Crick and G-U wobble pair geometries. H and S in the top panel indicate the Hoogsteen and sugar edges.

most commonly observed base pairs (A-U, G-C, and G-U) are nearly identical, several base pairs can be neatly placed on top of each other to continuously stack and form a rigid A-type helix with 11 base pairs per turn. Unlike the B-form of the DNA helix, the A-form RNA helix is characterized by a deep, narrow major groove and a relatively shallow minor groove (Fig. 2.4a) [65].

Whereas the canonical A-U and G-C pairs and the G-U wobble pair engage through their Watson–Crick edges, other planar base interactions occur through the Hoogsteen edge. This edge is defined by purine positions 6, 7, and 8 and pyrimidine positions 4 and 5. Similarly, the sugar edge is formed by the 2′ hydroxyl group of the ribose with positions 2 and 3 on purine and the oxygen atom at position 2 of the pyrimidine. The Hoogsteen and sugar edges are labeled H and S in Fig. 2.2 (top panel). Given that the glycosidic bonds can be oriented either in *cis* or *trans*, 12 principal geometric types with at least 2 hydrogen bonds connecting the bases are possible [42].

The ability of the bases and the ribose to form hydrogen bonds in numerous combinations and occasionally engage more than two ribonucleosides is largely responsible for the formidable structural and functional versatility of RNA. Unlike in DNA, the 2′ hydroxyl group of the ribose contributes to the capacity of the RNA to interact in versatile ways with itself or a variety of ligands. For example, antiparallel, consecutive hydrogen bonding between the ribose 2′-hydroxyls from two regions of an RNA chain form a ribose zipper [10].

The base pairings in RNA molecules are conveniently displayed in their secondary structure diagrams to indicate the locations of stems, bulges, and loops (Fig. 2.3). Paired residues are shown as letters connected by lines, dots, circles, or other symbols to signify which edges are hydrogen bonded [42]. Unpaired stacked bases can be depicted by placing the letters closer to each other without connecting symbols [39]. Similar conventions are used to indicate tertiary interactions. These are sites that connect distant primary and secondary structure elements through stacking and hydrogen bonding.

Water, metal, or other ions contribute significantly to the folding, conformation, and stability of the predominantly negatively charged RNA. These small molecules may bind at specific sites or may

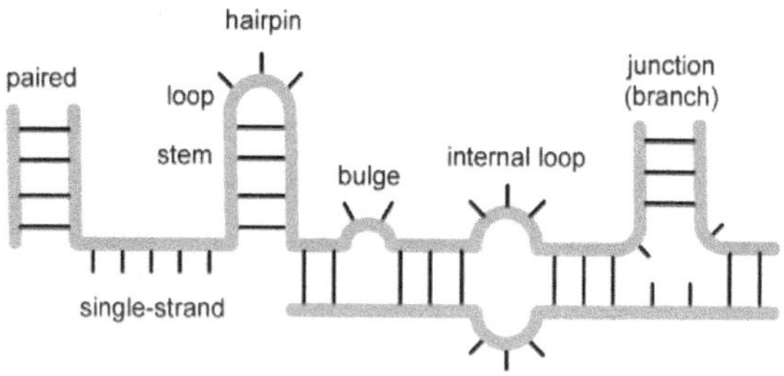

Figure 2.3 Naming conventions for features of the RNA shown in its secondary structure.

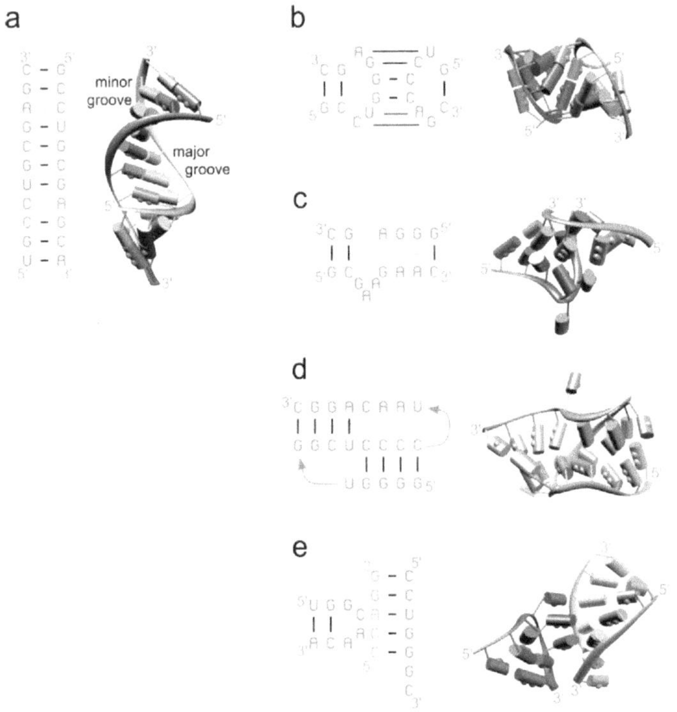

Figure 2.4 RNA motifs and RNA–RNA interactions. Helix (a), kissing hairpin loops (b), kink-turn (c), pseudoknot (d), and an example of a tetraloop–tetraloop receptor complex. Coordinates were chosen from the RCSB PDB files 1KFO (a), 2JLT (b), 1FFK (c), 2RP0, (d) and 2JYJ (e) [6] and displayed with UCSF Chimera [53]. *Abbreviations*: RSCB, Research Collaboratory for Structural Bioinformatics; PDB, Protein Data Bank; UCSF, University of California, San Francisco.

be more delocalized [13]. For example, magnesium ions and their functional analog spermidine are known to bind in the major groove of $tRNA^{Phe}$ [55].

2.3.2 *High-Resolution RNA Structure Milestones*

RNA molecules adopt sophisticated and precisely defined three-dimensional structures that rival in their complexity the structures of proteins. Solving the molecular structure of yeast $tRNA^{Phe}$ at atomic resolution was key to an understanding of certain architectural principles that govern the folded RNA [29, 59]. Subsequent milestones were the determination of the molecular structures of the hammerhead ribozyme [54, 56], the P4–P6 domain of the group I intron RNA [10], and the hepatitis delta ribozyme [17]. Elucidating the structures of ribosomes and their subunits [4, 67, 71] remains the high point of these accomplishments. These studies demonstrated that during evolution, RNA molecules are able to preserve their overall three-dimensional shape and function despite substantial differences in their sequences. The structures continue to be a treasure drove for deciphering how RNA folds, how it binds other molecules, and how it dynamically changes its structure in response to ligands.

2.3.3 *Predicting the RNA Structure*

In the current vibrant genome-sequencing era where sequences are abundant and obtaining structures using biophysical approaches is still an elaborate undertaking, efforts to determine the RNA structure from the sequences alone has become an increasingly useful approach (see Chapters 3 and 4).

Observing covariances and compensating bases changes (e.g., changes from a G-C pair to a C-G or to an A-U pair) in an alignment of phylogenetically related RNA sequences is the most reliable method for predicting the RNA structure [18]. The larger the number of such compensations, the stronger the support for the existence of a base pair because, during evolution, random mutations would not have been corrected to maintain the base pair unless it was required [39]. Semiautomated procedures have been developed to assist in the identification of covariations and compensatory base changes and calculate the level of support for each covariation [3]. Covariation analysis has the added advantage that it allows to identify tertiary structure elements and motifs that are particularly valuable for constraining the overall structure [40].

A quicker but necessarily less stringent approach for finding possible secondary structures is to calculate the overall low free energy of an RNA [62, 72]. This approach can provide secondary structures as a starting point for further experimental exploration.

Chemical and enzymatic probing of RNA molecules identifies the surface-exposed sites (see Chapter 4). Although these methods do not provide direct information about the existence of a particular base pair, they are nevertheless useful to support the prevalence of one structure over the alternative. These approaches also identify structural changes and protected regions caused by ligand binding.

Site-directed mutagenesis provides another experimental avenue for establishing the existence of a base pair even when it is protected from modifications inside the folded molecule. By introducing compensatory double mutations, the activity of the RNA is expected to be restored if the target base pair exists. Merging the computational and bench-derived data is particularly useful in the construction of high-quality three-dimensional models in order to gain insight into the structure and function of an RNA molecule [41, 49].

2.3.4 *Commonly Observed RNA Motifs*

Structures obtained by X-ray crystallography, nuclear magnetic resonance (NMR), or model building have revealed sites in RNA with repeatedly observed characteristic features. These motifs have been given descriptive names such as kissing hairpin, kink- or K-turn, pseudoknot, or the tetranucleotide loop (tetraloop) receptor complex (Fig. 2.4). More refined motif classifications using details of structure, function, and potential tertiary interactions are available [26, 34]. Progress has been made in identifying RNA motifs in the genome sequences using pattern matching [14]. These approaches fulfill an urgent need in the quest to identify and characterize the RNA-encoding regions in the large amount of the raw sequence data [5].

Like LEGO, the naturally occurring RNA motifs can be combined to assemble artificial nanoscale structures with potential use in synthetic biology, as sensors, and as delivery vehicles in nanomedicine [1, 2, 11, 27, 35, 57].

2.4 RNA Ligands

2.4.1 *RNA-Binding Proteins*

The majority of the RNA molecules form complexes with proteins through induced-fit structural changes. Although RNA may have appeared first, this suggests that RNA and proteins share a long coevolutionary history.

The protein domains that associate with the RNA are typically predominantly positively charged to counterbalance the negatively charged phosphates. The strategies for the formation of complexes employ to various degrees hydrogen bonding, stacking of aromatic residues, and recognition of intricate complementary surfaces. By applying the lessons learned from the structures of naturally occurring RNPs, nanotechnology has incorporated proteins into the design of structures that appear to form more readily than when using only RNA as building blocks [51].

2.4.2 *Drugs Targeting RNA*

Small molecules that bind to the naturally occurring RNAs either prevent the binding of an important macromolecule (protein or RNA), distort the RNA's active conformation, or bind in competition with a cofactor. Aminoglycosides, such as paromomycin and spectinomycin, are among the antibiotics that impede translation by binding to the bacterial rRNA [70]. The emergence of resistant strains and the formidable toxicity of the conventional antibiotics strongly motivate the development of new drugs with high affinity, superior specificity for the target RNA, and potentially fewer side effects. Besides the rRNA, other RNAs with essential cellular functions, including SRP RNA and tmRNA, have potential as not-yet-been-exploited targets.

In active development for the treatment of human disease are small molecule inhibitors directed, among others, toward the mRNA of a gene that causes myotonic dystrophy type 1 (DM1) [66], the hepatitis C virus (HCV) [52], and toward the transactivation response element (TAR) of human immunodeficiency virus (HIV) [37].

2.5 RNA Drugs

The development of RNA-based therapeutics is still at an early stage. Approaches that have been investigated are the use of antisense RNAs [7, 50], ribozymes [15, 60], small interfering RNA (siRNA) [69], and miRNA [36], as well as molecules that function as decoys and aptamers capable of binding to various targets [31]. In the future, RNA promises to reach its full potential as a new class of sophisticated drugs that will be able to treat a wide range of diseases.

Solid-phase chemistry and automated oligonucleotide synthesizers provide formidable flexibility in the incorporation of modified nucleotides into the desired RNA drug. The most commonly modified sites are the 2′ position of the ribose, the 5 positions of pyrimidines, and the 8 positions of purines (Fig. 2.1). The 2′-fluoro and 2′-*O*-methyl modifications render the RNA resistant to the degradation by RNases and provide a means to improve the pharmacokinetics of the effective drug.

Systematic evolution of ligands by exponential enrichment (SELEX) is a powerful approach that permits to isolate in vitro RNA molecules (aptamers) that bind specifically with micro- to nanomolar affinities to proteins, antibiotics, or small metabolites [44]. Effective selection has been demonstrated with nucleotides containing aromatic, alkyl, heterocyclic, amino acid, and other groups [16]. These extensively modified aptamers are characterized by enhanced affinity and specificity and are well suited to simultaneously monitor the levels of several proteins in human serum [23].

The energy that is liberated by the binding of an aptamer to its target can be used to switch structures. These riboswitches are natural regulators of gene expression and may be manipulated to sense substances within the body or in the environment [20].

2.6 Outlook

The field of RNA research continues its steady expansion. RNA-based therapeutics, the development of new antibiotics, and the use of RNA technology for improving crop yields promise to ultimately enable us to face the challenges associated with extended life spans, increasing global populations, and climate change. The recent discovery of the bacterial and archaeal CRISPR RNAs [28, 61] and of the Piwi-interacting RNAs (piRNAs) [22] signifies a trend that is expected to lead to many new insights into the role of RNA. Developing the tools for identifying additional RNAs in the ocean of sequences data will help to promote further discovery. Our ability to understand and take advantage of RNA's rich natural diversity will be crucial for expanding the repertoire of building blocks that can be used for the construction of powerful nanometer-scale devices.

Acknowledgments

I am indebted to Jacek Wower for critical reading of the manuscript.

References

1. Afonin, K.A., et al. (2010). In vitro assembly of cubic RNA-based scaffolds designed in silico. *Nat. Nanotechnol.*, **5**, 676–682.
2. Andersen, E.S., et al. (2009). Self-assembly of a nanoscale DNA box with a controllable lid. *Nature*, **459**, 73–76.
3. Andersen, E.S., et al. (2007). Semiautomated improvement of RNA alignments. *RNA*. **13**, 1850–1859.
4. Ban, N., et al. (2000). The complete atomic structure of the large ribosomal subunit at 2.4 A resolution. *Science*, **289**, 905–920.
5. Bateman, A., et al. (2011). RNAcentral: a vision for an international database of RNA sequences. *RNA*, **17**(11), 1941–1946.
6. Berman, H.M., et al. (2000). The Protein Data Bank. *Nucleic Acids Res.*, **28**, 235–242.
7. Brantl, S. (2002). Antisense-RNA regulation and RNA interference. *Biochim. Biophys. Acta*, **1575**, 15–25.
8. Brosius, J., et al. (1978). Complete nucleotide sequence of a 16S ribosomal RNA gene from Escherichia coli. *Proc. Natl. Acad. Sci. U S A*, **75**, 4801–4805.

9. Brown, C.J., et al. (1992). The human XIST gene: analysis of a 17 kb inactive X-specific RNA that contains conserved repeats and is highly localized within the nucleus. *Cell*, **71**, 527–542.
10. Cate, J.H., et al. (1996). Crystal structure of a group I ribozyme domain: principles of RNA packing. *Science*, **273**, 1678–1685.
11. Chworos, A., et al. (2004). Building programmable jigsaw puzzles with RNA. *Science*, **306**, 2068–2072.
12. Clark, B.F. (2006). The crystal structure of tRNA. *J. Biosci.*, **31**, 453–457.
13. Conn, G.L., and Draper, D.E. (1998). RNA structure. *Curr. Opin. Struct. Biol.*, **8**, 278–285.
14. Disz, T., et al. (2010). Accessing the SEED genome databases via web services API: tools for programmers. *BMC Bioinf.*, **11**, 319.
15. Doherty, E.A., and Doudna, J.A. (2001). Ribozyme structures and mechanisms. *Annu. Rev. Biophys. Biomol. Struct.*, **30**, 457–475.
16. Eaton, B.E. (1997). The joys of in vitro selection: chemically dressing oligonucleotides to satiate protein targets. *Curr. Opin. Chem. Biol.*, **1**, 10–16.
17. Ferre-D'Amare, A.R., Zhou, K., and Doudna, J.A. (1998). Crystal structure of a hepatitis delta virus ribozyme. *Nature*, **395**, 567–574.
18. Fox, G.E., and Woese, C. (1975). 5S RNA secondary structure. *Nature (London)*, **256**, 505–507.
19. Gardner, P.P., et al. (2011). Rfam: Wikipedia, clans and the0 "decimal" release. *Nucleic Acids Res.*, **39**, D141–D145.
20. Garst, A.D., Edwards, A.L., and Batey, R.T. (2011). Riboswitches: structures and mechanisms. *Cold Spring Harb. Perspect. Biol.*, **3**(6), a003533.
21. Giese, M.R., et al. (1998). Stability of RNA hairpins closed by wobble base pairs. *Biochemistry*, **37**, 1094–1100.
22. Girard, A., et al. (2006). A germline-specific class of small RNAs binds mammalian Piwi proteins. *Nature*, **442**, 199–202.
23. Gold, L., et al. (2012). Aptamers and the RNA world, past and present. *Cold Spring Harb. Perspect. Biol.*, **4**(3), pii: a003582.
24. Gott, J.M., and Emeson, R.B. (2000). Functions and mechanisms of RNA editing. *Annu. Rev. Genet.*, **34**, 499–531.
25. Green, R., and Noller, H.F. (1997). Ribosomes and translation. *Annu. Rev. Biochem.*, **66**, 679–716.
26. Hendrix, D.K., Brenner, S.E., and Holbrook, S.R. (2005). RNA structural motifs: building blocks of a modular biomolecule. *Q. Rev. Biophys.*, **38**, 221–243.

27. Horiya, S., et al. (2003). RNA LEGO: magnesium-dependent formation of specific RNA assemblies through kissing interactions. *Chem. Biol.*, **10**, 645–654.
28. Jansen, R., et al. (2002). Identification of a novel family of sequence repeats among prokaryotes. *OMICS*, **6**, 23–33.
29. Jovine, L., Djordjevic, S., and Rhodes, D. (2000). The crystal structure of yeast phenylalanine tRNA at 2.0 A resolution: cleavage by Mg(2+) in 15-year old crystals. *J. Mol. Biol.*, **301**, 401–414.
30. Kapranov, P., et al. (2005). Examples of the complex architecture of the human transcriptome revealed by RACE and high-density tiling arrays. *Genome Res.*, **15**, 987–997.
31. Keefe, A.D., Pai, S., and Ellington, A. (2010). Aptamers as therapeutics. *Nat. Rev. Drug Discovery*, **9**, 537–550.
32. Keiler, K.C., Waller, P.R., and Sauer, R.T. (1996). Role of a peptide tagging system in degradation of proteins synthesized from damaged messenger RNA. *Science*, **271**, 990–993.
33. Kirsebom, L.A., and Trobro, S. (2009). RNase P RNA-mediated cleavage. *IUBMB Life*, **61**, 189–200.
34. Klosterman, P.S., et al. (2002). SCOR: a structural classification of RNA database. *Nucleic Acids Res.*, **30**, 392–394.
35. Ko, S.H., et al. (2008). Reversible switching of pRNA activity on the DNA packaging motor of bacteriophage phi29. *J. Am. Chem. Soc.*, **130**, 17684–17687.
36. Krutzfeldt, J., et al. (2005). Silencing of microRNAs in vivo with "antagomirs." *Nature*, **438**, 685–689.
37. Kumar, S., and Arya, D.P. (2011). Recognition of HIV TAR RNA by triazole linked neomycin dimers. *Bioorg. Med. Chem. Lett.*, **21**, 4788–4792.
38. Lander, E.S., et al. (2001). Initial sequencing and analysis of the human genome. *Nature*, **409**, 860–921.
39. Larsen, N., and Zwieb, C. (1991). SRP-RNA sequence alignment and secondary structure. *Nucleic Acids Res.*, **19**, 209–215.
40. Larsen, N. (1992). Higher order interactions in 23s rRNA. *Proc. Natl. Acad. Sci. U S A*, **89**, 5044–5048.
41. Lavender, C.A., et al. (2010). Robust and generic RNA modeling using inferred constraints: a structure for the hepatitis C virus IRES pseudoknot domain. *Biochemistry*, **49**, 4931–4933.
42. Leontis, N.B., and Westhof, E. (2001). Geometric nomenclature and classification of RNA base pairs. *RNA*, **7**, 499–512.

43. Lerner, M.R., et al. (1980). Are snRNPs involved in splicing. *Nature*, **283**, 220–224.
44. Lorsch, J.R., and Szostak, J.W. (1994). In vitro selection of RNA aptamers specific for cyanocobalamin. *Biochemistry*, **33**, 973–982.
45. Lund, E., and Dahlberg, J.E. (2006). Substrate selectivity of exportin 5 and Dicer in the biogenesis of microRNAs. *Cold Spring Harb. Symp. Quant. Biol.*, **71**, 59–66.
46. Maeda, N., et al. (2006). Transcript annotation in FANTOM3: mouse gene catalog based on physical cDNAs. *PLoS Genet.*, **2**, e62.
47. Mituyama, T., et al. (2009). The Functional RNA Database 3.0: databases to support mining and annotation of functional RNAs. *Nucleic Acids Res.*, **37**, D89–D92.
48. Mortazavi, A., et al. (2008). Mapping and quantifying mammalian transcriptomes by RNA-Seq. *Nat. Methods*, **5**, 621–628.
49. Mueller, F., et al. (1997). A new model for the three-dimensional folding of Escherichia coli 16 S ribosomal RNA. III. The topography of the functional center. *J. Mol. Biol.*, **271**, 566–587.
50. Oberstrass, J. and Nellen, W. (1997). Genes regulating with antisense RNA, in *Antisense Oligodeoxynucleotides and Antisense RNA*, 171–195, CRC Press.
51. Ohno, H., et al. (2011). Synthetic RNA-protein complex shaped like an equilateral triangle. *Nat. Nanotechnol.*, **6**, 116–120.
52. Parker, M.A., et al. (2011). An efficient new route to dihydropyranobenzimidazole inhibitors of HCV replication. *Molecules*, **16**, 281–290.
53. Pettersen, E.F., et al. (2004). UCSF Chimera—a visualization system for exploratory research and analysis. *J. Comput. Chem.*, **25**, 1605–1612.
54. Pley, H.W., Flaherty, K.M., and McKay, D.B. (1994). Three-dimensional structure of a hammerhead ribozyme. *Nature*, **372**, 68–74.
55. Quigley, G.J., Teeter, M.M., and Rich, A. (1978). Structural analysis of spermine and magnesium ion binding to yeast phenylalanine transfer RNA. *Proc. Natl. Acad. Sci. U S A*, **75**, 64–68.
56. Scott, W.G., Finch, J.T., and Klug, A. (1995). The crystal structure of an all-RNA hammerhead ribozyme: a proposed mechanism for RNA catalytic cleavage. *Cell*, **81**, 991–1002.
57. Severcan, I., et al. (2010). A polyhedron made of tRNAs. *Nat. Chem.*, **2**, 772–779.
58. Sharp, P.A. (2005). The discovery of split genes and RNA splicing. *Trends Biochem. Sci.*, **30**, 279–281.

59. Shi, H., and Moore, P.B. (2000). The crystal structure of yeast phenylalanine tRNA at 1.93 A resolution: a classic structure revisited. *RNA*, **6**, 1091–1105.

60. Stull, R.A., and Szoka, F.C.J. (1995). Antigene, ribozyme and aptamer nucleic acid drugs: progress and prospects. *Pharm. Res.*, **12**, 465–483.

61. Tang, T.H., et al. (2002). Identification of 86 candidates for small non-messenger RNAs from the archaeon Archaeoglobus fulgidus. *Proc. Natl. Acad. Sci. U S A*, **99**, 7536–7541.

62. Turner, D.H., Sugimoto, N., and Freier, S.M. (1988). RNA structure prediction. *Annu. Rev. Biophys. Biophys. Chem.*, **17**, 167–192.

63. Walter, P., and Blobel, G. (1982). Signal recognition particle contains a 7S RNA essential for protein translocation across the endoplasmic reticulum. *Nature*, **299**, 691–698.

64. Watson, J.D., and Crick, F.H. (1953). A structure for deoxyribose nucleic acid. *Nature*, **171**, 737–738.

65. Weeks, K.M., and Crothers, D.M. (1993). Major groove accessibility of RNA. *Science*, **261**, 1574–1577.

66. Wheeler, T.M., et al. (2009). Reversal of RNA dominance by displacement of protein sequestered on triplet repeat RNA. *Science*, **325**, 336–339.

67. Wimberly, B.T., et al. (2000). Structure of the 30S ribosomal subunit. *Nature*, **407**, 327–339.

68. Woese, C. (1968). *The Genetic Code*, Harper & Row.

69. Xie, F.Y., Woodle, M.C., and Lu, P.Y. (2006). Harnessing in vivo siRNA delivery for drug discovery and therapeutic development. *Drug Discovery Today*, **11**, 67–73.

70. Yonath, A. (2005). Antibiotics targeting ribosomes: resistance, selectivity, synergism and cellular regulation. *Annu. Rev. Biochem.*, **74**, 649–679.

71. Yusupov, M.M., et al. (2001). Crystal structure of the ribosome at 5.5 A resolution. *Science*, **292**, 883–896.

72. Zuker, M. (2003). Mfold web server for nucleic acid folding and hybridization prediction. *Nucleic Acids Res.*, **31**, 3406–3415.

Chapter 3

Computational Methods for RNA Structure Prediction and Analysis

David Dufour,[a,b] Emidio Capriotti,[c] and Marc A. Marti-Renom[a,b,*]

[a] *Genome Biology Group, Structural Genomics Team, Centre Nacional d'Anàlisi Genòmic (CNAG), Barcelona, Spain*
[b] *Structural Genomics Group, Centre de Regulació Genòmica (CRG), Barcelona, Spain*
[c] *Department of Mathematics and Computer Science, University of Balearic Islands, Palma de Mallorca, Spain*
mmarti@pcb.ub.cat

RNA cannot be considered anymore as a simple transfer molecule. On the contrary, a plethora of noncoding RNA molecules are being discovered, which is transforming our thinking about how the cell is regulated. Large and small RNAs carry now a large repertory of diverse biological functions within cells. Altogether, RNA is now considered as a major player in the molecular regulation of essential cellular processes. Similar to proteins, RNAs adopt three-dimensional (3D) structures that are necessary for performing their functional roles. Unfortunately, despite advances in understanding the folding and unfolding of RNA molecules, our knowledge of the atomic mechanism by which RNA molecules adopt their biologically active structures is still limited. Moreover, experimental

*Parc Científic de Barcelona—Torre I, Baldiri Reixac, 4, 08028 Barcelona, Spain

RNA Nanotechnology
Edited by Bin Wang

ISBN 978-981-4411-64-6 (Hardcover), 978-981-4411-65-3 (eBook)
www.panstanford.com

determination of RNA structures either by X-ray crystallography or nuclear magnetic resonance is challenging, given the instability of RNA molecules. Therefore, computational approaches for predicting the 3D structure of RNAs are becoming essential in the study of the molecular mechanisms of RNA function. Here we start by outlining the general principles of the RNA structure, and then we describe the databases and algorithms for analyzing and predicting RNA secondary and 3D structures.

3.1 Introduction

Ribonucleic acid (RNA) molecules are now known to act upon enzymatic functions [1], gene transcriptional regulation [1–3], protein biosynthesis regulation [4], development [5], and disease [6]. Characterizing the molecular details of such a diverse repertory of functions requires the knowledge of the three-dimensional (3D) structure of the RNA molecules as well as their interaction with other biomolecules in the cell. Since the seventies, when the first RNA structure was determined (i.e., the yeast phenylalanine transfer RNA (tRNA) [7]), the number of known RNA structures has steadily grown, and only recently the increase of new structures has been exponential (Fig. 3.1). However, the first computational algorithms to predict base pairing from an RNA sequence were developed in the late seventies/early eighties [8–10]. Another 10 years passed by before the first RNA 3D structure (i.e., a conserved core of group I introns) was predicted by the Westhof group [11].

Nowadays, only a limited number of automatic predictions of the 3D structures of large RNA molecules has so far been accomplished (for some examples see [12–14]). However, and given the limited number and diversity of known RNA structures, computational algorithms for RNA structure prediction has been one of the sources for characterizing the structural diversity in RNA molecules and its relationship to function [15]. Most of the existing algorithms rely in the principle that RNA folding is a hierarchical process and that knowledge of its secondary structure (i.e., the determination of all base pairing in an RNA sequence) may improve the prediction of its 3D conformation. Consequently, in recent years, several computational programs have been developed, which try to predict the base pair interactions in RNA from its sequence (see, e.g., [16–18]). However, the growing amount of available structural data for RNA molecules and the initial attempts for classifying their motifs [19, 20] open the possibility for applying comparative approaches previously used for protein structure prediction [21–23]. It is known that in general, it will be more difficult to predict large RNA 3D structures based on comparative approaches than predicting protein structures [24]. Such a statement relies on (at least) two properties of RNA: (i) Its folding is essentially driven by its base pair and its regular motifs [25] (in contrast to the hydrophobic forces that drive protein folding [26]), and (ii) within the same functional family, RNA sequence conservation is usually limited to very short fragments of nucleotides, while still maintaining a substantial conservation of their secondary structure [25]. Both principles make necessary that the base pairs of an RNA molecule need be determined (or predicted) before attempting to predict its 3D structure and that reliable comparative approaches be limited to RNA sequences that align with more than 60% sequence identity to a known structure [15].

We begin this chapter by describing the RNA structure and the initial attempts for classifying of the RNA structural space. We continue by outlining recent developments and methods for secondary and tertiary structure prediction from sequences. We then conclude by discussing possible implications of the use of comparative approaches to predict the 3D conformation of RNA sequences based on existing known structures.

3.2 RNA Structure

Structurally, RNA is composed of a combination of riboses, phosphates, and aromatic bases. Riboses and phosphates are connected through the phosphodiester bond, forming a backbone from which the aromatic bases are attached in a regular fashion through the C1′ atom of the ribose moiety. As they are being synthesized, RNA molecules fold mainly by the driving force of hydrogen bonding and stacking interactions between bases. Long stretches of canonical

Watson–Crick (WC) base pairs result in an antiparallel double helix. However, each base has three sides (i.e., WC, Hoogsteen, and sugar sides) that can interact resulting in 28 different base-pairing possibilities between two nucleotides [19]. Nevertheless, canonical helices are maintained by WC base pairing. Additionally, nucleotide bases may also interact with the ribose or phosphate atoms as well as noncanonical base–base interactions, which may result in special RNA geometries. In contrast to deoxyribonucleic acid (DNA), this plethora of possible pairwise interactions between bases results in RNA adopting complex 3D structures. A base pair in RNA is maintained by a minimum of two hydrogen bonds between the paired nucleotide bases. The combination of base pairs in an RNA molecule defines the so-called secondary structure, which is composed of stems (double helices), loops, bulges, stem junctions, and pseudoknots. The final 3D RNA structure is maintained by tertiary interactions, including loop–loop interactions, stem–loop interactions, coaxial stacking, and triple and quadruple helices.

3.2.1 *RNA Base Pairs*

Over the last few years there has been a rapid growth in the number of RNA structures made available through the Protein Data Bank (PDB) (Fig. 3.1) [27, 28]. This increment is mostly due to the recent structural determination of ribosome machineries [29–32]. Thus, the availability of such data has allowed the application of a more robust classification of base pair interactions in RNA molecules. Although there are differences in the interaction of two RNA bases, a stable classification depending on the edges involved in the interaction (i.e., WC, Hoogsteen, or sugar edges) has already been proposed [33, 34]. In such a classification, each base can form several nonbonded interactions that involve different types of atoms: (i) phosphate-phosphate interaction mediated by water molecules, (ii) phosphate–sugar interaction, (iii) sugar–sugar interaction, (iv) base–phosphate interaction, (v) base–sugar interaction, and (vi) base–base interaction. Moreover, these six different interaction types can be formed in either a *cis* or a *trans* state, resulting in 12 possible different conformations (Fig. 3.2A,B) [35]. Only about 60% of the base pairs in known RNA structures adopt the canonical WC–WC interaction in *cis* conformation.

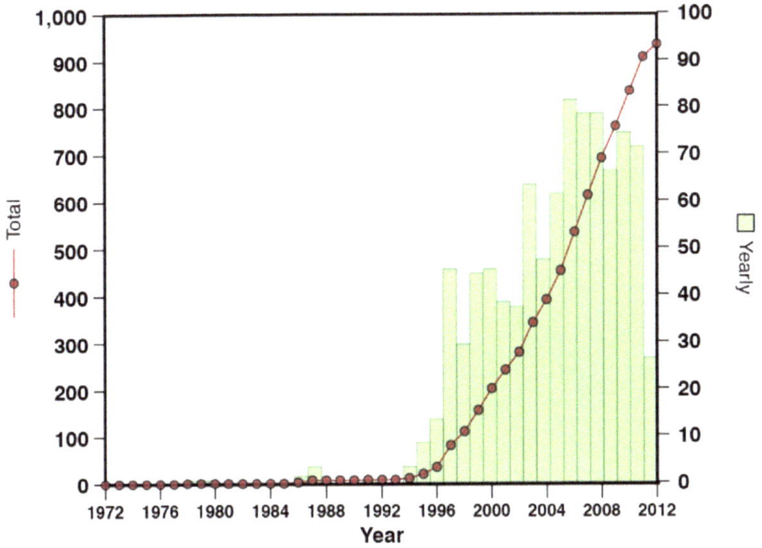

Figure 3.1 RNA structure deposition in the PDB database. Green bars (right y-axis) indicate yearly new PDB entries, and the red line (left y-axis) represents the total number of RNA structures in the PDB database. The data ends in July 2012.

3.2.2 *RNA Backbone*

Differently from proteins, RNA molecules are characterized by well-packed side chains stabilized by hydrogen bonds and a flexible backbone. RNA backbone conformation can be described in terms of the torsion angles α, β, γ, δ, ε, and ζ, while the ribose conformation is determined by the endocyclic torsion angles ν_0 to ν_4 (Fig. 3.2C). Richardson *et al.* have analyzed a set of RNA structures with crystallographic resolutions higher than 3 Å and no atom clashes, identifying 42 discrete RNA backbone conformers [36]. Different studies concluded that the RNA backbone is rotameric and can be classified into discrete conformers [37, 38]. These types of analyses have been possible because the quality and amount of determined RNA structures have considerably grown over the last few years [39–41]. However, most large RNA structures can only be determined at low resolutions. At this resolution, the phosphate and base plane can be accurately positioned but the sugar ring and the rest of the

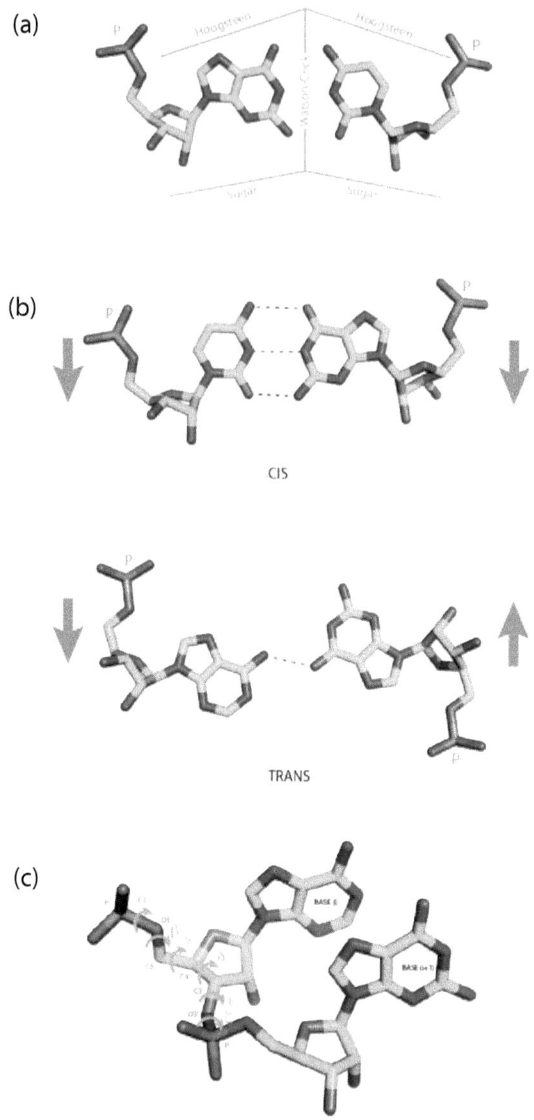

Figure 3.2 Base pair interactions. (A) WC, Hoogsteen, and sugar edges for a base pair interaction. (B) *cis* and *trans* states of a base pair interaction. (C) RNA backbone torsion angles.

backbone atoms may contain errors. Indeed, on average, one error can occur every two bases based on the analysis and classification of RNA torsion angles from the RNABase database [39].

3.2.3 *RNA Motifs*

RNA motifs correspond to recurrent RNA structural elements subject to 3D spatial constraints [20, 42]. This broad definition of RNA motifs already indicates the difficulty in uniquely describing or classifying them. The RNA secondary structure partially explains some of the known RNA motifs such as bulges, hairpins, internal loops, and multihelical motifs. However, the prediction of pseudoknots is a more challenging task in secondary structure prediction programs. Pseudoknots contain two stem–loop motifs in which the first stem's loop forms part of the second stem. Structural data indicates that the final 3D RNA structure is mostly determined by base pair stacking (i.e., WC base pairs) and non-WC interactions. Thus, characterizing, analyzing, and ultimately predicting the stacking of those bases will help the goal of classifying complex RNA motifs.

3.3 RNA Structural Databases

Since the seventies, when the first RNA structures became available [7], there has been an attempt to store, organize, and classify the RNA structural space. Next, we briefly describe available databases that classify RNA structures.

The Nucleic Acid Database (**NDB**) [27] stores all molecules containing nucleic acids and complements them with additional information such as classification of nucleic acids and their interaction with proteins, backbone conformation angles, and base pair classification. The Structural Classification of RNA (**SCOR**) database [41] organizes RNA motifs in a hierarchical classification system similar to the SCOP database for protein domains [43]. SCOR classifies RNA structures from three properties: first, the RNA structural classification describes RNA motifs according to the number of strands connecting double helices; second, the RNA functional classification divides each entry by the biological function of their molecule, motif, and structural model; and third, the RNA tertiary interaction group classifies RNA structures by their inter- and intramolecular interactions differing from WC and non-WC base pairs. The SCOR database stores 8,270 structural motifs (October 2004), some of which are further classified into functional and RNA tertiary interaction classes (Table 3.1). The RNA family database (**Rfam**) [44, 45] classifies noncoding RNA molecules into families of members that conserve sequence and secondary structure. The conservation of RNA secondary structure implies a degree of conservation of its function [15]. The Metal ion binding sites in RNA structures database (**MeRNA**) [46] was manually curated by analyzing each RNA structure and comparing them to previously described binding motifs and includes eight well-characterized metal ion–binding motifs. **PseudoBase** [47], a searchable database

Table 3.1 RNAbase classification. The number of RNA structure entries stored in RNAbase classified by their functional categories

Category	Entries
tRNAs	217
rRNAs	283
mRNAs	126
Transcription-related RNAs	86
Introns	26
Splicing-related RNAs	59
Signal recognition particle RNAs	22
Ribozymes	115
RNase P	21
Aptamers	30
Pseudoknots	31
Tetraloops	81
Bulges	69
DNA–RNA hybrids	115
PNA–RNA hybrids	1
Drug–RNA complexes	137
Viral and phage RNAs	221

Abbreviations: mRNA, messenger RNA; RNase P, ribonuclease P.

of pseudoknot secondary structures, contains over 250 records of pseudoknots determined by crystallography, nuclear magnetic resonance (NMR), mutational experiments, and sequence comparisons. **PseudoBase++** [48] is an extension of PseudoBase for searching, formatting, and visualization of pseudoknots. PseudoBase++ links each pseudoknot in PseudoBase to the GenBank record of the corresponding nucleotide sequence and allows scientists to automatically visualize RNA secondary structures with PseudoViewer. It also includes the capabilities of fine-grained reference searching and collecting of new pseudoknot information. The **RNAJunction** database [49] contains structure and sequence information for RNA structural elements such as helical junctions, kissing loops, internal loops, bulges, and loop–loop interactions. The database can be searched using the PDB code, structural classification, sequence, and interhelix angles. The RNA secondary STRucture and statistical ANalysis Database (**STRAND**) [50] provides a collection of known RNA secondary structures drawn from diverse public databases. The database is searchable based on one or various criteria defined by the user, like RNA type, organism of origin, external source, length, the number of molecules in the complex, and other features. Moreover, the RNA Secondary Structure Analyzer, a tool developed by the same group to analyze RNA secondary structures, provides comprehensive statistical information on the secondary structures in the database. **MODOMICS** [51] is a database devoted to the systems biology of RNA modification. It provides information on the chemical structure of modified nucleosides, pathways of their biosynthesis, sequences of RNAs containing these modifications, and RNA-modifying enzymes. It contains curated tRNA and ribosomal RNA (rRNA) sequences and all RNAs with 3D structures in the NDB database for which modified nucleosides are known. Database of Base-Pair Structures (**BPS**) is a database [52] that stores RNA base pair structures with quantitative information on the spatial arrangements of interacting bases, including higher-order base associations and isosteric pairs. The structures are taken from the NDB database, and the base pairs are identified and characterized with the 3DNA software package [53]. The interactions are classified in terms of residue identities, base pair positioning, and hydrogen-bonding patterns and related to the structural context in which they occur.

Table 3.2 SCOR classification. Number of RNA structure motifs stored in the SCOR database classified by structural, functional, and RNA tertiary interaction categories

Classification	Subclasses	RNA motifs
Structural classification	Internal loops	5,350
	Hairpin loops	2,920
Functional classification	Molecular function	480
	Motif function	179
	Structural models	137
RNA tertiary interaction	Coaxial helices	7
	Tetraloop-receptor	1
	A-minor motif	240
	Kissing-hairpin loops	32
	tRNA D-loop:T-loop	7
	Pseudoknots	17
	Ribose zipper	657

BPS also includes an atlas with representative images of the various base pairs, higher-order base interactions and isosteric pairs, and links to statistical information about these groups of structures. The RNA FRAgments search engine & dataBASE (**FRABASE**) [54] allows for the automatic search of 3D RNA fragments within a set of RNA structures by the input of either RNA sequence(s) and/or secondary structure(s). The database contains RNA sequences and secondary structures in the 'dot bracket' notation derived from the PDB, a collection of atom coordinates of unmodified and modified nucleotide residues occurring in RNA structures, calculated RNA torsion angles and sugar pucker parameters, and information about base pairs. The Characterization of Secondary Structure Motifs (**CoSSMos**) database [55] is an online database of 3D characteristics of internal, bulge, and hairpin loops. It contains each loop's structural information, including sugar pucker, glycosidic linkage, hydrogen-bonding patterns, and stacking interaction. Users can search via general PDB information, experimental parameters, sequences, and specific motifs and by specific structural parameters in the subquery page after the initial search.

3.4 RNA Secondary Structure Prediction

Predicting the secondary structure of an RNA sequence can prove very useful for gaining insight into its tertiary structure and its function [56]. The RNA-folding process is hierarchical [57], which means that local interactions occur first and are energetically stronger than tertiary interactions [58]. Therefore, the RNA secondary structure provides a scaffold to its native 3D structure. This property already indicates that the RNA secondary structure can be predicted without the knowledge of tertiary interactions (to the exception of the so-called pseudoknots). The first methods for predicting the secondary structure of RNA molecules were developed assuming that the minimum free-energy conformation for the native state could be searched by dynamic programming [8, 9] and the Nussinov's algorithm [10]. The scoring functions for such approaches were based on free-energy parameters from physics, which were derived from empirical calorimetric experiments [59] or from known RNA structures deposited in the PDB [17]. Unfortunately, the minimum free-energy (MFE) approach does not guarantee that the selected or predicted final structure will be the native structure and typically corresponds to a near-native conformation [60]. Other implementations of the MFE principle include the use of a heuristic search for suboptimal secondary structures [8, 59, 61], the computation of all suboptimal alignments near the optimal folding space [62], and the selection of suboptimal solutions based on RNA shape analysis [63].

In the 1990, McCaskill first implemented a method based on the equilibrium partition function for secondary structure and associated probabilities of various substructures [64]. Such a method allowed the statistical characterization of the equilibrium ensemble of RNA secondary structures. It has been noted that higher base pair probabilities, computed by the partition function approach, correspond to higher predictive reliability when considering structures determined by comparative sequence analysis [61]. More recently, new computational approaches based on the statistical sampling of known RNA secondary structures [17] or genetic algorithms [65–67] have also been implemented for secondary structure prediction. However, most of the methods described so far are not suitable for predicting RNA pseudoknots, because they are based on the recursive approach. It has been demonstrated that the prediction of secondary structure motifs with pseudoknots is a NP-complete problem making it computational intractable [68]. To address this problem, modified dynamic programming and stochastic context-free grammar algorithms [69] have been recently introduced. For example, the **PKNOTS** program implements thermodynamic folding in a rather large subclass of pseudoknots on $O(N^4)$ and $O(N^6)$ time space, which makes it only usable for short sequences [70]. The partition function approach implemented by Dirks [71, 72] has an $O(N^5)$ complexity. Despite this computational complexity, the accuracy for pseudoknots prediction has significantly increased by using an innovative dynamic partner sequence-stacking algorithm [73].

RNA secondary structure prediction from a single sequence somehow neglects the evolutionary forces acting upon RNA sequence variation. Therefore, the inclusion of multiple sequences for predicting the RNA secondary structure allows the incorporation of constraints based on the commonalities of the compared sequences [74]. Evolution tends to conserve the RNA secondary structure more than sequence [15]. Indeed, it is known that a mutation in an RNA molecule is usually compensated by a second mutation that restores base pair interaction [75, 76]. Several methods for secondary structure prediction use this principle by attempting to detect such covariance between different positions in the multiple sequence alignment. An initial implementation of such an approach used mutual information theory to extract the covariance between bases [77, 78]. However, these approaches resulted in limited accuracy [79] and have been replaced by more recent implementations such as the **RNAalifold** program [80], which scores possible solutions by combining free energy with a covariance term; the **Pfold** program [81], which uses an evolutionary SCFG approach; or the Iterated Loop Matching (**ILM**) program [82, 83], which combines thermodynamic and mutual information in a single score. The **Foldalign** program [84, 85] heuristically considers local sequence alignments and the maximum number of base pairs at the same time. The **Dynalign** program [86] is a pairwise alignment method that searches for common low-energy structures between two sequences. The algorithm complexity is reduced considering a maximum value of the sequence distance between two aligned residues and by limiting the size of any internal loop. Finally, the Comparative Analysis of RNA by Cofolding (**Carnac**) program [87, 88], which is not a strict implementation of a simultaneous align-and-fold approach, relies on a thermodynamic model with energy minimization by combining information from locally conserved elements and mutual information between sequences. Finally, the **RNAforester** [89] and Multiple Alignment and consensus structure prediction of RNAs based on sequence structure comparisons (**MARNA**) [90] programs first fold RNA sequences using single-sequence secondary structure prediction methods and then align the resulting structures using tree-based methods.

3.5 RNA Tertiary Structure Analysis and Prediction

The increase over the last decade of the number of available structures deposited in the PDB, including X-ray and NMR models (Fig. 3.1), has stimulated the structural biology community to develop computational tools for analyzing the RNA structural space. Next, we outline some of the existing methods for RNA structure analysis and prediction.

3.5.1 *RNA 3D Structure Analysis*

The Probing RNA structures to Identify Motifs and Overall Structural changes (**PRIMOS**) program [91] describes an RNA structure with pseudo torsion angles η ($C4'_{i-1}$-P_i-$C4'_i$-P_{i+1}) and θ (P_i-$C4'_i$-P_{i+1}-$C4'_{i+1}$) obtained using the **AMIGOS** program [92]. Then the search comparison is done over such simplified version of the RNA structure, allowing the identification of common small motifs between two RNAs or an RNA structural motif and a database of RNA structures. The Nucleic Acids Search for Substructures and Motifs (**NASSAM**) program [93], which was designed for identifying common substructural motifs between two RNA structures, implements a simplified vector representation of each nucleic acid base with respect to its position in the structure. Then the vectors and their edges are transformed into a graph connecting the bases and compared using the Ullman subgraph isomorphism algorithm. The Alignment of RNA Tertiary Structures (**ARTS**) [94] and DIhedral ALignment (**DIAL**) [95] programs for structural comparison of RNA structures were developed to overcome the limitation of sequence continuity. The ARTS program describes RNA structures by a set of contiguous *quadrats* (i.e., four phosphate atoms located in two successive base pairs). The program then identifies similar *quadrats* between two RNA structures and uses them as seeds for the final alignment. Finally, the algorithm finds the maximal matching in a bipartite graph between the two structures by extending the structure alignment that maximizes the number of aligned bases and base pairs. The DIAL program uses a dynamic programming algorithm to align two RNA structures based on a scoring function that combines a base, a dihedral angle, and a base-paring similarity measure. DIAL can be run as a web server and provides the user with the option of producing global (Needleman–Wunsch), local (Smith–Waterman), or global/semiglobal (motif search) alignments. The Structural Alignment of Ribonucleic Acids (**SARA**) program [96] applies a unit-vector root mean square approach to pairwise structural alignment. It can also assign RNA structures to functional classes as defined in the Structural Classification of RNA (SCOR) database [97]. The SARA has been used within the Tree-based Consistency Objective Function for alignment Evaluation (T-Coffee) framework as SARA-Coffee for the multiple sequence alignment of RNAs based on their tertiary structure, which is helpful for the alignment of low-conservation homologs [98]. The Structural Alignment of RNA using a Structural Alphabet (**SARSA**) program [99] uses vectors to obtain a structural alphabet of RNA backbone conformations. The input structures are represented using such an alphabet, and the two RNA structures are finally aligned using dynamic programming based on the alphabet sequence. The Find RNA 3D (**FR3D**) program [100], and its web-based interface **WebFR3D** [101], identifies recurrent motifs in a base-centered approach using geometric, symbolic, or sequence information. To score and rank candidate motifs, FR3D calculates a geometric discrepancy by rigidly rotating candidates to optimally align with the query motif and then comparing the relative orientations of the corresponding bases in the query and candidate motifs. The Fast and Accurate Search Tool for RNA 3D structures (**FASTR3D**) program [102] allows users to specify a range of nucleotides from a PDB file as a query to look for similar structures in a list of PDB files using the secondary structure information and backbone torsion angles of the query structure. Alternatively, it can take primary and/or secondary structures as an input. The **RNAMotifScan** program [103] detects similar RNA structure motifs based on the two-dimensional (2D) alignments. It was observed that many noncanonical base pairs in RNA structural motifs are isosteric, and these base pairs can interchange with each other without affecting the overall RNA structure, so RNAMotifScan takes into account isosteric base pairs and multipairings. The **FRASS** program [104] is capable of handling large RNA fragments and is designed for global similarity searching. The user can select an entire chain from a PDB file or upload a structure to the server. The searching method is based on Gauss integrals that are used to compare the shapes of backbones of RNA molecules. The Web RNA 3D Align (**WebR3DAlign**) server [105] identifies all motifs conserved in the 3D structures of two possibly homologous RNA molecules. It produces a nucleotide-to-nucleotide alignment of two 3D structures to identify conserved motifs, while allowing for differences in the global structure of the molecules, like domain motions.

3.5.2 *RNA 3D Structure Prediction*

Predicting the 3D structure of an RNA molecule is straightforward and usually requires human intervention [106]. In contrast to the current status of protein structure prediction, a fully automated approach is not able to reliably predict a large RNA 3D structure from its sequence [24]. However, over the last few years, a plethora of methods has been developed that aid the manual or semiautomatic prediction of RNA structures. For example, the Editor for RNA in 3D (**ERNA-3D**) program [12] automatically generates an RNA 3D structure, starting from its secondary structure. ERNA-3D, which has successfully been used to model the structure of transfer-messenger RNA molecules [13], is able to model RNA motifs by using high-resolution structural information from the

SCOR database. The **MANIP** program [107] builds complete RNA structural models based on the assembly of RNA motifs or fragments from a selected library. The final refinement protocol combines canonical as well as noncanonical base-pairing constraints with restraints imposed by covalent geometry, stereochemistry, and van der Waals contacts. The Macromolecular Conformations by SYMbolic programming (**MC-Sym**) [108] builds RNA 3D structures using the coordinates and relations between bases from known RNA structures. Additional constraints can be applied to the model during the building procedure to ensure the conservation of particular structural features. Mc-Sym uses molecular dynamic simulations to minimize the energy of the predicted structure. The **RNA2D3D** program [109] builds RNA structural models by first spacing the atoms of a nucleotide along a fixed backbone and then predicting the final structure of the model by an helical winding procedure. The model is further refined by interactively moving groups of nucleotides to better-fit, known structural information or by minimizing it using molecular dynamics simulations. The Interactive RNA folding simulations (**iFoldRNA**) program [110] uses coarse-grained structural models to perform molecular dynamics simulations of RNA structures. iFoldRNA has been used to predict the structure of RNA molecules smaller than 50 nucleotides to near-atomic resolution (i.e., 2 to 5 Å Root Mean Square Deviation (RMSD) to its native structure). The Nucleic Acid Simulation Tool (**NAST**) program [14] uses a knowledge-based, coarse-grained dynamics engine for modeling RNA structures. NAST allows the end user to provide secondary or tertiary experimentally derived restraints to filter the predicted 3D models. The BAyesian network model of RNA using Circular distributions and maximum Likelihood Estimation (**BARNACLE**) program [111], which introduced a probabilistic model of the RNA structure, allows an efficient sampling of RNA conformations in continuous space. The Fragment Assembly of RNA with Full-Atom Refinement (**Rosetta/FARFAR**) program [112, 113], inspired by the protein structure prediction method [114], has been applied to predict the 3D structure of 20 RNA sequences of ~30 nucleotides. The authors report that their method is able to correctly predict the native conformation for ~90% of WC and about one-third of non-WC base pairs. Their results also suggest that improvements in the energy function together with the use of predictions from phylogenetic approaches are necessary for an accurate structure prediction of more complex RNA molecules. The **RNAmoIP** framework [115, 116] uses integer programming to refine predicted or known secondary structures to accommodate the insertion of RNA 3D motifs. Then, the predictions are used as templates to generate complete 3D structures with the MC-Sym program. Integer programming techniques have gained a lot of interest recently as they provide state-of-the-art methods for predicting RNA secondary structures with pseudoknots. The **RNABuilder** program [117] is a software package that generates model RNA structures by treating the kinematics and forces at separate multiple levels of resolution. Kinematically, bonds in bases, certain stretches of residues, and some entire molecules are rigid, while other bonds remain flexible. Forces act on the rigid bases and selected individual atoms. The **Assemble** program [118], a graphical user interface (GUI) semiautomated modeling program that can be performed by homology and ab initio with or without electron density maps, allows interactive editing of the secondary structure and the use of a library of annotated tertiary structures. It combines automated and manual protocols within an iterative modeling process, where the user can insert 3D motifs and modify backbone angles of a coarse-grained input structure. **ModeRNA** [119] is a program for comparative modeling of RNA 3D structures. It requires a pairwise sequence alignment and a structural template to generate a 3D structural model of the target RNA sequence via either fully automated or script-based approaches. ModeRNA is capable of handling 115 different nucleotide modifications and bridging gaps using fragments derived from an extensive fragment library. The **Vfold** program [120] evaluates the stable structures and the folding free energies for RNA secondary structures and pseudoknotted structures. The program predicts the secondary structure of the RNA sequence allowing the build of a coarse-grained structure, which is later refined with an all-atom representation that relies in a library of PDB fragments. RNA-As-Graph-Pools (**RAGPOOLS**) is a recent application of a graph theoretic approach to represent RNA molecules [121, 122], which allows for a simplified representation of RNA structural motifs that are then used for predicting and designing new complex structures of RNA molecules. Finally, the **RSIM** program [123] provides a fully automated application predicting RNA tertiary structures using fragment assembly from secondary structure constraints predicted by the ViennaRNA package. These tertiary structures are further refined with Monte Carlo simulations utilizing a novel sampling method, an expanded statistical potential, and a diverse fragment library. Finally, RSIM stores the refinement paths, which allows the representation of the predicted RNA conformational space as a graph with secondary structures as nodes and simulation paths as edges.

3.5.3 *RNA 3D Structure Assessment*

The large number of new tools for predicting the RNA 3D structure is likely to result in an increasing number of predicted structures that will need to be assessed. According to this, the scientific community is now using known 3D structures deposited in the PDB to develop knowledge-based potentials of mean force to assess the accuracy of predicted RNA structures. In the 2009, Jonikas *et al.* presented the first coarse-grained knowledge-based function to select native-like structures [14]. The statistical potential is embedded in the **NAST** algorithm and assesses the RNA structure using a representation based on C3′ atoms. The method has been validated against thousands of decoys and assessing how well it samples RNA structures close to the solved crystal structures. Unfortunately, the NAST knowledge-based potential provides a nucleotide-based score that cannot distinguish between models with different atom positions except for C3′. The NAST algorithm was successfully used to generate and properly assess the accuracy of a modeled tRNA and P4–P6 structures. Later, the **FARFAR** method [113] implemented a statistical potential developed to score the structures predicted using the FARNA algorithm. The scoring function in the FARFAR algorithm is composed of several terms, including pairwise distance-dependent potentials, hydrogen bonds, hydrophobic and hydrophilic contributions, and a term describing the screened electrostatic interactions between phosphates. FARFAR was derived using a dataset of 32 motifs automatically extracted from a set of high-resolution crystallographic structures of ribozymes, riboswitches, and other noncoding RNAs. The Ribonucleic Acids Statistical Potential (**RASP**) [124] is based on a distance-dependent scoring function including four different types of RNA molecule representations with increasing level of complexity. The simplest RASP potential included four atom types corresponding to the four C3′ atoms of each nucleotide. Other representations consisted of 28 atom types for the backbone atoms, 44 atom types for the backbone and sugar ring, and 23 for all atoms without taking into consideration the differences between nucleotides. All the RASP knowledge-based potentials were tested using a leave-one-out procedure over a set of 85 nonidentical RNA structures and their associated decoys sets composed of 500 structures with different RMSD and Global Distance Test - Total Score (GDT-TS) values. The RASP tool was favorably compared with previously existing methods such as NAST, FARFAR and Assisted Model Building with Energy Refinement (AMBER) [125]. The results showed that the RASP full atom was the most accurate method in the ranking native structures. More recently, Bernauer *et al.* have developed another distance-dependent statistical potential based on the Dirichlet process mixture model [126]. This procedure allows us to obtain an analytical form of the potential as a sum of Gaussian functions that makes the scoring function fully differentiable and suitable for energy minimization or molecular dynamics. The method has been trained using a set of 77 nonredundant and high-quality RNA structures, and its performances have been evaluated over a set of decoys generated by molecular dynamics simulations and a normal-mode perturbation method. The method has been compared with FARFAR, resulting in similar accuracies.

3.6 Perspectives

The increase in the number of known RNA structures in the PDB clearly shows the existence of regular and recurrent 3D RNA motifs. Thus, the next logical step for structural biologists is to detect, store, analyze, and classify such structural motifs to aid in ab initio and/or knowledge-based structural prediction of whole RNA sequences [19]. The current amount and diversity of known structures of RNA molecules has allowed the development of a plethora of approaches for RNA structure prediction, which have been briefly outlined in this chapter. However, it is difficult to predict whether such methods will be readily applicable to RNA and, more importantly, will result in reliable models. The first collective blind test experiment in RNA 3D structure prediction (called RNA-Puzzles) was recently organized. RNA-Puzzles is a Critical Assessment of protein Structure Prediction (CASP)-like experiment that aims at evaluating the accuracy of both manual and automatic methods for RNA structure prediction [24]. The results from the RNA-Puzzles experiment provide deeper insights into the accuracy of available methods for different applications at the same time that stimulate the RNA structure prediction community for its ongoing efforts to improve its tools. In its first edition, seven different research groups that tried to predict the structure of several RNA molecules within three different types of scenarios participated in the RNA-Puzzles experiment. Overall, the methods implemented by the Bujnicki, Chen, and Das groups were scored among the top methods [24]. The website for automatic RNA model evaluation and additional information about the RNA-Puzzles experiment is http://paradise-ibmc.u-strasbg.fr/rnapuzzles/.

Acknowledgments

MAM-R acknowledges support from the Spanish Ministry of Economy and Competitively (BFU2010-19310). EC acknowledges support from the European Community through the Marie Curie International Outgoing Fellowship program (PIOF-GA-2009-237225). This chapter is an updated version of our previous review [127].

References

1. Staple, D.W., and Butcher, S.E. (2005). Pseudoknots: RNA structures with diverse functions. *PLoS Biol.*, **3**, e213.
2. Bartel, D.P. (2004). MicroRNAs: genomics, biogenesis, mechanism, and function. *Cell*, **116**, 281–297.

3. Dorsett, Y., and Tuschl, T. (2004). siRNAs: applications in functional genomics and potential as therapeutics. *Nat. Rev. Drug Discovery*, **3**, 318–329.
4. Doudna, J.A. (2000). Structural genomics of RNA. *Nat. Struct. Biol.*, **7**(Suppl.), 954–956.
5. Aravin, A.A., *et al.* (2003). The small RNA profile during Drosophila melanogaster development. *Dev. Cell*, **5**, 337–350.
6. Lu, J., *et al.* (2005). MicroRNA expression profiles classify human cancers. *Nature*, **435**, 834–838.
7. Kim, S.H., *et al.* (1974). Three-dimensional tertiary structure of yeast phenylalanine transfer RNA. *Science*, **185**, 435–440.
8. Zuker, M., and Stiegler, P. (1981). Optimal computer folding of large RNA sequences using thermodynamics and auxiliary information. *Nucleic Acids Res.*, **9**, 133–148.
9. Zuker, M., and Sankoff, D. (1984). RNA secondary structure and their prediction. *Bull. Math. Biol.*, **46**, 591–621.
10. Nussinov, R., and Jacobson, A.B. (1980). Fast algorithm for predicting the secondary structure of single-stranded RNA. *Proc. Natl. Acad. Sci. U S A.*, **77**, 6309–6313.
11. Michel, F., and Westhof, E. (1990). Modelling of the three-dimensional architecture of group I catalytic introns based on comparative sequence analysis. *J. Mol. Biol.*, **216**, 585–610.
12. Zwieb, C., and Muller, F. (1997). Three-dimensional comparative modeling of RNA. *Nucleic Acids Symp. Ser.*, **36**, 69–71.
13. Burks. J., Zwieb, C., Muller, F., Wower, I., and Wower, J. (2005). Comparative 3-D modeling of tmRNA. *BMC Mol. Biol.*, **6**, 14.
14. Jonikas, M.A., *et al.* (2009). Coarse-grained modeling of large RNA molecules with knowledge-based potentials and structural filters. *RNA*, **15**, 189–199.
15. Capriotti, E., and Marti-Renom, M.A. (2010). Quantifying the relationship between sequence and three-dimensional structure conservation in RNA. *BMC Bioinf.*, **11**, 322.
16. Hofacker, I.L. (2003). Vienna RNA secondary structure server. *Nucleic Acids Res.*, **31**, 3429–3431.
17. Do, C.B., Woods, D.A., and Batzoglou, S. (2006). CONTRAfold: RNA secondary structure prediction without physics-based models. *Bioinformatics*, **22**, e90–98.

18. Dowell, R.D., and Eddy, S.R. (2006). Efficient pairwise RNA structure prediction and alignment using sequence alignment constraints. *BMC Bioinf.*, **7**, 400.
19. Leontis, N.B., *et al.* (2006). The RNA Ontology Consortium: an open invitation to the RNA community. *RNA*, **12**, 533–541.
20. Leontis, N.B., Lescoute, A., and Westhof, E. (2006). The building blocks and motifs of RNA architecture. *Curr. Opin. Struct. Biol.*, **16**, 279–287.
21. Baker, D., and Sali, A. (2001). Protein structure prediction and structural genomics. *Science*, **294**, 93–96.
22. Marti-Renom, M.A., *et al.* (2000). Comparative protein structure modeling of genes and genomes. *Annu. Rev. Biophys. Biomol. Struct.*, **29**, 291–325.
23. Sali, A., and Blundell, T.L. (1993). Comparative protein modelling by satisfaction of spatial restraints. *J. Mol. Biol.*, **234**, 779–815.
24. Cruz, J.A., *et al.* (2012). RNA-Puzzles: a CASP-like evaluation of RNA three-dimensional structure prediction. *RNA*, **18**, 610–625.
25. Schuster, P., Stadler, P.F., and Renner, A. (1997). RNA structures and folding: from conventional to new issues in structure predictions. *Curr. Opin. Struct. Biol.*, **7**, 229–235.
26. Dill, K.A. (1990). Dominant forces in protein folding. *Biochemistry*, **29**, 7133–7155.
27. Berman, H.M., *et al.* (1992). The nucleic acid database. A comprehensive relational database of three-dimensional structures of nucleic acids. *Biophys. J.*, **63**, 751–759.
28. Berman, H.M., *et al.* (2002). The Protein Data Bank. *Acta Crystallogr. D Biol. Crystallogr.*, **58**, 899–907.
29. Schuwirth, B.S., *et al.* (2005). Structures of the bacterial ribosome at 3.5 A resolution. *Science*, **310**, 827–834.
30. Gao, H., Ayub, M.J., Levin, M.J., and Frank, J. (2005). The structure of the 80S ribosome from Trypanosoma cruzi reveals unique rRNA components. *Proc. Natl. Acad. Sci. U S A*, **102**, 10206–10211.
31. Noller, H.F. (2005). RNA structure: reading the ribosome. *Science*, **309**, 1508–1514.
32. Felden, B. (2007). RNA structure: experimental analysis. *Curr. Opin. Microbiol.*, **10**, 286–291.
33. Leontis, N.B., and Westhof, E. (2001). Geometric nomenclature and classification of RNA base pairs. *RNA*, **7**, 499–512.

34. Leontis, N.B., Stombaugh, J., and Westhof, E. (2002). The non-Watson-Crick base pairs and their associated isostericity matrices. *Nucleic Acids Res.*, **30**, 3497–3531.

35. Yang, H., *et al.* (2003). Tools for the automatic identification and classification of RNA base pairs. *Nucleic Acids Res.*, **31**, 3450–3460.

36. Murray, L.J., Arendall, W.B., 3rd, Richardson, D.C., and Richardson, J.S. (2003). RNA backbone is rotameric. *Proc. Natl. Acad. Sci. U S A*, **100**, 13904–13909.

37. Schneider, B., Moravek, Z., and Berman, H.M. (2004). RNA conformational classes. *Nucleic Acids Res.*, **32**, 1666–1677.

38. Hershkovitz, E., Sapiro, G., Tannenbaum, A., and Williams, L.D. (2006). Statistical analysis of RNA backbone. *IEEE/ACM Trans. Comput. Biol. Bioinform.*, **3**, 33–46.

39. Murthy, V.L., and Rose, G.D. (2003). RNABase: an annotated database of RNA structures. *Nucleic Acids Res.*, **31**, 502–504.

40. Griffiths-Jones, S., *et al.* (2005). Rfam: annotating non-coding RNAs in complete genomes. *Nucleic Acids Res.*, **33**, D121–D124.

41. Klosterman, P.S., Tamura, M., Holbrook, S.R., and Brenner, S.E. (2002). SCOR: a structural classification of RNA database. *Nucleic Acids Res.*, **30**, 392–394.

42. Leontis, N.B., and Westhof, E. (2003). Analysis of RNA motifs. *Curr. Opin. Struct. Biol.*, **13**, 300–308.

43. Andreeva, A., *et al.* (2004). SCOP database in 2004: refinements integrate structure and sequence family data. *Nucleic Acids Res.*, **32**, D226–D229.

44. Griffiths-Jones, S., *et al.* (2003). Rfam: an RNA family database. *Nucleic Acids Res.*, **31**, 439–441.

45. Gardner, P.P., *et al.* (2011). Rfam: Wikipedia, clans and the "decimal" release. *Nucleic Acids Res.*, **39**, D141–D145.

46. Stefan, LR., *et al.* (2006) MeRNA: a database of metal ion binding sites in RNA structures. *Nucleic Acids Res.*, **34**, D131–D134.

47. van Batenburg, F.H., *et al.* (2000). PseudoBase: a database with RNA pseudoknots. *Nucleic Acids Res.*, **28**, 201–204.

48. Taufer, M., *et al.* (2009). PseudoBase++: an extension of PseudoBase for easy searching, formatting and visualization of pseudoknots. *Nucleic Acids Res.*, **37**, D127–D135.

49. Bindewald, E., *et al.* (2008). RNAJunction: a database of RNA junctions and kissing loops for three-dimensional structural analysis and nanodesign. *Nucleic Acids Res.*, **36**, D392–D397.

50. Andronescu, M., Bereg, V., Hoos, H.H., and Condon, A. (2008). RNA STRAND: the RNA secondary structure and statistical analysis database. *BMC Bioinf.*, **9**, 340.

51. Czerwoniec, A., *et al.* (2009). MODOMICS: a database of RNA modification pathways. 2008 update. *Nucleic Acids Res.*, **37**, D118–D121.

52. Xin, Y., and Olson, W.K. (2009). BPS: a database of RNA base-pair structures. *Nucleic Acids Res.*, **37**, D83–D88.

53. Lu, X.J., and Olson, W.K. (2003). 3DNA: a software package for the analysis, rebuilding and visualization of three-dimensional nucleic acid structures. *Nucleic Acids Res.*, **31**, 5108–5121.

54. Popenda, M., *et al.* (2010). RNA FRABASE 2.0: an advanced web-accessible database with the capacity to search the three-dimensional fragments within RNA structures. *BMC Bioinf.*, **11**, 231.

55. Vanegas, P.L., *et al.* (2012). RNA CoSSMos: characterization of secondary structure motifs—a searchable database of secondary structure motifs in RNA three-dimensional structures. *Nucleic Acids Res.*, **40**, D439–D444.

56. Mathews, D.H., Moss, W.N., and Turner, D.H. (2010). Folding and finding RNA secondary structure. *Cold Spring Harb. Perspect. Biol.*, **2**, a003665.

57. Tinoco, I., Jr., and Bustamante, C. (1999). How RNA folds. *J. Mol. Biol.*, **293**, 271–281.

58. Mathews, D.H., and Turner, D.H. (2006). Prediction of RNA secondary structure by free energy minimization. *Curr. Opin. Struct. Biol.*, **16**, 270–278.

59. Mathews, D.H., Sabina, J., Zuker, M., and Turner, D.H. (1999). Expanded sequence dependence of thermodynamic parameters improves prediction of RNA secondary structure. *J. Mol. Biol.*, **288**, 911–940.

60. Reeder, J., *et al.* (2006). Beyond Mfold: recent advances in RNA bioinformatics. *J. Biotechnol.*, **124**, 41–55.

61. Mathews, D.H. (2004). Using an RNA secondary structure partition function to determine confidence in base pairs predicted by free energy minimization. *RNA*, **10**, 1178–1190.

62. Wuchty, S., Fontana, W., Hofacker, I.L., and Schuster, P. (1999). Complete suboptimal folding of RNA and the stability of secondary structures. *Biopolymers*, **49**, 145–165.

63. Giegerich, R., Voss, B., and Rehmsmeier, M. (2004). Abstract shapes of RNA. *Nucleic Acids Res.*, **32**, 4843–4851.

64. McCaskill, J.S. (1990). The equilibrium partition function and base pair binding probabilities for RNA secondary structure. *Biopolymers*, **29**, 1105–1119.
65. van Batenburg, F.H., Gultyaev, A.P., and Pleij, C.W. (1995). An APL-programmed genetic algorithm for the prediction of RNA secondary structure. *J. Theor. Biol.*, **174**, 269–280.
66. Gultyaev, A.P., van Batenburg, F.H., and Pleij, C.W. (1995). The computer simulation of RNA folding pathways using a genetic algorithm. *J. Mol. Biol.*, **250**, 37–51.
67. Shapiro, B.A., Wu, J.C., Bengali, D., and Potts, M.J. (2001). The massively parallel genetic algorithm for RNA folding: MIMD implementation and population variation. *Bioinformatics*, **17**, 137–148.
68. Lyngso, R.B., and Pedersen, C.N. (2000). RNA pseudoknot prediction in energy-based models. *J. Comput. Biol.*, **7**, 409–427.
69. Cai, L., Malmberg, R.L., and Wu, Y. (2003). Stochastic modeling of RNA pseudoknotted structures: a grammatical approach. *Bioinformatics*, **19**(Suppl 1), i66–73.
70. Rivas, E., and Eddy, S.R. (1999). A dynamic programming algorithm for RNA structure prediction including pseudoknots. *J. Mol. Biol.*, **285**, 2053–2068.
71. Dirks, R.M., and Pierce, N.A. (2003). A partition function algorithm for nucleic acid secondary structure including pseudoknots. *J. Comput. Chem.*, **24**, 1664–1677.
72. Dirks, R.M., and Pierce, N.A. (2004). An algorithm for computing nucleic acid base-pairing probabilities including pseudoknots. *J. Comput. Chem.*, **25**, 1295–1304.
73. Huang, X., and Ali, H. (2007). High sensitivity RNA pseudoknot prediction. *Nucleic Acids Res.*, **35**, 656–663.
74. Gaspin, C., and Westhof, E. (1995). An interactive framework for RNA secondary structure prediction with a dynamical treatment of constraints. *J. Mol. Biol.*, **254**, 163–174.
75. Rousset, F., Pelandakis, M., and Solignac, M. (1991). Evolution of compensatory substitutions through G.U intermediate state in Drosophila rRNA. *Proc. Natl. Acad. Sci. U S A*, **88**, 10032–10036.
76. Kirby, D.A., Muse, S.V., and Stephan, W. (1995). Maintenance of pre-mRNA secondary structure by epistatic selection. *Proc. Natl. Acad. Sci. U S A*, **92**, 9047–9051.
77. Chiu, D.K., and Kolodziejczak, T. (1991). Inferring consensus structure from nucleic acid sequences. Comput. *Appl. Biosci.*, **7**, 347–352.

78. Gutell, R.R., *et al.* (1992). Identifying constraints on the higher-order structure of RNA: continued development and application of comparative sequence analysis methods. *Nucleic Acids Res.*, **20**, 5785–5795.
79. Lindgreen, S., Gardner, P.P., and Krogh, A. (2006). Measuring co-variation in RNA alignments: physical realism improves information measures. *Bioinformatics*, **22**, 2988–2995.
80. Hofacker, I.L., Fekete, M., and Stadler, P.F. (2002). Secondary structure prediction for aligned RNA sequences. *J. Mol. Biol.*, **319**, 1059–1066.
81. Knudsen, B., and Hein, J. (2003). Pfold: RNA secondary structure prediction using stochastic context-free grammars. *Nucleic Acids Res.*, **31**, 3423–3428.
82. Ruan, J., Stormo, G.D., and Zhang, W. (2004). ILM: a web server for predicting RNA secondary structures with pseudoknots. *Nucleic Acids Res.*, **32**, W146–W149.
83. Ruan, J., Stormo, G.D., and Zhang, W. (2004). An iterated loop matching approach to the prediction of RNA secondary structures with pseudoknots. *Bioinformatics*, **20**, 58–66.
84. Havgaard, J.H., Lyngso, R.B., and Gorodkin, J. (2005). The FOLDALIGN web server for pairwise structural RNA alignment and mutual motif search. *Nucleic Acids Res.*, **33**, W650–W653.
85. Torarinsson, E., Havgaard, J.H., and Gorodkin, J. (2007). Multiple structural alignment and clustering of RNA sequences. *Bioinformatics*, **23**, 926–932.
86. Mathews, D.H., and Turner, D.H. (2002). Dynalign: an algorithm for finding the secondary structure common to two RNA sequences. *J. Mol. Biol.*, **317**, 191–203.
87. Perriquet, O., Touzet, H., and Dauchet, M. (2003). Finding the common structure shared by two homologous RNAs. *Bioinformatics*, **19**, 108–116.
88. Touzet, H., and Perriquet, O. (2004). CARNAC: folding families of related RNAs. *Nucleic Acids Res.*, **32**, W142–W145.
89. Hochsmann, M., Toller, T., Giegerich, R., and Kurtz, S. (2003). Local similarity in RNA secondary structures. *Proc. IEEE Comput. Soc. Bioinf. Conf.*, **2**, 159–168.
90. Siebert, S., and Backofen, R. (2005). MARNA: multiple alignment and consensus structure prediction of RNAs based on sequence structure comparisons. *Bioinformatics*, **21**, 3352–3359.

91. Duarte, C.M., Wadley, L.M., and Pyle, A.M. (2003). RNA structure comparison, motif search and discovery using a reduced representation of RNA conformational space. *Nucleic Acids Res.*, **31**, 4755–4761.
92. Duarte, C.M., and Pyle, A.M. (1998). Stepping through an RNA structure: A novel approach to conformational analysis. *J. Mol. Biol.*, **284**, 1465–1478.
93. Harrison, A., *et al.* (2003). Recognizing the fold of a protein structure. *Bioinformatics*, **19**, 1748–1759.
94. Dror, O., Nussinov, R., and Wolfson, H. (2005). ARTS: alignment of RNA tertiary structures. *Bioinformatics*, **21**(Suppl 2), ii47–ii53.
95. Ferre, F., Ponty, Y., Lorenz, W.A., and Clote, P. (2007). DIAL: a web server for the pairwise alignment of two RNA three-dimensional structures using nucleotide, dihedral angle and base-pairing similarities. *Nucleic Acids Res.*, **35**, W659–W668.
96. Capriotti, E., and Marti-Renom, M.A. (2008). RNA structure alignment by a unit-vector approach. *Bioinformatics*, **24**, i112–118.
97. Capriotti, E., and Marti-Renom, M.A. (2009). SARA: a server for function annotation of RNA structures. *Nucleic Acids Res.*, **37**, W260–W265.
98. Kemena, C., Bussotti, G., Capriotti, E., Marti-Renom, M.A., and Notredame, C. (2013). Using tertiary structure for the computation of highly accurate multiple RNA alignments with the SARA-Coffee package. *Bioinformatics*, **29**(9), 1112–1119.
99. Chang, Y.F., Huang, Y.L., and Lu, C.L. (2008). SARSA: a web tool for structural alignment of RNA using a structural alphabet. *Nucleic Acids Res.*, **36**, W19–W24.
100. Sarver, M., *et al.* (2008). FR3D: finding local and composite recurrent structural motifs in RNA 3D structures. *J. Math. Biol.*, **56**, 215–252.
101. Petrov, A.I., Zirbel, C.L., and Leontis, N.B. (2011). WebFR3D—a server for finding, aligning and analyzing recurrent RNA 3D motifs. *Nucleic Acids Res.*, **39**, W50–W55.
102. Lai, C.E., *et al.* (2009). FASTR3D: a fast and accurate search tool for similar RNA 3D structures. *Nucleic Acids Res.*, **37**, W287–W295.
103. Zhong, C., Tang, H., and Zhang, S. (2010). RNAMotifScan: automatic identification of RNA structural motifs using secondary structural alignment. *Nucleic Acids Res.*, **38**, e176.
104. Kirillova, S., Tosatto, S.C., and Carugo, O. (2010). FRASS: the web-server for RNA structural comparison. *BMC Bioinf.*, **11**, 327.

105. Rahrig, R.R., Leontis, N.B., and Zirbel, C.L. (2010). R3D Align: global pairwise alignment of RNA 3D structures using local superpositions. *Bioinformatics*, **26**, 2689–2697.

106. Shapiro, B.A., Yingling, Y.G., Kasprzak, W., and Bindewald, E. (2007). Bridging the gap in RNA structure prediction. *Curr. Opin. Struct. Biol.*, **17**, 157–165.

107. Massire, C., and Westhof, E. (1998). MANIP: an interactive tool for modelling RNA. *J. Mol. Graph Model*, **16**, 197–205, 255–197.

108. Major, F. (2003). Building three-dimensional ribonucleic acid structures. *Comput. Sci. Eng.*, **5**, 44–53.

109. Martinez, H.M., Maizel, J.V., Jr., and Shapiro, B.A. (2008). RNA2D3D: a program for generating, viewing, and comparing 3-dimensional models of RNA. *J. Biomol. Struct. Dyn.*, **25**(6), 669–683.

110. Sharma, S., Ding, F., and Dokholyan, N.V. (2008). iFoldRNA: three-dimensional RNA structure prediction and folding. *Bioinformatics*, **24**, 1951–1952.

111. Frellsen, J., *et al.* (2009). A probabilistic model of RNA conformational space. *PLoS Comput. Biol.*, **5**, e1000406.

112. Das, R., and Baker, D. (2007). Automated de novo prediction of native-like RNA tertiary structures. *Proc. Natl. Acad. Sci. U S A*, **104**, 14664–14669.

113. Das, R., Karanicolas, J., and Baker, D. (2010). Atomic accuracy in predicting and designing noncanonical RNA structure. *Nat. Methods*, **7**, 291–294.

114. Bradley, P., *et al.* (2005). Free modeling with Rosetta in CASP6. *Proteins*, **61**(Suppl 7), 128–134.

115. Poolsap, U., Kato, Y., and Akutsu, T. (2009). Prediction of RNA secondary structure with pseudoknots using integer programming. *BMC Bioinf.*, **10**(Suppl 1), S38.

116. Sato, K., Kato, Y., Hamada, M., Akutsu, T., and Asai, K. (2011). IPknot: fast and accurate prediction of RNA secondary structures with pseudoknots using integer programming. *Bioinformatics*, **27**, i85–93.

117. Flores, S.C., Wan, Y., Russell, R., and Altman, R.B. (2010). Predicting RNA structure by multiple template homology modeling. *Pac. Symp. Biocomput.*, 216–227.

118. Jossinet, F., Ludwig, T.E., and Westhof, E. (2010). Assemble: an interactive graphical tool to analyze and build RNA architectures at the 2D and 3D levels. *Bioinformatics*, **26**, 2057–2059.

119. Rother, M., Rother, K., Puton, T., and Bujnicki, J.M. (2011). ModeRNA: a tool for comparative modeling of RNA 3D structure. *Nucleic Acids Res.*, **39**, 4007–4022.

120. Cao, S., and Chen, S.J. (2011). Physics-based de novo prediction of RNA 3D structures. *Journal of Physical Chemistry B*, **115**, 4216–4226.

121. Izzo, J.A., Kim, N., Elmetwaly, S., and Schlick, T. (2011). RAG: an update to the RNA-As-Graphs resource. *BMC Bioinf.*, **12**, 219.

122. Kim, N., *et al.* (2007). RagPools: RNA-As-Graph-Pools–a web server for assisting the design of structured RNA pools for in vitro selection. *Bioinformatics*, **23**, 2959–2960.

123. Bida, J.P., and Maher, L.J., 3rd (2012). Improved prediction of RNA tertiary structure with insights into native state dynamics. *RNA*, **18**, 385–393.

124. Capriotti, E., Norambuena, T., Marti-Renom, M.A., and Melo, F. (2011). All-atom knowledge-based potential for RNA structure prediction and assessment. *Bioinformatics*, **27**, 1086–1093.

125. Case, D.A., *et al.* (2005). The Amber biomolecular simulation programs. *J. Comput. Chem.*, **26**, 1668–1688.

126. Bernauer, J., Huang, X., Sim, A.Y., and Levitt, M. (2011). Fully differentiable coarse-grained and all-atom knowledge-based potentials for RNA structure evaluation. *RNA*, **17**, 1066–1075.

127. Capriotti, E., and Marti-Renom, M.A. (2008). Computational RNA structure prediction. *Curr. Bioinf.*, **3**, 32–45.

Chapter 4

Computational Modeling for RNA Structure Discovery and Characterization

Brian J. Parker and Thomas Preiss
Genome Biology Department, The John Curtin School of Medical Research, The Australian National University, Acton (Canberra), ACT 0200, Australia
brian.parker@anu.edu.au, thomas.preiss@anu.edu.au

4.1 Introduction

Ribonucleic acid (RNA) structures are key building blocks in both natural molecular-scale mechanisms and in artificial nanotechnologies [3, 24, 42], and thus computational RNA structural modeling can be an important tool in both searching for natural examples of RNA mechanisms as well as in deducing the putative structure of experimentally determined and designed structured RNA sequences, as a first step in fully characterizing their function. In this chapter we discuss applying computational and probabilistic modeling of RNA structures, with a particular focus on applying these models to RNA structure characterization, comparison, search and discovery.

The RNA structure can be described at three levels (see Fig. 4.1): (i) primary sequence of the A, G, C, and U bases (with uracil, U,

RNA Nanotechnology
Edited by Bin Wang

ISBN 978-981-4411-64-6 (Hardcover), 978-981-4411-65-3 (eBook)
www.panstanford.com

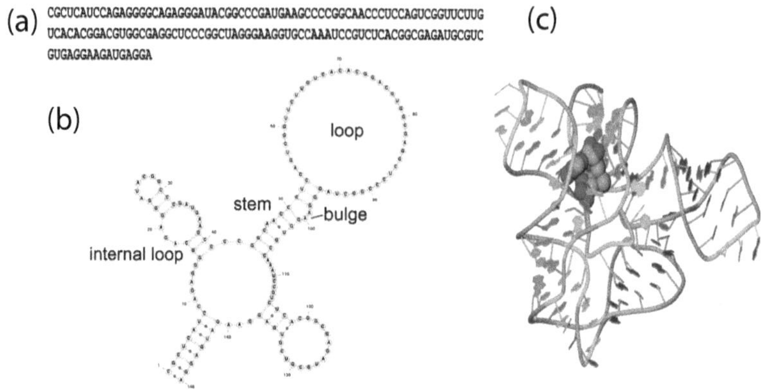

Figure 4.1 RNA structure description levels. Example of a SAM riboswitch (a) primary sequence; (b) secondary structure (secondary structure subtypes shown: stem, loop, bulge, internal loop); and (c) tertiary structure of a SAM riboswitch with bound metabolite (source: wikimedia commons). *Abbreviation*: SAM, *S*-adenosyl methionine.

replacing thymine, T, of the template deoxyribonucleic acid (DNA)) (Fig. 4.1a); (ii) secondary structure (Fig. 4.1b), which schematically describes the nested interactions of bases; and (iii) tertiary structure (Fig. 4.1c), which describes the precise three-dimensional (3D) positioning of RNA segments; (see Bevilacqua and Blose [2] for details).

Computational RNA structure prediction research has primarily focused on the secondary structure, and this is the focus in this chapter. The secondary structure is the level that often provides the key mechanistic description for the function of an RNA structure (see [46] for examples) and provides the main structural features to enable probabilistic models to search and compare RNA structural motifs. Tertiary structure determination can be essential for precise analysis of the function of a structural RNA, and algorithms exist for computational estimation of the tertiary structure, either de novo (see Shapiro et al. [44]) or in combination with experimental data, for example, low-resolution small-angle X-ray scattering (SAXS) data [14, 34, 58]; however, a detailed description of these algorithms is outside of the scope of this chapter.

RNA has different base-pairing rules to DNA, and in particular non-Watson–Crick G/U wobble pairs bind almost as stably as the canonical Watson–Crick G/C and A/U pairs. This can lead to strand asymmetries in structure stability, and thus predicted RNA structures are typically specific to a particular transcribed strand. In terms of binding energy, G/C is the most stable base pair, followed by A/U, and with G/U wobble pairs being the least stable. Note that other noncanonical base pairing can occur in particular structures and can affect the tertiary structure [57], but this is typically ignored in a first-order approximation of the secondary structure.

4.1.1 *RNA Structure Prediction*

There are two broad approaches to predicting and modeling RNA structures, (i) free-energy minimization approaches and (ii) comparative genomics approaches utilizing the covariance signal of compensatory mutations between corresponding positions in homologous examples of an RNA structure.

4.1.1.1 Energy-based folding of a single sequence

Energy-based methods attempt to find the most stable, minimum free-energy (MFE) secondary structure or structures. This approach can either find a single, predicted, most stable structure or return a range of predicted near-optimum stable structures. The initial efficient dynamic programming solutions to find the most energetically stable structure were given by Zuker and Stiegler [62], which determined a single optimum structure. The later Mfold program by Zuker [61] returns the energetically most favorable as well as other near-optimal solutions within a defined energy.

A dynamic programming approach (see Fig. 4.2 and [13]) is used to efficiently find the most energetically stable base pairing by recursively extending optimal partial solutions with the next most stable pairing (Fig. 4.2a). By saving an array of estimated free energy of all smaller partial solutions (i.e., a dynamic programming algorithm), this can be computed efficiently (Fig. 4.2b). The change in energy of extending a partial solution is based on estimates of stacking energies between adjacent base pairs in double-stranded

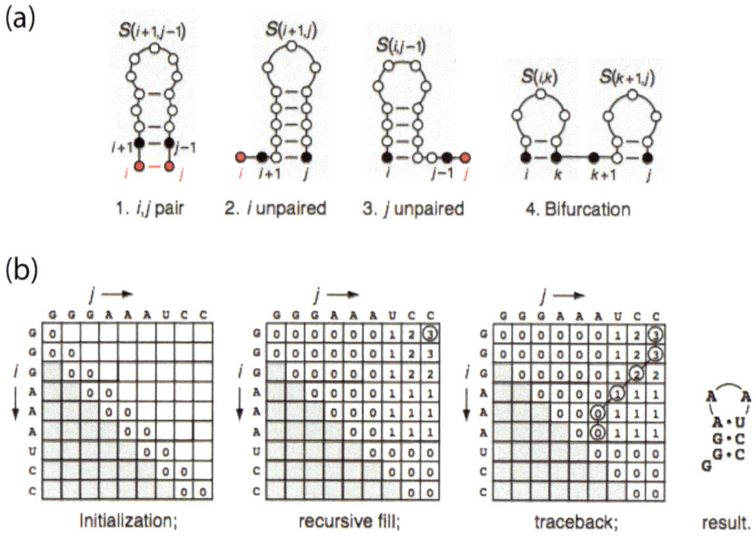

Figure 4.2 Dynamic programming algorithm for efficiently determining the optimal structure. In this simplified example from [13], a structure with the maximum number of base pairs is computed, but the same approach is used to compute a minimum free-energy solution. (a) Assuming that optimal structures have already been computed for all smaller subsequences, the structure with optimal number of base pairs for a subsequence from base i to j can be computed recursively as the best of (i) adding a base pair (red) to the optimal subsequence from $i+1$ to $j-1$, (ii) adding an unpaired base (red) to the optimal subsequence from $i+1$ to j, (iii) adding an unpaired base (red) to the optimal subsequence from i to $j-1$, or (iv) forming the structure for i to j from two smaller optimal substructures from i to k and k to j. (b) A dynamic programming algorithm can execute this recursive scheme efficiently by computing all partial results for all i to j subsequences and saving them in an array indexed by i (row) and j (column). This is done recursively by (i) starting with the trivial single base cases on the diagonal; (ii) recursively extending the table for increasing i and j by choosing the best combination of smaller subsequences (which have been previously computed), as described in (a); and (iii) when the table is complete, thus storing the final optimal number of base pairs in array position $(1, N)$, where N is the length of the sequence. The optimum structure can then be found by tracing back through the array over the steps used to compute this optimum value (figure from [13], with permission).

regions and destabilizing effects of single-stranded regions, based on previous experimental melting studies [48]. Note that the strength of base pairing is affected by neighboring bases, and so such dinucleotide base stacking needs to be incorporated into accurate free-energy models.

A connection between the free energy of a potential structure and its probability of occurrence is given by the Boltzmann distribution: the probability of forming a substructure with free energy ΔG is given as $\exp(-\Delta G/kT)$ divided by the partition function Q, where Q is the sum of probabilities of all structures S: $Q = \sum_S \exp(-\Delta G_S/kT)$. A dynamic programming approach is used to compute the partition function efficiently: the RNAFold program of the Vienna RNA package [26] uses this Boltzmann/partition method to predict the most probable and stable suboptimal structures and also returns base-pairing probabilities, which are useful in visualizing and analyzing the structures.

An advantage of these free energy–based methods is that they can be applied to a single example RNA sequence, as opposed to the solutions utilizing comparative genomic information (discussed in the sequel), which require multiple examples of the RNA structure. By contrast, however, such predictions based on a single RNA sequence, and particularly a single optimal structure prediction, can be highly unstable and not representative of the true in vivo structure. Such methods are most accurate for short sequences and become less reliable with longer sequences. Thus, producing an estimate of several near-optimal structures can be more meaningfully interpreted and can be especially useful when modeling, for example, RNA switch mechanisms (see below). In the limit, the approach of Ding [10] is to return a sampling of structures weighted by the entire Boltzmann distribution of energies (as implemented in his Sfold program or Lfold of the Vienna toolkit). The RNAshapes approach of Steffen et al. [45] similarly returns a set of possible near-optimal structures sampled from the Boltzmann distribution but with structures limited to representatives of alternative distinct structure classes. The centroid of this Boltzmann ensemble of structures has been proposed as a more robust estimate of the RNA structure [5, 43], and features based on the Boltzmann ensemble of structures have been suggested to be useful to distinguish classes of structured and nonstructured RNAs [6, 31]. These energy-based methods can also be extended to model general RNA–RNA interactions [32].

4.1.1.2 Comparative genomics approaches based on covariation of mutations across multiple sequences

Comparative genomics approaches utilize the covariance signal of compensatory mutations between corresponding paired positions in homologous examples of an RNA structure (see Fig. 4.3). This signal can be captured by (generative) probabilistic models, with the basic nested structure of RNA folding captured by, for example, stochastic context-free grammar (SCFG) models.

SCFG models use nested grammar production rules, with probabilistic weighting of the transitions between states of the model, to give a probability distribution over (possibly nested) RNA structures; the model is fitted by utilizing the covarying compensatory mutations across multiple sequence alignments (msas) of homologous examples of the RNA sequence across (i.e., orthologs) or within (i.e., paralogs) species. Note that SCFGs cannot directly model non-nested structures such as pseudoknots. Such general probabilistic models can be used for de novo genome-wide structure prediction.

In particular, a conserved secondary structure would be expected to show correlated variation with an increased incidence of compensatory double substitutions (i.e., both bases of a pair mutate near simultaneously in evolutionary time so as to maintain correct base pairing, for example, a C/G pair may mutate to a U/A pair) or compatible single substitutions (e.g., C/G mutating to U/G, which still maintains a sequence compatible with the structure as the U/G wobble pair is energetically stable (Fig. 4.3a)).

The general parameters of the model are trained against known annotated structures, for example, RFAM data. The optimal matching structure for this SCFG class of models against input data can then be found using the CYK algorithm [12], which is a dynamic programming solution closely related to the dynamic programming solution of minimum energy discussed above. Note that this

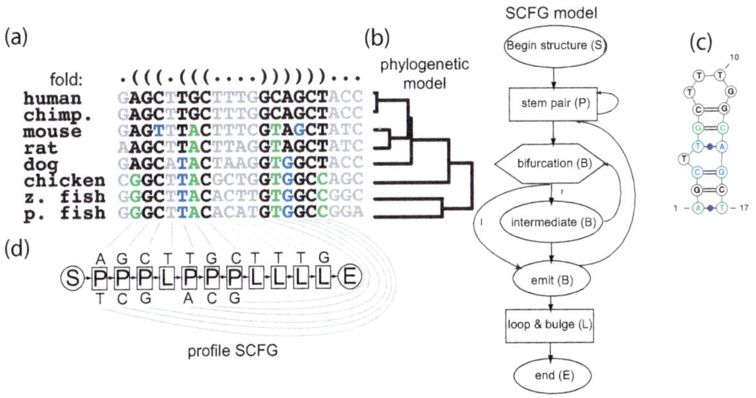

Figure 4.3 Covariance of mutations maintaining the RNA structure across evolutionary time can be captured by phylo-SCFG models. (a) Input is an msa of orthologous regions across species with a known phylogenetic relationship. Compensatory double mutations (green) and compatible single substitutions (blue) maintain the functional structure across evolutionary time. (b) A phylo-SCFG model can detect and be fitted to arbitrarily nested RNA structures, with the best-fitting structure estimated using the CYK, or related, dynamic programming algorithm. Such general models have distinct states representing paired (P) and unpaired (L) regions, with additional states to process bifurcations (B). The probability of transition between these states (arrows), and the emission probabilities for paired or unpaired regions, are determined from known examples of aligned RNA structures (EvoFold-structured SCFG model shown [37]). (c) The output of this initial structured RNA search and fitting stage is the detected structured region(s) with an estimated optimal secondary structure. (d) A profile SCFG model (aka covariance model) can then be trained to model an entire specific family of (orthologous or paralogous) examples of structural RNAs with similar structure and sequence. In contrast to the general SCFG of (b), such profile models have multiple states each representing a particular matching base pair (P) or unpaired base (L and R) in the sequence alignment for the given predicted optimal RNA structure (as well as bifurcation (B) states—not shown). The matching bases of the human sequence are shown above and below (for paired states) each model state; gray lines show correspondence of states and alignment columns. Note that each such state itself has additional internal states (not shown) that are used to handle sequence mismatches or indels (see the Infernal user guide Nawrocki et al. [33] for full details).

algorithm is relatively slow for long sequences, with time complexity increasing as the third power of the sequence length.

If the msa is across multiple species with a known (or estimated) phylogenetic relationship, then a phylogenetic model can be combined with the SCFG model to give a more accurate model of expected mutations, giving a phylo-SCFG model (Fig. 4.3b). The phylogenetic model can estimate the probabilities of different substitution patterns in paired and unpaired regions, including an increase in the overall mutation rate outside of stem regions, as well as compensatory and compatible mutations. Approaches using this model include Pfold [21] and EvoFold [37]. EvoFold uses a phylo-SCFG formulation and a sliding-window approach to detect and predict structures genome-wide: a comparison of the likelihoods of such a model of a structured RNA region compared with an alternative unstructured model can be used to detect and model structured regions. Strand asymmetries due to G/U pairings can be used to give a partial strand estimation.

The energy-based methods discussed previously can also be extended to incorporate comparative genomic compensatory mutation information if suitable alignments across species are available: an example is RNAz [50].

These models start with an msa generated without using structural information, which is a suitable approach for conserved structures that do not themselves evolve substantially over time. Another approach is to start with unaligned sequences and combine the structure modeling with the multiple alignment stage. This approach would be expected to be more sensitive to structures that have poor sequence conservation and are not stable across evolutionary time: examples include RNApromo [38] and CMFinder [59]; CMFinder is based on an SCFG model. These approaches essentially find the maximum likelihood structural motifs similarly to sequence motif finders such as Meme [1].

Several other issues may need to be considered when modeling the RNA structure in some cases. Comparative genomics methods predicting RNA structures in coding regions need to account for and incorporate the codon biases into the model: see [36] and [25] for examples. In vivo RNA folding may differ from a single predicted stable structure, due to effects such as cotranscriptional folding, which can potentially limit the possible RNA structures that can fold; Transat [54] attempts to detect alternative potential functional folds using a comparative genomics approach.

The WAR server [47] is a web service that can apply multiple structure prediction methods, including several of the methods discussed above, to uploaded RNA sequence examples to compare the robustness of the independent predictions across methods and to produce a consensus structure prediction.

4.1.1.3 RNA structure-specific profile SCFG models for model comparison and database search

Once a structured RNA region is detected de novo and its secondary structure estimated using the methods described above, and if alignments of other homologous examples of the structure are available, then a more specialized probabilistic model can be built that captures the structure and sequence variation of the particular RNA structural family.

A common probabilistic model used for this purpose is a *profile* SCFG, also known as a *covariance model*. This model is a specialization of the general SCFG model discussed in the previous section and is the SCFG equivalent of a profile hidden Markov model (pHMM) often used for modeling linear (often protein or DNA) sequence motifs, extended to also be suitable for modeling nested structures [12].

SCFG models can model a particular family of RNA structures, incorporating both sequence and structural identity. This model can be used (i) as a basis for an intermodel distance measure to cluster structures into paralogous families, for example, EvoFam [35], (ii) for homology search against a database of similar structures (see RFAM below), and (iii) for search against a related genome to detect similar paralogous or orthologous RNA family examples.

A profile SCFG differs from the general SCFG model for arbitrary structures discussed in the previous section in that it has states representing each sequence position and so is specific to a given structural family defined by an msa (Fig. 4.3c).

A profile SCFG has a tree-like graph of nodes, based on a parse tree for the nested RNA structure (Fig. 4.3d). Similarly to a pHMM, used to model linear sequences, each node has three states of *match*, *insert*, and *delete* for each column in the msa, allowing mismatches and indels to be modeled.

The Infernal toolkit [33] can be used to train and search with profile SCFG/covariance models.

Similarity measures between RNA structural models Once such profile models have been trained, distance measures between the probability distribution over RNA sequences induced by the models can be defined. For example, a modified form of Kullback–Leibler divergence between the probability distributions given by the models can be used to identify families of homologous structures [35], such as a family of hairpins discovered in the 3′UTR of *MAT2A* (Fig. 4.5a).

4.1.2 *RFAM: A Database of Probabilistic Models of Structural RNA Families*

The RFAM database [15] is a database of alignments and profile SCFG models built to model numerous families of RNA structures based on deep sequence alignments, including both structured cis-regulatory elements and structured noncoding RNAs (ncRNAs). This is a valuable resource for predicting the function of a newly detected structure, as the similarity in a search with all known RFAM structures may reveal a strongly homologous match. This can be used to help annotate genome-wide structural RNA predictions in newly sequenced genomes, or metagenomic samples.

4.1.3 *Visualizing the RNA Secondary Structure*

The secondary RNA structure can be visualized in several ways (see Fig. 4.4). The most basic approach is the dot–bracket notation (Fig. 4.4a), which gives a simple linear representation showing nested stems through matching parentheses. Different forms of a nested structure are sometimes distinguished by different bracket types. An energy dot plot, or base-pairing probability dot plot, for the Boltzmann/partition method (Fig. 4.4b), has the RNA sequence listed along the *x*-axis and *y*-axis, with each base pair position at the intersection of a row and a column having intensity proportional

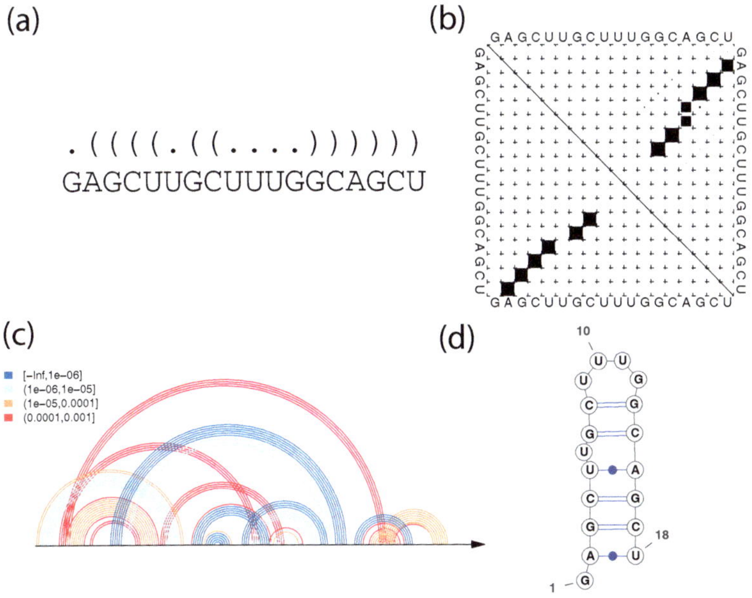

Figure 4.4 Visualization of the RNA structure. (a) Dot–bracket notation; (b) base-pairing probability dot plot; (c) arc plot, for a structure with a non-nested structure (colors indicate evidence of base pairing; drawn with R-chie); (d) secondary structure diagram (drawn with Varna).

to its free-energy decrease or base-pairing probability, respectively (Lorenz et al. [26]). Various forms of arc plots (including circle plots) (Fig. 4.4c) can highlight non-nested structures (see [22]). A general secondary structure diagram (Fig. 4.4d) can show complex structures such as riboswitches more perspicuously. VARNA [9] is a flexible tool for visualizing the RNA secondary structure with such diagrams.

4.1.4 *Experimental Data for Computationally Predicted RNA Secondary Structure*

Once a structural RNA has been identified and its structure computationally predicted, further validation is needed to confirm the structure is biologically active. Even state-of-the-art computational predictions of the RNA structure may not be fully accurate, and

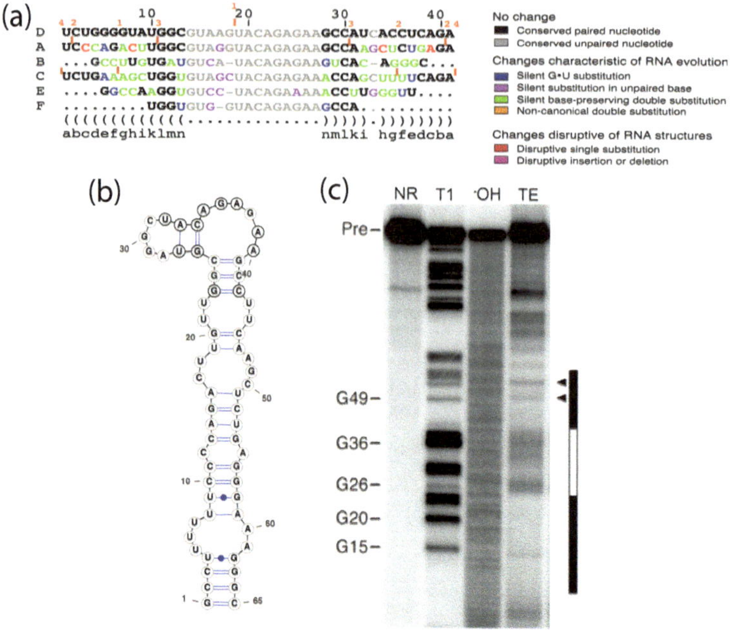

Figure 4.5 Applied example of a family of conserved hairpins in the human gene *MAT2A* (Parker et al. [35]) found by SCFG modeling, demonstrating (a) an msa across the six paralogous family members, with substitution evidence for the conserved RNA structure across hairpins highlighted in green for compensatory double substitutions and blue for compatible single substitutions; (b) computationally predicted RNA structure of hairpin A using a phylo-SCFG model; and (c) experimental confirmation of the RNA structure of hairpin A by inline probing analysis of a MAT2A construct. RNA cleavage products resulting from TE were resolved by denaturing 10% PAGE. (Other lanes: NR, no reaction; T1, partial digest with ribonuclease T1 (RNase T1); ($^{-}$OH) partial alkaline digest; Pre, precursor RNA). Selected bands in the T1 lane are labeled with the positions of the respective 39-terminal guanosyl residues according to the numbering used for hairpin A in panel (b). Filled bars correspond to the positions within hairpin A that are predicted to be largely base paired, while the open bar corresponds to positions within the putative loop sequence. Arrowheads indicate putative bulged nucleotides C50 and A55. *Abbreviation*: TE, transesterification.

in particular, predictions on long, single RNA sequences may differ markedly from the true in vivo structure. If additional independent sequence examples are available, in addition to those used to train the models used for prediction, then they can be used as an independent test set to show additional confirmatory evidence of compensatory mutations (e.g., see [35] where a Monte Carlo *p*-value estimate (*evoP*) is used for additional confirmation).

The gold-standard validation method is experimental, as discussed in detail in subsequent chapters. Probing assays can distinguish single- and double-stranded RNA regions. Inline probing uses the relative stability of double-stranded compared with single-stranded RNA regions to reveal loops and bulges of the RNA structure and has been used to identify RNA conformational changes following ligand binding of riboswitches [40, 51]. Figure 4.5c shows an inline probing result compared with the structural prediction (Fig. 4.5b) of a member of a well-conserved family of 3′-UTR hairpins (aka *stem loops*) in the key methylation gene *MAT2A* (Fig. 4.5a). Other methods for confirming the functionality of a putative RNA structure include mutational analyses to disrupt the putative structure [19].

Recently, assays based on similar-RNA-probing methods have been applied genome-wide [29] in yeast [20], mouse [49], and fly and worm [23]. Such structural assays can be used to detect structured double-stranded RNA regions genome-wide and combined with computational models to increase specificity of structural RNA detection and structure prediction. Research on combining forms of probing data and computational secondary structure prediction is ongoing, for example, the selective 2′-hydroxyl acylation analyzed by primer extension (SHAPE) assay [27].

4.1.5 *Examples of Computational Analysis of Natural RNA Mechanisms*

Natural RNA structures occur in both cis-regulatory elements [16] in protein-coding genes, such as riboswitches and RNA switches, and in structured noncoding RNAs, for example, transfer RNAs (tRNAs), to effect their respective functions. In the following section, we give examples of the application of the above computational and statistical methods in identifying and analyzing natural examples of structural RNA mechanisms (see Fig. 4.6).

4.1.5.1 RNA structures in autoregulatory control of gene expression

One area of nanotechnology is research into approaches to control gene expression [8, 17]. A repeating theme in naturally evolved gene regulation of endonucleases that cleave specific RNA structures is autoregulation by cleavage of a target structure in the gene transcript itself.

Validated examples include a microRNA precursor stem loop-like structure in the messenger RNA (mRNA) of the gene *DGCR8*, which is involved in processing of precursor microRNA stem loops. This was discovered using a phylo-SCFG model (Pedersen et al. [37]) (Fig. 4.6a) and was shown to be cleaved by the associated Drosha component of the microRNA precursor hairpin-processing machinery [18]. Another suggestive example, though not experimentally validated, may be a tRNA-like structure in the transcript of *POP1*, discovered by a profile SCFG–based structural RNA family detection screen [35]: POP1 is a component of RNase P, which cleaves tRNAs, and a possible role for the structure could be such feedback regulation. Another example is bacterial RNase E, which involves feedback regulation by cleavage of a hairpin in its transcript [11].

Interestingly, a nanotechnology with a similar approach of utilizing natural RNases for gene knockdown had been developed by Altman et al. [60] and others: an engineered external guide sequence (EGS) RNA is transfected to induce a tRNA-like structure on a target transcript, which is then cleaved as a target of RNase P, effecting gene knockdown (Fig. 4.6b).

4.1.5.2 RNA switch controlling miRNA action in human *p27*

RNA can form bistable switch structures that can be designed to control various biological processes and are a current focus in nanotechnology research [7]. Natural examples can be found in all kingdoms of life, for example, riboswitches in bacteria or RNA switch control of *VEGFA* expression in humans [39]. One interesting example is the interaction with RNA-binding proteins (RBPs) to control microRNA targeting of mRNAs. It has been noted that adjacent binding of the RBP Pum2 next to a microRNA site in the 3′UTR of *p27* [19] acts as an RNA switch to obstruct access to the microRNA site, leading to coordinate regulation between the RBP and microRNA. In [19], a computational energy-based method

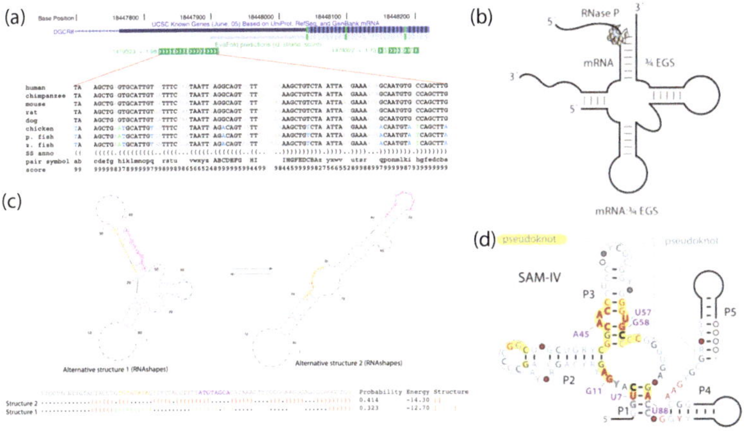

Figure 4.6 Selected examples of RNA regulatory mechanisms detected and analyzed by the computational methods discussed in this chapter. (a) A microRNA precursor-like stem loop detected in the microRNA-processing gene *DGCR8* autoregulates *DGCR8* expression through cleavage by the microRNA-processing machinery (Pedersen et al. [37]). A UCSC browser screenshot showing 3′UTR of *DGCR8* with detected hairpin location in green; the msa across species shows compensatory double (green) and compatible (blue) mutations (figure from [37]). (b) An EGS RNA can be used to induce a tRNA-like structure in a transcript [60], causing RNase P cleavage of the transcript. A tRNA-like structure in the RNase P gene *POP1* (Parker et al. [35]) may be similarly processed (figure from [28]). (c) An RNA switch in the *p27* transcript RNA is activated by the RBP PUM2 binding to make an adjacent microRNA target accessible. The most stable RNA structure shows the microRNA target (purple) inaccessible in a stem region, adjacent to a PUM2-binding site (orange). The top two most energetically stable configurations have high probability, suggesting a two-state RNA switch mechanism. (d) SAM IV bacterial riboswitch—a novel, naturally evolved motif that binds the metabolite SAM, detected using an SCFG model. Purple labels are bases thought to directly contact metabolite (figure from [53]).

(Mfold) was used to determine the likely RNA structure before experimental validation by mutation.

RNA structure prediction methods that predict near-optimal structures are potentially useful in detecting and modeling such switch mechanisms, for example, Fig. 4.6c shows the top predictions of the Boltzmann distribution–based RNAShapes method, which suggests that as well as the published structure (1), it appears to act as a two-state switch with a second stable RNA structure position (2) (unpublished observation—J. Wen, 2011).

4.1.5.3 Riboswitches

Riboswitches have aptamers that directly bind metabolites [41]. Aptamer design is an important issue in nanotechnological applications of RNA [30, 55], and thus natural examples are of interest. As in Systematic Evolution of Ligands by Exponential Enrichment (SELEX) [4], the aptamers in the riboswitches have been evolutionarily optimized to bind a specific metabolite [56]. Large-scale searches of metagenomic samples using SCFG models have revealed entirely new aptamers that bind SAM (Fig. 4.6d), providing additional insights into the design of such aptamers [52, 53].

4.2 Conclusion

Modeling and prediction of the RNA structure, and in particular the secondary structure, are well-studied problems in computational biology and can be a valuable tool in discovering or characterizing both natural and artificially designed examples of RNA structures and mechanisms. In this chapter we have given an overview of the main computational approaches used for studying the RNA structure and have highlighted a selected set of computational tools implementing these methods, including: Vienna RNA package, PFold, EvoFold/EvoFam, RNAz, Transat, RNAShapes, WAR server, Infernal, RFAM, and VARNA. We have given examples of their use in identifying and analyzing naturally evolved examples of RNA-based mechanisms that are of interest to practitioners in the field of nanotechnology, and the mathematical and computational principles discussed will be similarly applicable to the design of artificial constructs as well.

Acknowledgments

TP is supported by the John Curtin School of Medical Research, as well as grants from the National Health & Medical Research Council and the Australian Research Council. BJP is supported by the John Curtin School of Medical Research, and the reported research was partly funded by a Statistics Network Fellowship, Department of Mathematical Sciences, University of Copenhagen.

References

1. Bailey, T. L. and Elkan, C. (1994). Fitting a mixture model by expectation maximization to discover motifs in biopolymers. *Proceedings of the International Conference on Intelligent Systems for Molecular Biology*, **2**, 28–36.
2. Bevilacqua, P. C. and Blose, J. M. (2008). Structures, kinetics, thermodynamics, and biological functions of RNA hairpins, *Annual Review of Physical Chemistry*, **59**(1), 79–103.
3. Bhatia, D., Sharma, S. and Krishnan, Y. (2011). Synthetic, biofunctional nucleic acid-based molecular devices. *Current Opinion in Biotechnology*, **22**(4), 475–484.
4. Burke, D. H. and Gold, L. (1997). RNA aptamers to the adenosine moiety of S-adenosyl methionine: structural inferences from variations on a theme and the reproducibility of SELEX. *Nucleic Acids Research*, **25**(10), 2020–2024.
5. Carvalho, L. E. and Lawrence, C. E. (2008). Centroid estimation in discrete high-dimensional spaces with applications in biology. *Proceedings of the National Academy of Sciences*, **105**(9), 3209–3214.
6. Chan, C. Y. and Ding, Y. (2007). Boltzmann ensemble features of RNA secondary structures: a comparative analysis of biological RNA sequences and random shuffles, *Journal of Mathematical Biology*, **56**(1-2), 93–105.

7. Chang, A. L., Wolf, J. J. and Smolke, C. D. (2012). Synthetic RNA switches as a tool for temporal and spatial control over gene expression. *Current Opinion in Biotechnology*, **23**(5), 679–688.
8. Chin, J. W. (2006). Programming and engineering biological networks. *Current Opinion in Structural Biology*, **16**(4), 551–556.
9. Darty, K., Denise, A. and Ponty, Y. (2009). VARNA: Interactive drawing and editing of the RNA secondary structure. *Bioinformatics*, **25**(15), 1974–1975.
10. Ding, Y. (2003). A statistical sampling algorithm for RNA secondary structure prediction, *Nucleic Acids Research*, **31**(24), 7280–7301.
11. Diwa, A., Bricker, A. L., Jain, C. and Belasco, J. G. (2000). An evolutionarily conserved RNA stem-loop functions as a sensor that directs feedback regulation of RNase E gene expression. *Genes & Development*, **14**(10), 1249–1260.
12. Durbin, R., Eddy, S. R., Krogh, A. and Mitchison, G. (1998). *Biological Sequence Analysis: Probabilistic Models of Proteins and Nucleic Acids* (Cambridge University Press).
13. Eddy, S. R. (2004). How do RNA folding algorithms work? *Nature Biotechnology*, **22**(11), 1457–1458.
14. Flores, S. C. and Altman, R. B. (2010). Turning limited experimental information into 3D models of RNA. *RNA*, **16**(9), 1769–1778.
15. Gardner, P. P., Daub, J., Tate, J., Moore, B. L., Osuch, I. H., Griffiths-Jones, S., Finn, R. D., Nawrocki, E. P., Kolbe, D. L., Eddy, S. R. and Bateman, A. (2011). Rfam: Wikipedia, clans and the "decimal" release. *Nucleic Acids Research*, **39**, Database issue, D141–5.
16. Gebauer, F., Preiss, T. and Hentze, M. W. (2012). From cis-regulatory elements to complex RNPs and back. *Cold Spring Harbor Perspectives in Biology* **4**, 7.
17. Greber, D. and Fussenegger, M. (2007). Mammalian synthetic biology: engineering of sophisticated gene networks. *Journal of Biotechnology*, **130**(4), 329–345.
18. Han, J., Pedersen, J. S., Kwon, S. C., Belair, C. D., Kim, Y.-K., Yeom, K.-H., Yang, W.-Y., Haussler, D., Blelloch, R. and Kim, V. N. (2009). Posttranscriptional crossregulation between Drosha and DGCR8. *Cell*, **136**(1), 75–84.
19. Kedde, M., van Kouwenhove, M., Zwart, W., Vrielink, J. A. F. O., Elkon, R. and Agami, R. (2010). A Pumilio-induced RNA structure switch in p27-3′ UTR controls miR-221 and miR-222 accessibility. *Nature*, **12**(10), 1014–1020.

20. Kertesz, M., Wan, Y., Mazor, E., Rinn, J. L., Nutter, R. C., Chang, H. Y. and Segal, E. (2010). Genome-wide measurement of RNA secondary structure in yeast. *Nature*, **467**(7311), 103–107.
21. Knudsen, B. (2003). Pfold: RNA secondary structure prediction using stochastic context-free grammars. *Nucleic Acids Research*, **31**(13), 3423–3428.
22. Lai, D., Proctor, J. R., Zhu, J. Y. A. and Meyer, I. M. (2012). R-CHIE: a web server and R package for visualizing RNA secondary structures. *Nucleic Acids Research*, **40**(12), e95.
23. Li, F., Zheng, Q., Ryvkin, P., Dragomir, I., Desai, Y., Aiyer, S., Valladares, O., Yang, J., Bambina, S., Sabin, L. R., Murray, J. I., Lamitina, T., Raj, A., Cherry, S., Wang, L.-S. and Gregory, B. D. (2012). Global analysis of RNA secondary structure in two metazoans. *Cell Reports*, **1**(1), 69–82.
24. Liang, J. C., Bloom, R. J. and Smolke, C. D. (2011). Engineering biological systems with synthetic RNA molecules. *Molecular Cell*, **43**(6), 915–926.
25. Lin, M. F., Kheradpour, P., Washietl, S., Parker, B. J., Pedersen, J. S. and Kellis, M. (2011). Locating protein-coding sequences under selection for additional, overlapping functions in 29 mammalian genomes. *Genome Research*, **21**(11), 1916–1928.
26. Lorenz, R., Bernhart, S. H., Höner Zu Siederdissen, C., Tafer, H., Flamm, C., Stadler, P. F. and Hofacker, I. L. (2011). ViennaRNA Package 2.0. *Algorithms for Molecular Biology*, **6**, 26.
27. Low, J. T. and Weeks, K. M. (2010). SHAPE-directed RNA secondary structure prediction. *Methods*, **52**(2), 150–158.
28. Lundblad, E. W. and Altman, S. (2010). Inhibition of gene expression by RNase P. *New Biotechnology*, **27**(3), 212–221.
29. Mauger, D. M. and Weeks, K. M. (2010). Toward global RNA structure analysis. *Nature*, **28**(11), 1178–1179.
30. McDaniel, R. and Weiss, R. (2005). Advances in synthetic biology: on the path from prototypes to applications. *Current Opinion in Biotechnology*, **16**(4), 476–483.
31. Miklos, I., Meyer, I. and Nagy, B. (2005). Moments of the Boltzmann distribution for RNA secondary structures. *Bulletin of Mathematical Biology*, **67**(5), 1031–1047.
32. Muckstein, U., Tafer, H., Hackermuller, J., Bernhart, S. H., Stadler, P. F. and Hofacker, I. L. (2006). Thermodynamics of RNA-RNA binding. *Bioinformatics*, **22**(10), 1177–1182.

33. Nawrocki, E. P., Kolbe, D. L. and Eddy, S. R. (2009). Infernal 1.0: inference of RNA alignments. *Bioinformatics*, **25**(10), 1335–1337.
34. Parisien, M. and Major, F. (2012). Determining RNA three-dimensional structures using low-resolution data. *Journal of Structural Biology*, **179**(3), 252–260.
35. Parker, B. J., Moltke, I., Roth, A., Washietl, S., Wen, J., Kellis, M., Breaker, R. and Pedersen, J. S. (2011). New families of human regulatory RNA structures identified by comparative analysis of vertebrate genomes. *Genome Research*, **21**(11), 1929–1943.
36. Pedersen, J. S. (2004). A comparative method for finding and folding RNA secondary structures within protein-coding regions. *Nucleic Acids Research*, **32**(16), 4925–4936.
37. Pedersen, J. S., Bejerano, G., Siepel, A., Rosenbloom, K., Lindblad-Toh, K., Lander, E. S., Kent, J., Miller, W. and Haussler, D. (2006). Identification and classification of conserved RNA secondary structures in the human genome. *PLoS Computational Biology*, **2**(4), e33.
38. Rabani, M., Kertesz, M. and Segal, E. (2008). Computational prediction of RNA structural motifs involved in posttranscriptional regulatory processes. *Proceedings of the National Academy of Sciences*, **105**(39), 14885–14890.
39. Ray, P. S., Jia, J., Yao, P., Majumder, M., Hatzoglou, M. and Fox, P. L. (2009). A stress-responsive RNA switch regulates VEGFA expression. *Nature*, **457**(7231), 915–919.
40. Regulski, E. E. and Breaker, R. R. (2008). In-line probing analysis of riboswitches. *Methods in Molecular Biology*, **419**, 53–67.
41. Roth, A. and Breaker, R. R. (2009). The structural and functional diversity of metabolite-binding riboswitches. *Annual Review of Biochemistry*, **78**(1), 305–334.
42. Saito, H. and Inoue, T. (2009). Synthetic biology with RNA motifs. *International Journal of Biochemistry and Cell Biology*, **41**(2), 398–404.
43. Sato, K., Hamada, M., Asai, K. and Mituyama, T. (2009). CENTROIDFOLD: a web server for RNA secondary structure prediction. *Nucleic Acids Research*, **37**, Web Server, W277–W280.
44. Shapiro, B. A., Yingling, Y. G., Kasprzak, W. and Bindewald, E. (2007). Bridging the gap in RNA structure prediction. *Current Opinion in Structural Biology*, **17**(2), 157–165.
45. Steffen, P., Voss, B., Rehmsmeier, M., Reeder, J. and Giegerich, R. (2006). RNAshapes: an integrated RNA analysis package based on abstract shapes. *Bioinformatics*, **22**(4), 500–503.

46. Svoboda, P. and Cara, A. D. (2006). Hairpin RNA: a secondary structure of primary importance. *Cellular and Molecular Life Sciences*, **63**(7-8), 901–908.

47. Torarinsson, E. and Lindgreen, S. (2008). WAR: Webserver for aligning structural RNAs. *Nucleic Acids Research*, **36**, Web Server, W79–W84.

48. Turner, D. H. and Mathews, D. H. (2010). NNDB: the nearest neighbor parameter database for predicting stability of nucleic acid secondary structure. *Nucleic Acids Research*, **38**, Database issue, D280–2.

49. Underwood, J. G., Uzilov, A. V., Katzman, S., Onodera, C. S., Mainzer, J. E., Mathews, D. H., Lowe, T. M., Salama, S. R. and Haussler, D. (2010). FragSeq: transcriptome-wide RNA structure probing using high-throughput sequencing. *Nature Methods*, **7**(12), 995–1001.

50. Washietl, S., Hofacker, I. L. and Stadler, P. F. (2005). Fast and reliable prediction of noncoding RNAs. *Proceedings of the National Academy of Sciences*, **102**(7), 2454–2459.

51. Weeks, K. M. (2010). Advances in RNA structure analysis by chemical probing. *Current Opinion in Structural Biology*, **20**(3), 295–304.

52. Weinberg, Z., Barrick, J. E., Yao, Z., Roth, A., Kim, J. N., Gore, J., Wang, J. X., Lee, E. R., Block, K. F., Sudarsan, N., Neph, S., Tompa, M., Ruzzo, W. L. and Breaker, R. R. (2007). Identification of 22 candidate structured RNAs in bacteria using the CMfinder comparative genomics pipeline. *Nucleic Acids Research*, **35**(14), 4809–4819.

53. Weinberg, Z., Regulski, E. E., Hammond, M. C., Barrick, J. E., Yao, Z., Ruzzo, W. L. and Breaker, R. R. (2008). The aptamer core of SAM-IV riboswitches mimics the ligand-binding site of SAM-I riboswitches. *RNA*, **14**(5), 822–828.

54. Wiebe, N. J. P. and Meyer, I. M. (2010). TRANSAT– method for detecting the conserved helices of functional RNA structures, including transient, pseudo-knotted and alternative structures. *PLoS Computational Biology*, **6**(6), e1000823.

55. Win, M. N. and Smolke, C. D. (2007). A modular and extensible RNA-based gene-regulatory platform for engineering cellular function. *Proceedings of the National Academy of Sciences*, **104**(36), 14283–14288.

56. Winkler, W. C., Nahvi, A., Sudarsan, N., Barrick, J. E. and Breaker, R. R. (2003). An mRNA structure that controls gene expression by binding S-adenosylmethionine. *Nature Structural Biology*, **10**(9), 701–707.

57. Xin, Y. and Olson, W. K. (2009). BPS: a database of RNA base-pair structures. *Nucleic Acids Research*, **37**, Database, D83–D88.

58. Yang, S., Parisien, M., Major, F. and Roux, B. (2010). RNA structure determination using SAXS data. *The Journal of Physical Chemistry B*, **114**(31), 10039–10048.
59. Yao, Z., Weinberg, Z. and Ruzzo, W. L. (2006). CMfinder–a covariance model based RNA motif finding algorithm. *Bioinformatics*, **22**(4), 445–452.
60. Yuan, Y., Hwang, E. and Altman, S. (1992). Targeted cleavage of mRNA by human RNase P. *Proceedings of the National Academy of Sciences*, **89**(17), 8006.
61. Zuker, M. (1989). On finding all suboptimal foldings of an RNA molecule. *Science*, **244**(4900), 48–52.
62. Zuker, M. and Stiegler, P. (1981). Optimal computer folding of large RNA sequences using thermodynamics and auxiliary information. *Nucleic Acids Research*, **9**(1), 133–148.

Chapter 5

Traditional Approaches to RNA Structure Analysis

Jacek Wower
Department of Animal Sciences, Cellular and Molecular Biosciences Program, Auburn University, 210 Upchurch Bldg, 361 Mell St. Auburn, Alabama 36849, USA
wowerja@auburn.edu

Our knowledge of RNA structures has advanced immeasurably since the elucidation of the first such structure, that of yeast tRNA^{Phe}, by X-ray crystallography. Recent advances in massively parallel cDNA sequencing (RNA-seq) have allowed new insights in transcriptomes and greatly expanded the RNA universe. X-ray crystallography and nuclear magnetic resonance (NMR) cannot cope with the avalanche of newly discovered RNA molecules. This chapter focuses on powerful well-tested methods for the analysis of the secondary and tertiary structures of RNA molecules both in vitro and in vivo.

5.1 X-Ray and NMR Studies of RNA Structures

In 1958, when Francis Crick proposed the central dogma of molecular biology, only three classes of RNA molecules were known to exist: transfer RNAs (tRNAs), ribosomal RNAs (rRNAs) and

RNA Nanotechnology
Edited by Bin Wang

ISBN 978-981-4411-64-6 (Hardcover), 978-981-4411-65-3 (eBook)
www.panstanford.com

messenger RNAs (mRNAs). Structural analysis of tRNAs provided the first insights into the complex nature of RNA. Soon after baker's yeast $tRNA^{Ala}$ had been sequenced by Robert W. Holley, several independent studies revealed that secondary structures of all tRNAs resemble the cloverleaf [36]. An attempt to integrate the UV-induced cross-link between the U8 and C13 bases in $tRNA^{Phe}$ yielded the first three-dimensional rendering of the cloverleaf structure [43]. Many aspects of this model were confirmed by X-ray analysis of yeast $tRNA^{Phe}$ crystals, which demonstrated that tRNA molecules fold into an L-shaped structure in which helical domains formed by the anticodon/D stems and the acceptor/T stems orient perpendicular to one another. Twenty years after the three-dimensional structure of tRNA was elucidated, the first high-resolution images of a complete ribosome were produced using new tools for X-ray analysis of macromolecular crystals [5, 16, 82].

Due to tremendous progress in X-ray analysis, the research collaboratory for structural bioinformatics (RCSB) Protein Data Bank stored 80,000 high-resolution structures in 2012. However, only a few hundred solved RNA structures exist in current databases because the difficulty in growing a single well-diffracting RNA crystal constitutes a major limitation to the study of RNA structure [27, 34, 39]. There are many reasons why RNAs do not crystallize as readily as proteins. First, RNA function depends heavily upon an ability to undergo dramatic conformational changes, which are often triggered by interactions with protein cofactors. Thus, because of this inherent flexibility, RNA molecules often exist in multiple isoforms at the high concentrations that are required for crystallographic studies [37]. Second, the functional species of an RNA molecule might not be the most stable and readily reproduced isoform in vitro, because many RNA molecules have to associate with proteins to assume their physiologically relevant structure. Third, it is not simple to force all RNA molecules into a single conformation in vitro and such efforts often yield RNAs in kinetically trapped conformations [66]. Furthermore, the chances of misfolding increase for large RNAs. Finally, RNA molecules with extensive self-complementarity often form multimers.

Although X-ray crystallography remains the method of choice for determining high-resolution structures of large RNAs, valuable structural information about the structure of small RNAs can also be derived from NMR studies. Because molecules are investigated in solution, NMR analysis provides insights not only into the overall structure of RNA molecules but also into RNA dynamics. Most RNA structures deposited in the RCSB Protein Data Bank come from NMR studies of RNA molecules with molecular masses of 15 kDa. Only four structures of RNAs with molecular weights raging from 75 to 111 nt were determined by NMR [18, 20, 47]. Although the use of NMR to study RNA is currently limited by tremendous technical challenges, combining NMR techniques currently available promises to provide new insights into the solution properties of RNAs with molecular masses up to 100 kDa. Recent and anticipated developments in NMR methods for the characterization of structure and dynamics of large RNAs are insightfully described by Tzakos [65].

The explosion of interest in RNA, which stems from the discovery of catalytic RNAs, led to discovery of a vast universe of non-coding RNAs (ncRNAs). Based on their origin, functions, and properties, ncRNAs may be classified broadly as DNA markers, gene regulators, abiotic stress signals and biotic stress signals [25]. High-throughput sequencing technologies allow systematic searches for ncRNAs and every year thousands of new sequences are added to databases. As many of the newly discovered ncRNAs do not fit into the above listed ncRNA classifications, one can expect new exciting discoveries in this area of RNA research.

Addition of SELEX (systematic evolution of ligands by exponential enrichment) to the molecular biology toolbox produced aptamers that usually consist of short strands of oligoribonucleotides [64, 71]. These RNAs are engineered through repeated rounds of in vitro selection to bind various molecular targets from large libraries of sequences. Structures of many published RNA aptamers are very poorly characterized. Aptamers are currently being used in many diagnostic, biosensing and biochemical screening applications. Several clinical trials are underway using aptamer-based therapeutics for treating neovascularization, intravascular thrombus, diabetes, von-Hippel-Lindau syndrome, angiomas and myeloid leukemia (clinicaltrials.gov, National Institutes of Health (NIH)). In 2004, the US Food and Drug Administration approved

Macugen (Pegaptanib), the first aptamer based therapeutic for the treatment of age-related macular degeneration [51]. The Apamer Database is not comprehensive and lists only several hundred RNA aptamers [42].

Along with increased interest in RNA molecules has come the recognition that understanding their functions will require knowledge of their structures and conformational dynamics. Despite recent advances in X-ray crystallography and NMR of RNA molecules, these advanced methods will not be able to cope with the fast growing universe of natural and man-made RNAs. However, elucidation of RNA structure may be accelerated by the combination of traditional approaches for probing RNA structure in solution with the new tools provided by computer sciences.

5.2 Enzymatic Footprinting

The simplest method, at least conceptually, to map nucleotides forming single- and double-stranded regions in the RNA molecule involves ribonucleases and nucleases. Such an approach is usually used in investigations of small molecules composed of up to 350 nucleotides. A typical footprinting experiment is executed in several stages. First, as in all structural studies, researchers test the integrity, homogeneity and folding of the RNA molecule. The integrity and homogeneity of RNA samples usually is established using denaturing gel electrophoresis, which allows separation of small RNA molecules differing by a single nucleotide. Whether RNA molecules form multimers or folding isoforms can often be detected on the native polyacrylamide gel. The best insurance that the investigated RNA molecule is correctly folded is to test its functional activities. In the second step, the RNA molecule of interest is ^{32}P-labeled either at its 5' or 3' end as bacteriophage T4 polynucleotide kinase is able to transfer the γ phosphate from [γ-^{32}P]ATP to the 5' hydroxyl group or exchange it with a single 5' phosphate group present on the RNA substrate [57, 58]. RNA transcripts have 5' triphosphates that can be removed with a bacterial alkaline phosphatase. A simpler method of generating 5'-^{32}P-labeled RNA transcripts is to synthesize them in vitro with T7 RNA polymerase and [γ-^{32}P]GTP [49]. Both natural and in vitro transcribed RNAs can be ^{32}P-labeled at their 3' ends by bacteriophage T4 RNA ligase that catalyzes phosphodiester formation between the 3' hydroxyl of RNA and cytidine 3',5'-[5'-^{32}P]bisphosphate [13]. In the third step, the accessibility of particular nucleotides in the ^{32}P-labeled RNA molecule is interrogated with a number of enzymes that cleave the RNA backbone at specific sites. These reactions are performed under conditions where RNA is cleaved only once per molecule. In the fourth step, the RNA fragments are resolved on denaturing polyacrylamide gels to identify the positions of cleavage sites in the investigated RNA molecule. The latter analysis is facilitated by running partial alkaline hydrolysis reaction products and ribonuclease T1 (RNase T1) digests of the same ^{32}P-labeled RNA sample alongside the cleavage reaction products.

Table 5.1 lists enzymes that have been use thus far in such RNA footprinting experiments. Almost all enzymes listed specifically recognize nucleotides not involved in Watson-Crick base pairing. The exception is cobra venom ribonuclease, also known as RNase V1 [14]. Many RNases commonly used for RNA footprinting in 1980's, are now scarcely available [24]. Owing to their commercial availability and cleaving properties, endonucleases T1, A, V1 and nuclease S1 belong to enzymatic probes that are presently most frequently used.

RNases T1 and *A* specifically split the RNA backbone adjacent to the 3'-phosphate of unpaired guanosines and pyrimidines, respectively [32]. This reaction produces nucleoside 2',3'-cyclic phosphate, which is readily hydrolyzed to yield RNA fragments with a 3'-terminal phosphate.

RNase V1 recognizes three nucleotides on either side of an RNA helix and cuts double-stranded regions of RNA without any base specificity to produce RNA fragments with a 5'-terminal phosphate [14]. RNase V1 is not exceptionally specific and frequently cleaves phosphodiester bonds adjacent to double-stranded regions. Moreover, Lockard and Kumar demonstrated that RNase V1 is able to cleave both wobble base pairs and some nucleotides known to be involved in tertiary interactions [46].

Nuclease S1 has been extensively used for probing single-stranded regions in RNA [79]. It produces fragments that have a 3'-phosphate. Nuclease S1 by itself is insufficient for discerning all single-stranded regions because some of them are held tightly

Table 5.1 Enzymatic probes

Enzyme	Specificity	Commercial availability
RNase A	ss Cp/N and Up/N	+
RNase CL3	ss Cp/N	**
RNase I	ss Np/N	+
RNase J1	ss Np/N	-
RNase PhyM	ss Ap/N and Up/N	**
RNase T1	ss Gp/N	+
RNase T2	ss Np/N	+
RNase U2	ss Ap/N>Gp/N	**
RNase V1*	ds N/pN	+
Mung bean nuclease	ss N/pN	**
Neurospora crassa nuclease	ss N/pN	**
S1 nuclease	ss N/pN	+

*Also known as cobra venom RNase; **Contact the RNA Society

together by tertiary interactions [78]. Recent studies suggest that RNase J1 may also be used for mapping single-stranded regions in RNA molecules [19]. Since the 5'-exonuclease activity of this enzyme is inhibited by 5' triphosphates, RNase J1 is likely to be useful for structural investigation of RNA transcripts labeled with [γ-^{32}P]GTP [49].

5.3 Chemical Probing of RNA Secondary Structure

Probing RNA structure with enzymatic probes yields low-resolution images of RNA structure because the steric radii of these proteins do not allow access to all solvated sites in the RNA molecule. Thus, interpretation of results derived from enzymatic footprinting is greatly facilitated by using chemical probes. The list of chemical probes available for interrogation of RNA structure has steadily increased over the past two decades and now includes reagents suitable for probing individual nucleotides in the secondary structure, detecting specific features of the tertiary structure and monitoring changes in the global structure. Chemical reagents allow good access to all regions of RNA molecule, are suitable for analyzing RNAs of any size and can often be used both in vitro and in vivo.

Because chemical probing does not always induce cleavages of the RNA backbone, primer extension is a method of choice for detecting modified sites [54]. In this method, a DNA oligomer acting as a primer is annealed to the investigated RNA and extended by reverse transcriptase to a point of cleavage or base modification. This reaction produces DNA fragments corresponding to the length from the 5' end of the primer to the termination site. These DNA fragments are labeled using either of two schemes. The primer is usually ^{32}P-labeled at its 5' end by bacteriophage T4 polynucleotide kinase and [γ-^{32}P]ATP [57, 58]. Alternatively, an extension reaction is carried out in the presence of either [α-^{32}P]dATP or [α-^{35}S]dATP. When radioactively labeled reverse transcripts are resolved by gel electrophoresis, the termination sites are visualized on an autoradiogram as bands. The reactivity of each site is reflected by the band intensity, and exact locations of modifications and cleavages can be determined by co-electrophoresing dideoxy-sequencing reactions carried out on an unmodified sample of investigated RNA. The use of ^{35}S-labeling in combination with electrophoresis on wedge-shaped denaturing polyacrylamide gels allows the analysis of up to 300 nucleotides from the primer. Complete analysis of primer extension patterns is not always possible because termination of reverse transcription by post-transcriptional modifications or at highly structured sites (e.g. tetraloops) often obscure signals produced by chemical probing.

Four different chemical reagents are routinely used to analyze base-pairings in RNA molecules. They are dimethyl sulphate (DMS), 1,1-dihydroxy-3-ethoxy-2-butanone (kethoxal), 1-cyclohexyl-3-[2-morpholinoethyl]carbodiimide metho-*p*-toluene sulphonate (CMTC) and 4'-(hydroxymethyl)-4,5', 8-trimethylpsoralen (HMT). Properties of these four and other less frequently used reagents are listed in Table 5.2.

DMS methylates primarily N7, N3 and N1 positions on single-stranded guanines, cytosines and adenines, respectively [11, 41]. Modifications on cytosines and adenines can be readily detected by primer extension. To detect DMS-modified guanines by reverse transcription, it is necessary to reduce the 7,8 bond of the alkylated ring with sodium borohydrate and then use aniline to cleave the RNA molecule at the modified site. The DMS reaction often detects

Table 5.2 Chemical probes

Reagent	Target	Detection	Source
CMTC	N3-U N1-G	B B	Aldrich
DEPC	N7-A	A*; B	Sigma
DMS	N1-A N3-C N7-G	B A*; B A*; B	Aldrich
Ethylnitrosourea	Phosphate oxygens	A; B*	Sigma
Kethoxal	N1-G and N2-G	B	Research Organics
Pb^{2+}	ss phosphates metal-binding sites	A; B	Sigma
Tb^{3+}	ss phosphates metal-binding sites	A; B	Sigma
Fe-bleomycin	Antibiotic-binding site	A; B	Meyne Pharma

*Additional chemical treatment necessary.

single-stranded guanines inaccessible to RNase T1 as this enzyme requires simultaneous access to N1 and 06 positions of guanine to express its cleaving activity [32]. Besides unpaired guanines, cytosines and adenines, DMS can also be used to detect G:A and Hoogsteen A:U base pairs, because their adenines are in the *syn* conformation. Unpaired guanines could be further explored with kethoxal, which modifies them at the N1 and N2 positions [56]. It has to be noted that kethoxal specificity is not absolute as it often exhibits reactivity towards G:C base pairs at the end of helical segments of RNA molecule.

CMTC primarily reacts with the N3 of uracil and to lesser extent with N1 of guanine and therefore is employed to map unpaired uracils and guanines. Uracils are frequently post-transcriptionally modified to form thymine (T) or pseudouracil (Ψ). CMTC modifies T residues. CMTC-modified U, G and T can be readily identified by the reverse transcription approach. To map Ψ residues in RNA molecules, Bakin and Ofengand developed two complementary procedures [3,4]. The first procedure involves reaction of all U and G residues with *N*-cyclohexyl-*N'*-*β*-(4-methylmorpholinium)ethylcarbodiimide *p*-tosylate (CMC), followed by alkaline removal of all CMC groups except those linked to the N3 of Ψ. CMC-modified Ψ are readily identified by reverse transcription. The second procedure involves uridine-specific hydrazinolysis. Because Ψ is resistant to hydrazinolysis, RNA is not cleaved adjacent to Ψ-occupied positions. This is reflected in the absence of bands representing Ψ residues in an autoradiogram illustrating primer extension analysis.

In contrast to the above-mentioned reagents, the water-soluble HMT intercalates between base pairs in the helical regions of RNA and upon UV irradiation (320–380 nm) forms well-characterized adducts with pyrimidine bases [1,2]. The first step of photoreaction yields either a furan-side or a pyrone-side monoadduct. The absorption of a second photon yields an interstrand cross-link with pyrimidine located on the other strand. Since HMT strongly reacts with uracil, most cross-links are between two uracils (U-U). However, C-U and C-C cross-links have also been observed [29]. The sites of cross-linking can readily be identified by reverse transcription [44]. Because of its high-specificity, the use of HMT is without parallel as a method for identification of uridines in helical regions of RNA [15]

5.4 Mapping Metal Ion-Binding Pockets

Mg^{2+} ions interact electrostatically with the phosphate backbone of RNA and are particularly effective in stabilizing the RNA double helix [53]. Frequently, when RNA molecules adopt a tertiary structure, distinct metal-binding "pockets" are created. In these pockets Mg^{2+} ions bind more tightly than to RNA helices. X-ray crystallography provided high-resolution images of four Mg^{2+} ion-binding sites in yeast $tRNA^{Phe}$ Two of them bind in the corner of the L-shaped structure of tRNA to stabilize the tertiary interactions between D and T loops [35, 55]. Using X-ray crystallography researchers also identified binding sites for Pb^{2+} ions and lanthanide ions such Sm^{2+} and Lu^{3+} [12, 38] Because lead and most lanthanide metals can cleave the RNA backbone in the presence of Mg^{2+}, they have been extensively employed for probing RNA structures both in vitro and in vivo [45,84]. The best-characterized probes are lead and terbium (Tb^{3+}). Cleavages induced by both metals result in the formation of 2',3'-cyclic phosphates and 5'-hydroxyl termini. A number of studies demonstrated that Tb^{3+} ions mimic Mg^{2+} binding more closely than Pb^{2+} ions. Because the binding constants for lanthanide metals is 600- to 10,000-fold higher than those for magnesium, low concentrations of Tb^{3+} ions (e.g. 40 uM) can be used for probing metal-binding pockets in RNA [23, 72]. At higher concentrations (>100 uM), Tb^{3+} ions cleave not only the metal-binding pockets, but also non-Watson-Crick base-paired and single-stranded regions in RNA molecules [68].

5.5 RNA Cross-Linking

Cross-linking represents one of the most successful strategies to investigate tertiary structure of RNA molecules as it rapidly generates a set of constraints useful for the development of three-dimensional models of structurally complex RNAs. When used appropriately, cross-linking provides detailed information about the proximity of distant parts of the secondary structure of RNA. Over the last forty years a large number of such approaches have been developed. While most of cross-linking experiments are carried out in vitro, many methods have also been used to investigate free RNAs and ribonucleoprotein complexes in vivo.

Cross-linking approaches can be divided into two groups. The most valuable are those producing "zero-length" cross-links [76]. Because strict geometric constraints are required for covalent bond formation, these approaches generate more accurate structural information than cross-links produced by the use of cross-linking reagents. At the early stages of structural analysis of RNA, cross-linking approaches that produce cross-links randomly are particularly useful. As structural information accumulates, site-specific cross-linking is usually used to systematically investigate the structurally and functionally interesting elements of the RNA molecule.

Irradiation with far-UV light (~254 nm) has been frequently used to induce covalent cross-links within unmodified RNA molecules and nucleic acid-protein complexes both in vitro and in vivo [10, 69]. This simple approach produced a cross-link between U8 and C13 residues in $tRNA^{Phe}$, which was instrumental for developing the first almost correct three-dimensional model of tRNA molecule [8]. Brimacombe and his coworkers used UV- induced cross-links for the construction of the first three-dimensional model of ribosomal subunits [9]. Over the period of twenty years, this group identified nucleotides involved in more than 1000 intra-RNA cross-links that are available at the Database of Ribosomal Cross-links [6]. More recently, UV irradiation of U2 and U6 snRNAs produced a cross-link, which identified the catalytic site of the spliceosome [59].

However, since far-UV light is structurally damaging to nucleic acids, this approach sometimes produces cross-links that lack biological relevance. Many natural RNAs possess a variety of modified nucleotides that are inherently photoreactive when irradiated with near-UV light (300–360 nm) [8]. Utility of these naturally photoreactive nucleotides for cross-linking is limited by their scarcity, preordained location in RNA sequences and a high specificity of cross-linking. Some of these shortcomings can be alleviated by derivatization of RNA with cross-linking reagents. Most of these reagents contain aryl azide groups that are positioned 10–30 Å away from the derivatized base. One of the better reagents of this type is *p*-azidophenacyl bromide that can be used for derivatization of thiolated bases [33]. This reagent is highly photoreactive and forms a broad range of covalent bonds upon generation of the reactive nitrenes upon irradiation with the near UV light. Because the cross-linked nucleotides are ~11 Å away from the derivatized base, data obtained with this and other cross-linking reagents do not reflect the most immediate environment of the base.

Over the past two decades a number of reagents and methods that allow for very short-range cross-links from both random and defined positions in the RNA molecule have been developed. The most interesting group involves azidonucleosides. Consistent with their classification as aryl azides, 2- and 8-azidoadenosine, 5-azidouridine and 2,6-diazidopurine are readily photolyzed with near–UV light [61]. To be incorporated into the RNA molecule, azidonucleosides have to be prepared either as 3',5'-bisphosphates or 5'-triphosphates [62, 74]. Numerous methods have been described for nucleoside phosphorylation. The simplest of them involves pyrophosphoryl chloride which produces a mixture of 2',5'- and 3',5'-bisphosphates [7]. This nucleotide mixture can be used in conjunction with RNA ligase, which recognizes only nucleoside 3',5'-bisphosphates and is able to attach them to the

3'-terminal hydroxyl group of single-stranded RNA molecules. When treated with nuclease P1, 3',5'-bisphosphates are converted to 5'-monophosphates which can be converted to 5'-triphosphates with the help of nucleoside monophosphate and nucleoside diphosphate kinases [63,67]. 5'-^{32}P-labeled 3',5'-bisphosphates can be prepared by exchanging the 5'-phosphate of a nucleoside 3',5'-bisphosphate with the γ phosphate of [γ-^{32}P]ATP using the 3'-phosphatase-free T4 polynucleotide kinase [62]. [5'-^{32}P]-labeled nucleoside 3',5'-bisphosphates can then be used for the synthesis of [α-^{32}P]-labeled nucleoside triphosphates using the procedure described above.

For site-specific incorporation of azidonucleotides into RNA molecules, a "cut-and-paste" approach is often utilyzed. A universal method for site-directed cleavage of RNA involves the use of RNase H which cleaves the RNA component within an RNA/DNA duplex [21]. Ohtsuka and co-workers demonstrated that the site cleaved by RNase H could be selected very precisely using oligodeoxyribonucleotides flanked on either side with a 2'-O-methyloligoribonucleotide. The pasting steps usually involve T4 RNA ligase which incorporates [5'-^{32}P]-labeled nucleoside 3',5'-bisphosphates between two oligonucleotides to form a photoreactive recombinant RNA derivative. The pasting steps can also be carried out using T4 DNA ligase [50]. This approach uses a complementary deoxyribonucleotide as a "splint" to bring the 5' and 3' termini of two RNA molecules in close juxtaposition for ligation with T4 DNA ligase. The "cut-and-paste" approach has been used to construct many photoreactive recombinant RNA molecules [31, 60, 75, 77].

Random incorporation of azidonucleotides into RNA molecules can be carried out by in vitro transcription catalyzed by T7 RNA polymerase. At the present time, it appears that only $5N_3UTP$ and $8N_3ATP$ can act as substrates for this enzyme [22, 26]. However, to incorporate $8N_3AMP$ into RNA, a T7 RNA polymerase-catalyzed reaction has to be carried out in a buffer containing a mixture of Mg^{2+} and Mn^{2+} ions [30]. Since azidoadenosine 3',5'-bisphosphates can be readily attached to the 3' ends of RNA, they may also be placed at different positions within molecules by synthesizing a series of circularly permuted RNAs [62]. A good alternative to azidonucleotides are 4-thiouridine (s^4U) and 5-bromouridine

(BrU). s^4U residues occur naturally in many tRNAs. Irradiation of s^4U with UV light at 330–360 nm yields unstable intermediates which readily react with neighboring nucleophiles [28]. Unlike s^4U, BrU is not a natural component of nucleic acids and is usually used for photocross-linking RNA molecules to their protein ligands. BrU-mediated photocross-linking is significantly increased when irradiation is carried out at 313 nm rather than 254 nm [70]. The potential utility of 2-thiothymidine (s^2T), 4-thiothymidine (s^4T) and 6-thioguanosine (s^6G) for photocross-linking has been already recognized [17, 52]. They are very stable and can be readily synthesized as [α-^{32}P]-labeled triphosphates using the above-described methods.

The usefulness of the cross-linking data is proportional to the degree of precision with which the sites of cross-linking can be localized. Analysis of cross-link sites at the nucleotide level is very challenging. In the first step, cross-linked RNAs are separated from non-cross-linked RNAs on polyacrylamide gels. This is facilitated by reduced mobility of cross-linked RNAs as they adopt "octopus-like" structures in the presence of urea. Many cross-linked nucleotides can be identified via primer extension mapping [54, 73]. A very efficient but labor-intensive method has been developed by Brimacombe and co-workers who used uniformly ^{32}P-labeled substrates for intra-RNA cross-linking [10]. The crosslinked species were partially digested with a suitable RNase (e.g. RNase V1). The digest was then fractionated by two-dimensional gel electrophoresis and the isolated intra-RNA cross-linked complexes were "fingerprinted" on polyethyleneimine cellulose thin-layer plates. Recent advances in mass spectrometry suggest that this approach can be used to accelerate identification of cross-linked nucleotides [40, 80, 81].

5.6 Outlook

The structural analysis of nucleic acids has become a very rich field. Hopefully, this chapter will give the reader a sense of the wide range of methods that are now available to individual researchers for analysis of RNA structure. Data generated by these methods are most effective when used in reference to secondary and tertiary models for RNA. The rapidly developing field of bioinformatics has already produced many effective tools for predicting RNA structure from both a single RNA sequence and phylogenetic sequence comparisons [48]. The development of better techniques for chemical and enzymatic RNA synthesis and innovative approaches for the introduction of site-specific modifications into functional RNA molecules are also likely to accelerate the structural analysis of RNA molecules [83].

Acknowledgements

I am indebted to Christian Zwieb for critical reading of the manuscript.

References

1. Bachellerie, J. P., and Hearst, J. E. (1982). Specificity of the photoreaction of 4'-(hydroxymethyl)-4,5',8-trimethylpsoralen with ribonucleic acid. Identification of reactive sites in Escherichia coli phenylalanine-accepting transfer ribonucleic acid, *Biochemistry*, **21**, pp. 1357–1363.
2. Bachellerie, J. P. et al. (1981). Identification of the modified nucleotides produced by covalent photoaddition of hydroxymethyltrimethylpsoralen to RNA, *Nucleic Acids Res*, **9**, pp. 2207–2222.
3. Bakin, A., and Ofengand, J. (1993). Four newly located pseudouridylate residues in Escherichia coli 23S ribosomal RNA are all at the peptidyl-transferase center: analysis by the application of a new sequencing technique, *Biochemistry*, **32**, pp. 9754–9762.
4. Bakin, A. V., and Ofengand, J. (1998). Mapping of pseudouridine residues in RNA to nucleotide resolution, *Methods Mol Biol*, **77**, pp. 297–309.
5. Ban, N. et al. (2000). The complete atomic structure of the large ribosomal subunit at 2.4 A resolution, *Science*, **289**, pp. 905–920.
6. Baranov, P. V. et al. (1999). The Database of Ribosomal Cross-links: an update, *Nucleic Acids Res*, **27**, pp. 184–185.
7. Barrio, J. R. et al. (1978). Synthesis of modified nucleoside 3',5'-bisphosphates and their incorporation into oligoribonucleotides with T4 RNA ligase, *Biochemistry*, **17**, pp. 2077–2081.

8. Bergstrom, D. E., Inoue, I., and Leonard, N. J. (1972). Synthesis of the 335-nm photoproduct of cytosine and 4-thiouracil, *J Org Chem*, **37**, pp. 3902–3905.
9. Brimacombe, R. (2000). The Bacterial Ribosome at Atomic Resolution, *Structure*, **8**, pp. R195–R200.
10. Brimacombe, R. et al. (1988). Intra-RNA and RNA-protein cross-linking techniques in Escherichia coli ribosomes, *Methods Enzymol*, **164**, pp. 287–309.
11. Brookes, P., and Lawley, P. D. (1961). The reaction of mono- and di-functional alkylating agents with nucleic acids, *Biochem J*, **80**, pp. 496–503.
12. Brown, R. S., Devan, J. C., and Klug, A. (1985). Crystallographic and biochemical investigation of the lead(II)-catalyzed hydrolysis of yeast phenylalanine tRNA, *Biochemistry*, **24**, pp. 4785–4801.
13. Bruce, A. G., and Uhlenbeck, O. C. (1978). Reactions at the termini of tRNA with T4 RNA ligase, *Nucleic Acids Res*, **5**, pp. 3665–3677.
14. Butorin, A. S. et al. (1982). Comparison of the hydrolysis patterns of several tRNAs by cobra venom ribonuclease in different steps of the aminoacylation reaction, *Eur J Biochem*, **121**, pp. 587–595.
15. Cimino, G. D. et al. (1985). Psoralens as photoactive probes of nucleic acid structure and function: organic chemistry, photochemistry, and biochemistry, *Annu Rev Biochem*, **54**, pp. 1151–1193.
16. Clemons, W. M. J. et al. (1999). Structure of a bacterial 30S ribosomal subunit at 5.5 A resolution, *Nature*, **400**, pp. 833–840.
17. Connolly, B. A., and Newman, P. C. (1989). Synthesis and properties of oligonucleotides containing 4-thiothymidine, 5-methyl-2-pyrimidinone-1-beta-D(2'-deoxyriboside) and 2-thiothymidine, *Nucleic Acids Res*, **17**, pp. 4957–4974.
18. D'Souza, V. et al. (2004). NMR structure of the 101-nucleotide core encapsidation signal of the Moloney murine leukemia virus, *J Mol Biol*, **337**, pp. 427–442.
19. Daou-Chabo, R., and Condon, C. (2009). RNase J1 endonuclease activity as a probe of RNA secondary structure, *RNA*, **15**, pp. 1417–1425.
20. Davis, J. H. et al. (2005). RNA helical packing in solution: NMR structure of a 30 kDa GAAA tetraloop-receptor complex, *J Mol Biol*, **351**, pp. 371–382.
21. Donis-Keller, H. (1979). Site specific enzymatic cleavage of RNA, *Nucleic Acids Res*, **7**, pp. 179–192.

22. Dontsova, O. A. et al. (1992). Identification of the Escherichia coli 30S ribosomal subunit protein neighboring mRNA during initiation of translation, *Biochimie*, **74**, pp. 363–371.
23. Draper, D. E. (1985). On the coordination properties of Eu3+ bound to tRNA, *Biophys Chem*, **21**, pp. 91–101.
24. Ehresmann, C. et al. (1987). Probing the structure of RNAs in solution, *Nucleic Acids Res*, **15**, pp. 9109–9128.
25. Erdmann, V. A. et al. (2001). Regulatory RNAs, *Cell Mol Life Sci*, **58**, Switzerland, pp. 960–977.
26. Evans, R. K., and Haley, B. E. (1987). Synthesis and biological properties of 5-azido-2'-deoxyuridine 5'-triphosphate: a photoreactive nucleotide suitable for making light-sensitive DNA, *Biochemistry*, **26**, pp. 269–276.
27. Ferre-D'Amare, A. R., Zhou, K., and Doudna, J. A. (1998). A general module for RNA crystallization, *J Mol Biol*, **279**, pp. 621–631.
28. Frischauf, A. M., and Scheit, K. H. (1973). Affinity labeling of E. coli RNA polymerase with substrate and template analogues, *Biochem Biophys Res Commun*, **53**, pp. 1227–1233.
29. Garrett-Wheeler, E., Lockard, R. E., and Kumar, A. (1984). Mapping of psoralen cross-linked nucleotides in RNA, *Nucleic Acids Res*, **12**, pp. 3405–3423.
30. Gopalakrishna, S. et al. (2004). Template-dependent incorporation of 8-N3AMP into RNA with bacteriophage T7 RNA polymerase, *RNA*, **10**, pp. 1820–1830.
31. Hayase, Y. et al. (1992). Recognition of bases in Escherichia coli tRNA(Gln) by glutaminyl-tRNA synthetase: a complete identity set, *EMBO J*, **11**, pp. 4159–4165.
32. Heinemann, U., and Saenger, W. (1983). Crystallographic study of mechanism of ribonuclease T1-catalysed specific RNA hydrolysis, *J Biomol Struct Dyn*, **1**, pp. 523–538.
33. Hixson, S. H., and Hixson, S. S. (1975). P-Azidophenacyl bromide, a versatile photolabile bifunctional reagent. Reaction with glyceraldehyde-3-phosphate dehydrogenase, *Biochemistry*, **14**, pp. 4251–4254.
34. Holbrook, S. R., Holbrook, E. L., and Walukiewicz, H. E. (2001). Crystallization of RNA, *Cell Mol Life Sci*, **58**, pp. 234–243.
35. Holbrook, S. R. et al. (1977). RNA-ligant interactions. (I) Magnesium binding sites in yeast tRNAPhe, *Nucleic Acids Res*, **4**, pp. 2811–2820.
36. Holley, R. et al. Structure of ribonucleic acid, *Science*, **147**, pp. 1462–1465.

37. Hyeon, C., Dima, R. I., and Thirumalai, D. (2006). Size, shape, and flexibility of RNA structures, *J Chem Phys*, **125**, pp. 194905.
38. Jack, A. et al. (1977). A crystallographic study of metal-binding to yeast phenylalanine transfer RNA, *J Mol Biol*, **111**, pp. 315–328.
39. Ke, A., and Doudna, J. A. (2004). Crystallization of RNA and RNA-protein complexes, *Methods*, **34**, pp. 408–414.
40. Kellersberger, K. A. et al. (2004). Top-down characterization of nucleic acids modified by structural probes using high-resolution tandem mass spectrometry and automated data interpretation, *Anal Chem*, **76**, pp. 2438–2445.
41. Lawley, P. D., and Brookes, P. (1961). Acidic dissociation of 7:9-dialkylguanines and its possible relation to mutagenic properties of alkylating agents, *Nature*, **192**, pp. 1081–1082.
42. Lee, J. F. et al. (2004). Aptamer database, *Nucleic Acids Res*, **32**, pp. D95–100.
43. Levitt, M. (1969). Detailed molecular model for transfer nucleic acid, *Nature*, **224**, pp. 759–763.
44. Liarakos, C. D. et al. (1988). Determination of secondary structure in the initiation region of ovalbumin mRNA, *Nucleic Acids Res*, **16**, pp. 11249–11265.
45. Lindell, M. et al. (2005). Lead(II) cleavage analysis of RNase P RNA in vivo, *RNA*, **11**, pp. 1348–1354.
46. Lockard, R. E., and Kumar, A. (1981). Mapping tRNA structure in solution using double-strand-specific ribonuclease V1 from cobra venom, *Nucleic Acids Res*, **9**, pp. 5125–5140.
47. Lukavsky, P. J. et al. (2003). Structure of HCV IRES domain II determined by NMR, *Nat Struct Biol*, **10**, pp. 1033–1038.
48. Marti-Renom, M. A., and Capriotti, E. (2008). Computational RNA structure prediction, *Current Bioinformatics*, **3**, pp. 32–45.
49. Milligan, J. F., and Uhlenbeck, O. C. (1989). Synthesis of small RNAs using T7 RNA polymerase, *Methods Enzymol*, **180**, pp. 51–62.
50. Moore, M. J., and Sharp, P. A. (1992). Site-specific modification of pre-mRNA: the 2'-hydroxyl groups at the splice sites, *Science*, **256**, pp. 992–997.
51. Ng, E. W. et al. (2006). Pegaptanib, a targeted anti-VEGF aptamer for ocular vascular disease, *Nat Rev Drug Discov*, **5**, pp. 123–132.
52. Nikiforov, T. T., and Connolly, B. A. (1992). Oligodeoxynucleotides containing 4-thiothymidine and 6-thiodeoxyguanosine as affinity labels

for the Eco RV restriction endonuclease and modification methylase, *Nucleic Acids Res*, **20**, pp. 1209–1214.

53. T. Pan, D. M. Long, and O. C. Uhlenbeck, Divalent Metal Ions in RNA Folding and Catalysis, in *The RNA World.*, Cold Spring Harbor Laboratory Press, 1993, pp. 271–302.
54. Qu, H. L., Michot, B., and Bachellerie, J. P. (1983). Improved methods for structure probing in large RNAs: a rapid 'heterologous' sequencing approach is coupled to the direct mapping of nuclease accessible sites. Application to the 5' terminal domain of eukaryotic 28S rRNA, *Nucleic Acids Res*, **11**, pp. 5903–5920.
55. Quigley, G. J., Teeter, M. M., and Rich, A. (1978). Structural analysis of spermine and magnesium ion binding to yeast phenylalanine transfer RNA, *Proc Natl Acad Sci U S A*, **75**, pp. 64–68.
56. Shapiro, R. et al. (1969). On the reaction of guanine with glyoxal, pyruvaldehyde, and kethoxal, and the structure of the acylguanines. A new synthesis of N2-alkylguanines, *Biochemistry*, **8**, pp. 238–245.
57. Silberklang, M., Gillum, A. M., and Raj Bhandary, U. L. (1977). The use of nuclease P1 in sequence analysis of end group labeled RNA, *Nucleic Acids Res*, **4**, pp. 4091–4108.
58. Silberklang, M., Gillum, A. M., and Raj Bhandary, U. L. (1979). Use of in vitro 32P labeling in the sequence analysis of nonradioactive tRNAs, *Methods Enzymol*, **59**, pp. 58–109.
59. Sun, J. S., Valadkhan, S., and Manley, J. L. (1998). A UV-crosslinkable interaction in human U6 snRNA, *RNA*, **4**, pp. 489–497.
60. Sylvers, L. A. et al. (1993). A 2-thiouridine derivative in tRNAGlu is a positive determinant for aminoacylation by Escherichia coli glutamyl-tRNA synthetase, *Biochemistry*, **32**, pp. 3836–3841.
61. Sylvers, L. A., and Wower, J. (1993). Nucleic acid-incorporated azidonucleotides: probes for studying the interaction of RNA or DNA with proteins and other nucleic acids, *Bioconjug Chem*, **4**, pp. 411–418.
62. Sylvers, L. A. et al. (1989). Preparation of 2-azidoadenosine 3',5'-[5'-32P]bisphosphate for incorporation into transfer RNA. Photoaffinity labeling of Escherichia coli ribosomes, *FEBS Lett*, **245**, pp. 9–13.
63. Symons, R. H. (1974). Synthesis of (alpha-32P) ribo- and deoxyribonucleoside 5'-triphosphates, *Methods Enzymol*, **29**, pp. 102–115.
64. Tuerk, C., and Gold, L. (1990). Systematic evolution of ligands by exponential enrichment: RNA ligands to bacteriophage T4 DNA polymerase, *Science*, **249**, pp. 505–510.

65. Tzakos, A. G. et al. (2006). NMR techniques for very large proteins and RNAs in solution, *Annu Rev Biophys Biomol Struct*, **35**, pp. 319–342.
66. Uhlenbeck, O. C. (1995). Keeping RNA happy, *RNA*, **1**, pp. 4–6.
67. Walseth, T. F., and Johnson, R. A. (1979). The enzymatic preparation of [alpha-(32)P]nucleoside triphosphates, cyclic [32P] AMP, and cyclic [32P] GMP, *Biochim Biophys Acta*, **562**, pp. 11–31.
68. Walter, N. G., Yang, N., and Burke, J. M. (2000). Probing non-selective cation binding in the hairpin ribozyme with Tb(III), *J Mol Biol*, **298**, pp. 539–555.
69. Watkins, K. P., and Agabian, N. (1991). In vivo UV cross-linking of U snRNAs that participate in trypanosome trans-splicing, *Genes Dev.*, **5**, pp. 1859–1869.
70. Weintraub, H. (1973). The assembly of newlyreplicated DNAinto chromatin, *Cold Spring Harbor Symp Quant Biol*, **38**, pp. 247–256.
71. Wilson, D. S., and Szostak, J. W. (1999). In vitro selection of functional nucleic acids, *Annu Rev Biochem*, **68**, pp. 611–647.
72. Wolfson, J. M., and Kearns, D. R. (1975). Europium as a fluorescent probe of transfer RNA structure, *Biochemistry*, **14**, pp. 1436–1444.
73. Wower, I. K. et al. (2000). Binding and cross-linking of tmRNA to ribosomal protein S1, on and off the Escherichia coli ribosome, *EMBO J*, **19**, pp. 6612–6621.
74. Wower, J. et al. (1994). Synthesis of 2,6-diazido-9-(beta-D-ribofuranosyl)purine 3',5'-bisphosphate: incorporation into transfer RNA and photochemical labeling of Escherichia coli ribosomes, *Bioconjug Chem*, **5**, pp. 158–161.
75. Wower, J. et al. (1990). Probing tRNA binding sites on the Escherichia coli 30 S ribosomal subunit with photoreactive analogs of the anticodon arm, *Biochim Biophys Acta*, **1050**, pp. 38–44.
76. Wower, J. et al. (1994). Recombinant photoreactive tRNA molecules as probes for cross-linking studies, *Biochimie*, **76**, pp. 1235–1246.
77. Wower, J. et al. (1993). Topography of the E site on the Escherichia coli ribosome, *EMBO J*, **12**, pp. 617–623.
78. Wrede, P., Woo, N. H., and Rich, A. (1979). Initiator tRNAs have a unique anticodon loop conformation, *Proc Natl Acad Sci U S A*, **76**, pp. 3289–3293.
79. Wurst, R. M., Vournakis, J. N., and Maxam, A. M. (1978). Structure mapping of 5'-32P-labeled RNA with S1 nuclease, *Biochemistry*, **17**, pp. 4493–4499.

80. Yu, E., and Fabris, D. (2003). Direct probing of RNA structures and RNA-protein interactions in the HIV-1 packaging signal by chemical modification and electrospray ionization fourier transform mass spectrometry, *J Mol Biol*, **330**, pp. 211–223.
81. Yu, E., and Fabris, D. (2004). Toward multiplexing the application of solvent accessibility probes for the investigation of RNA three-dimensional structures by electrospray ionization-Fourier transform mass spectrometry, *Anal Biochem*, **334**, pp. 356–366.
82. Yusupov, M. M. et al. (2001). Crystal structure of the ribosome at 5.5 A resolution, *Science*, **292**, pp. 883–896.
83. R. A. Zimmermann, M. J. Gait, and M. J. Mooer, *Incorporation of modified nucleotides into RNA for Studies on RNA Structure, Function and Intermolecular Interactions,* ASM Press, Washington, D.C., 1998.
84. Zito, K., Huttenhofer, A., and Pace, N. R. (1993). Lead-catalyzed cleavage of ribonuclease P RNA as a probe for integrity of tertiary structure, *Nucleic Acids Res*, **21**, pp. 5916–5920.

Chapter 6

RNA Structure: Probing Biochemical Analyses

Subash C.B. Gopinath
Biomedical Research Institute & Nanoelectronics Research Institute, National Institute of Advanced Industrial Science and Technology, 1-1-1 Higashi, Tsukuba City 305-8562, Ibaraki, Japan
gopi-subashchandrabose@aist.go.jp

RNA–protein interactions are essential events of biological functions, which undergo the conformational changes of interacting molecules. To pinpoint the residues involved in these interactions, several approaches have been proposed by biochemical and biophysical approaches. These studies were able to map the binding regions on the interactive molecules, to shorten the size of the RNA molecule and to analyze the structural changes upon binding. Herein, biochemical structural probing analyses of RNAs upon interaction with protein were discussed based on nucleotide analog interference mapping, in-line probing and selective 2′-hydroxyl acylation analysis by primer extension chemistries.

RNA Nanotechnology
Edited by Bin Wang

ISBN 978-981-4411-64-6 (Hardcover), 978-981-4411-65-3 (eBook)
www.panstanford.com

6.1 Introduction

Biological macromolecules including deoxyribonucleic acid (DNA) and ribonucleic acid (RNA) have intrinsic features and carry out various functions. DNA contains the genetic instructions to be used for the development and functions of all living things. RNA molecules are involved in most of the fundamental aspects and highly conserved cellular processes. Deemed as one of the most essential molecules in life, RNA participates in transferring genetic information to the translation machinery. Nucleic acids are made of nucleotides, which are a long chain of components consisting of nucleobases, sugars, and phosphate groups. With these components, both DNA and RNA fold to form a variety of helical structures in order to perform biological functions. RNA forms more diverse structures due to its flexibility and complexity. The diverse roles of RNA are determined by its folding fashion, which makes the proper structure. In this chapter, prior to elaboration on the description of structural probing aspects of RNA, it is vital to discuss some basic information about RNA. RNAs are composed of four nucleotides: adenosine (A), cystidine (C), guanosine (G), and uridine (U). In terms of chemistry, an –OH group (hydroxyl group) is at the 2′ position of the ribose sugar in RNA. In contrast, there is thymine (T) instead of the uracil base and a lack of oxygen (deoxy) at the 2′ position of the ribose sugar in DNA. Generally, RNA can be synthesized chemically (from 3′ to 5′) or enzymatically (from 5′ to 3′). In the enzymatic synthesis, usually a T7 promoter region is added at the 5′ end of the DNA template. This enzymatic process known as transcription involves the role of nucleoside triphosphates in linking the hydroxyl group of one nucleotide to the phosphate of another. The 5′ end of the resulting product will have one to three phosphates while the 3′ end has a free hydroxyl group. Thus, RNA molecules can be labeled at either the 5′ end following de-phosphorylation or the 3′ end by the ligation reaction.

Major subtypes of RNAs are categorized as ribosomal RNA (rRNA), messenger RNA (mRNA) and transfer RNA (tRNA). There are many RNA-nanotechnological applications characterized by different subtypes such as small interfering RNA (siRNA), RNA aptamer,

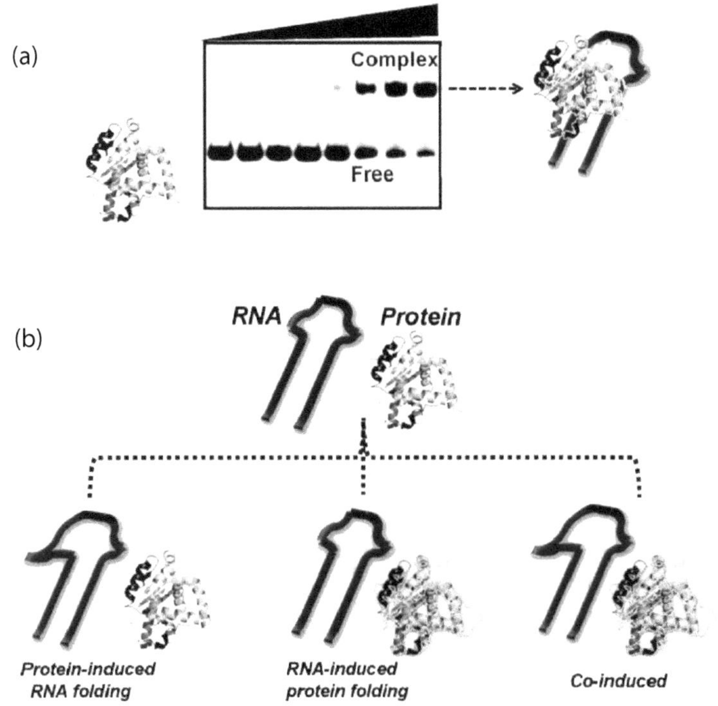

Figure 6.1 (a) RNA–protein complex formation is indicated by gel mobility shift. (b) Mechanisms involved in RNA–protein interactions [55].

antisense RNA, ribozymes, packaging RNA (pRNA), microRNA, small nuclear RNA (snRNA), small nucleolar RNA (snoRNA) and other subtypes. These RNAs play roles in the functional aspects by complex formation (Fig. 6.1a), characterized by RNA-induced ligand, ligand-induced RNA or coinduced conformational changes [55] (Fig. 6.1b). RNA undergoes self-folding to form two-dimensional (2D) or three-dimensional (3D) structures and contains a larger variety of single-stranded loops suitable for inter- and intramolecular interactions. Formation of secondary and tertiary structures with stems and loops of RNA can be predicted by computational methods. The occurrence of stem or loop structures also depends on stringent conditions (Fig. 6.2). To corroborate the structures predicted by the computational approach, structure chemical probing which is one of the common methods can be performed. In these

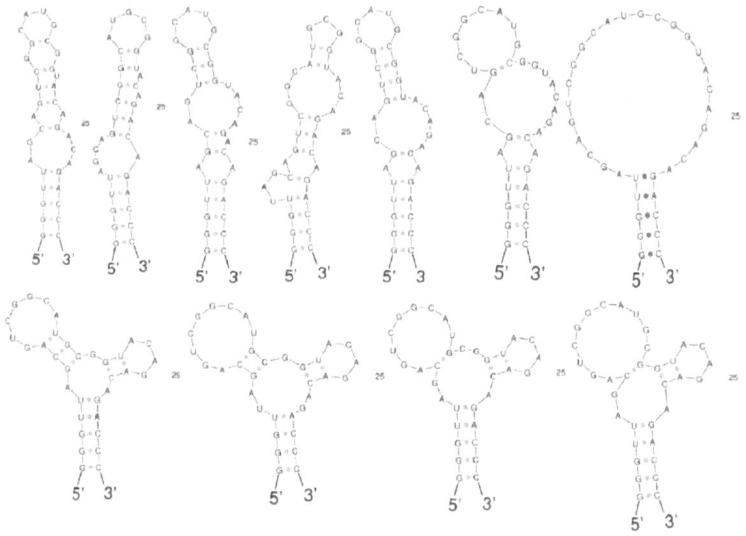

Figure 6.2 Predicted secondary structures of an anti-influenza aptamer [17]. The Sfold software was used. Single aptamer sequences may form different possible stem–loop structures.

methods biochemical mapping techniques are used to determine the arrangements and folding of the nucleic acids.

Results obtained from mapping studies are beneficial to further understand the conformational changes of the nucleic acid toward localizing the ligand-binding sites on nucleic acids—for example, footprinting, which generally refers to mapping of the backbone cleavage following the modification of the base or side chain. Footprinting of nucleic acids specifically refers to the cleavage of the phosphodiester backbone of the reaction products resolved on electrophoretic separation followed by autoradiography [13].

Patterns obtained from these analyses infer the molecular structure and are helpful to develop the shorter version of nucleic acid candidates (Fig. 6.3). Structure probing analysis can be carried out by many different methods, including nucleotide analog interference mapping (NAIM) in-line probing and selective 2′-hydroxyl acylation analyzed by primer extension (SHAPE). All these methods make the structural analyses of RNA molecules easier due to faster cleavage of the single-stranded loops. Chemical probing analyses

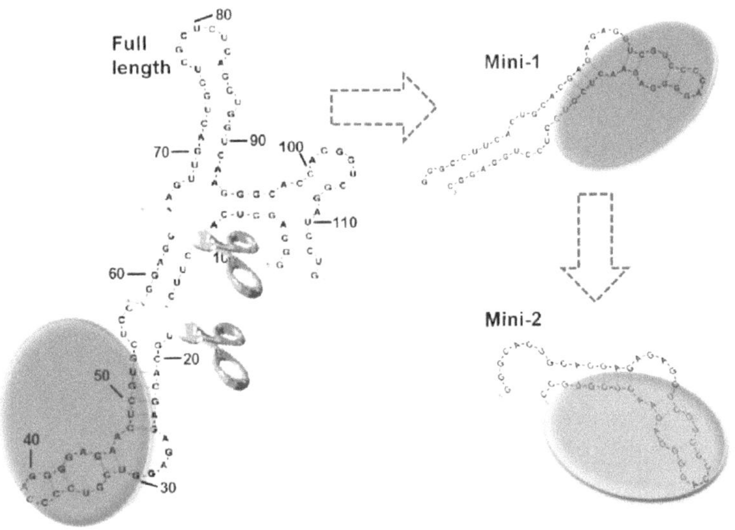

Figure 6.3 Illustration of making a shorter version from the full length. Shorter versions retain the important stem–loop structures. The protein-binding regions on the RNA aptamer are highlighted [15].

are in use for nearly 30 years; in these methods the end-labeled RNAs treated with chemical reagents (organic molecules, metal ions, enzymes) are subjected to interaction with the interacting partners before being resolved on urea-denaturing polyarylamide gel electrophoresis (PAGE) [52]. In this review, these biochemical probing analyses are discussed.

6.2 Nucleotide Analog Interference Mapping

NAIM is a chemogenetic approach that involves the identification of chemical groups that are important for RNA function. In this analysis a series of 5′*O*-(1-thio)nucleoside analog triphosphates are targeted in a modification procedure to probe the contribution of a functional group at every nucleotide position of an RNA molecule [37, 38]. NAIM is a straightforward approach and can be applicable for RNAs with any length, and analogs involved in this approach are lacking function. During in vitro transcription (using T7 RNA polymerase) random chemical substitutions to the base or sugar can be incorporated by α-phosphorothioate as the chemical tag. Subsequently, this phosphorothioate linkage can be cleaved by iodine (I_2), resulting in an RNA that is cleaved at the site of the nucleotide analog insertion. Upon cleavage, end-labeled RNA fragments can be visualized on urea-denaturing PAGE followed by autoradiography, which reveals the sites of analog incorporation. With these analyses, the active bases of the RNA population can be distinguished from the inactive bases. Plenty of nucleoside analogs are available as substrates for RNA polymerase in order to study the structure and function in biological systems [10]. It was reported that there are about 20 5′*O*-(1-thio)nucleoside triphosphates that are commercially available [46]. For the synthesis of phosphorothioate-tagged nucleoside triphosphates, the one-pot method developed by Arabshahi and Frey [1] is still in common use, even though several other highyielding methods were proposed [1, 46]. It was also found out that the addition of Mn^{2+} improved the incorporation efficiency [8, 40].

Using a NAIM strategy, several analyses on RNA and protein interactions have been performed [29, 35, 40, 46], and all these strategies are developed from the studies pioneered by Krupp et al. [8, 12, 36]. Overall steps involved in this strategy are displayed in Fig. 6.4, reproduced from guidelines prepared by Suydam and Strobel [46]. In the past, NAIM was also employed for studies on several functional aspects of RNA, which includes RNA folding [39, 50], catalytic activity [5, 20, 34, 37, 41, 44], protein binding [3], metal ions [2], small metabolites [19, 22], and in vivo studies [21, 47]. Using 5′*O*-(1-thio)inosine monophosphate in a NAIM assay, Strobel et al. have determined the importance of every N2 exocyclic amine of G of RNA for 5′ or 3′ exon ligation by the *Tetrahymena* group I intron [44]. Similarly, using NAIM Heide et al. [18] have identified the importance of guanosine 2-NH_2 groups of *Escherichia coli* ribonuclease (RNase P)-RNA in intramolecular tertiary contacts and direct interactions with tRNA. To characterize the ribonucleoprotein complexes, Buck et al. [6] carried out phosphorothioate footprinting techniques and identified the proteinbinding site and RNAfolding states of RNase P.

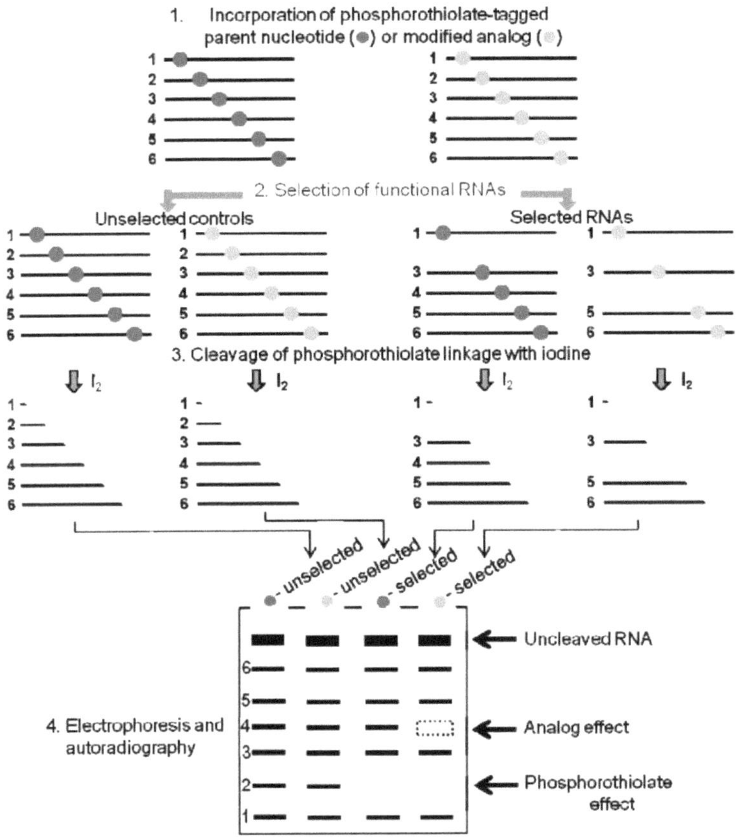

Figure 6.4 Schematic of the NAIM experiment described by Suydam and Strobel [46]. Parental and incorporated analog-modified nucleotides are shown. In this scheme phosphorothiorate interferes at position 2, whereas the modified analog interferes at position 4.

Quantitation of free-energy profiles in RNA–ligand interactions by NAIM (quantitative NAIM (QNAIM)) within the signal recognition particle (SRP), specifically the 4.5S RNA with the M domain of Ffh, was performed by Cochrane et al. [7]. Using NAIM, Sekiya et al. [40] have identified the involvement of functionally important nucleotides on the tertiary active structure of the RNA aptamer selected against hepatitis C Virus NS3 protease and found the tight interactions of conserved regions on the stem–loop aptamer structure. They quantified these interactions by the ratio: [Intensity of RNA bound with NS3(+)]/[Intensity of RNA untreated with NS3(−)] based on NAIM. Interference values were then normalized by dividing the values obtained at each position by the average values for all positions. The NAIM strategy can be applied to any analog that can be incorporated by an RNA polymerase to any RNA. Among several available biochemical RNA structure–probing analyses NAIM is considered as one of the powerful methods to substitute the specific nucleotide functional group(s) for rapid identification of single base-interactive site(s).

6.3 RNA In-Line Probing

RNA in-line probing assays rely on structure-dependent spontaneous cleavage and have been successfully used to map the binding sites of target molecules on RNA [27, 33]. In this assay, RNA phosphodiester linkages are cleaved spontaneously as the result of an S_N2 attack by the ribose 2′-oxygen on the adjacent phosphorus. As a result there will be fragments of 5′ with a 2′, 3′-cyclic phosphate and a 3′ with a 5′-hydroxyl terminus. The word "in-line" refers to the pseudobond angle [O2′(U2)–P(U3)–O5′(U3)]. The interatomic distance between the 2′-oxygen nucleophile and the phosphorus electrophile (degree of "in-line" positioning) is also expected to influence the rate of transesterification [26]. Maximal rates of spontaneous cleavage can be achieved when the position of the 2′-oxygen, the phosphorus center, and the 5′-oxygen-leaving group form a perfect 180° [49]. RNA involved in this assay does not require any type of chemicals or reagents to modify the RNA structures. In biochemical analyses an in-line probing assay has been used to determine changes in the structure upon ligand binding in a dosedependent manner, yielding different cleavage patterns with increasing order. In-line probing assays give information on the relative flexibility of specific RNA cleavages, greater cleavages occur in the unpaired regions rather than the paired regions, and these regions are identified by comparison. This assay also provides information on the sequence-dependent local structure changes [49].

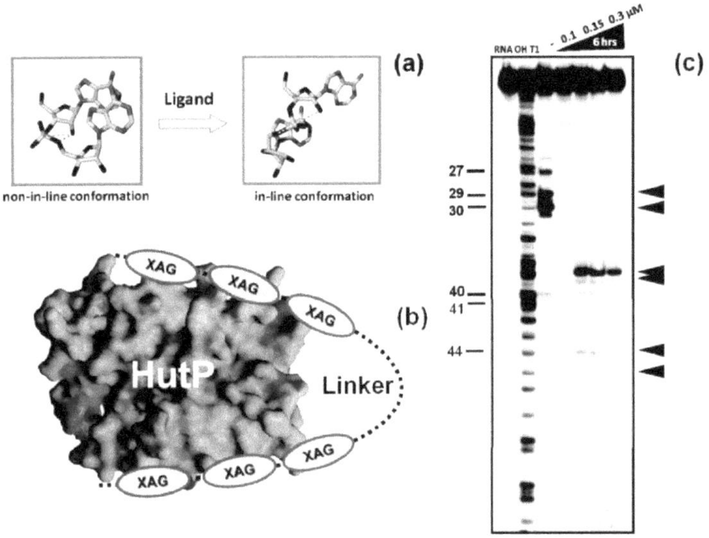

Figure 6.5 (a) Ligand-induced in-line conformation changes, (b) a HutP protein and RNA complex, and (c) an in-line probing assay demonstrated for an RNA aptamer and gD protein of herpes simplex virus-1. *Abbreviation*: HutP, histidine utilization protein.

In in-line probing, divalent cations (Mg^{2+} and Pb^{2+}) accelerate the cleavage of a 2′-hydroxyl-mediated transesterification reaction (Fig. 6.5a). The cleavage patterns obtained from several studies are mainly generated using magnesium ions [27, 33, 42]. In this assay the functional RNA molecules are incubated in the presence and absence of a ligand for different periods of intervals and the pattern of cleavages are visualized by resolving on the gel. These cleavages are structure dependent, and resultant band lengths are different. Experimental parameters such as temperature and incubation time and concentration of metal ions are also influential on the degree of cleavage. This technique can be useful to find the orientation changes in the structures. This method relies on a natural rate of spontaneous cleavage within RNA, which is helpful for accurate identification of the 5′ and 3′ ends of the functional RNA regions.

An in-line probing assay is commonly used in dynamic studies Major contributions to this technique were made by Breaker's lab with studies on riboswitches, and they demonstrated this assay with various ligands [4, 25–27, 33, 42, 45]. Using the small ligand, mitoxantrone, we have shown the sensitivity of this assay with vault noncoding RNAs (vRNAs) [16], which reside in the human vault particles. Vaults are the largest cytopalsmic ribonucleoproteins and are overexpressed in cancer cells. In our study, three different vRNAs were examined against three different drugs, and vRNA 1 and 2 were found to interact with the chemotherapeutic compound mitroxantrone. Similarly with HutP, an antitermination protein known to regulate the *hut* operon by an antitermination mechanism in *Bacillus subtilis*, we have demonstrated the interactions of this protein and RNA by an in-line probing assay [14]. In this study, to evaluate HutP interaction with RNA bearing multiple XAG motifs (3 motifs each in two binding sites), we chemically synthesized a model RNA (55 mer), in which two XAG-rich regions are joined by a linker region of 17 Us (Fig. 6.5b). In-line mapping experiments were performed by mixing 55-mer RNA with a HutP-L-histidine complex in the presence or absence of Mg^{2+} ions. Based on the resolved pattern on the gel, it was revealed that effective cleavage occurred in all six corresponding sites within the 55-mer RNA only in the presence of activated HutP (HutP-L-histidine–Mg^{2+}) [14].

Our recent studies performed using this assay on the RNA aptamer generated against the membrane glycoprotein (gD) of herpes simplex virus-1 [15], demonstrated a clear in-line pattern of cleavages in the presence of the target protein and Mg^{2+} ions. The generated RNA candidate has a full length of 113, bases and a shorter version with 57 mer (mini-1) was created based on preliminary analyses. To reduce the size further (mini-2), we performed in-line probing, and the cleavage pattern exhibited between 29 and 45 bases (Figs. 6.3 and 6.5c). Based on this assay, the length of the RNA aptamer could be reduced to produce the shortest derivate. We could also see the dose-dependent cleavages in the presence of different concentration of proteins. Among several available RNA structure probing analyses, in-line probing is considered as a good strategy to distinguish between base-paired and single-stranded regions of RNA in a sequence-dependent manner. However, this method shows only the regions that are undergoing conformational changes and does not provide any information about the involvement of other residues.

6.4 Selective 2′-Hydroxyl Acylation Analyzed by Primer Extension

SHAPE chemistry is a powerful approach to analyze nucleotide structure and dynamics and involves preferential modification on the backbone of RNA, especially in structurally flexible regions. Members of Kevin Weeks's originally developed SHAPE and most of the currently available procedures are originated on the basis of this team's work [31, 54] SHAPE has been used to analyze diverse RNA structures with unlimited length, including the entire genome. This strategy can identify the RNA structure participating in protein–RNA interactions at single-nucleotide resolution. Methods with SHAPE chemistry can identify mapping RNA conformations affected by intra or intermolecular interactions [48]. Similar to an in-line probing assay, SHAPE chemistry assists in monitoring the sequence-dependent local structure for every position [49].

Reagents for SHAPE are small-molecule electrophiles, such as *N*-methylisatoic anhydride (NMIA) and 1-methyl-7-nitroisatoic anhydride (1M7), and can selectively react with flexible RNA nucleotides at the 2′-hydroxyl group on the ribose sugar on the RNA backbone (Fig. 6.6), generating an adduct that can stop primer extension by reverse transcriptase [30]. The activity of the 2′-ribose of RNA is sensitive to the nucleotide conformation [30, 31]. There are several chemical reagents proposed for mapping analyses for RNA and protein interactions. However, compared to the other

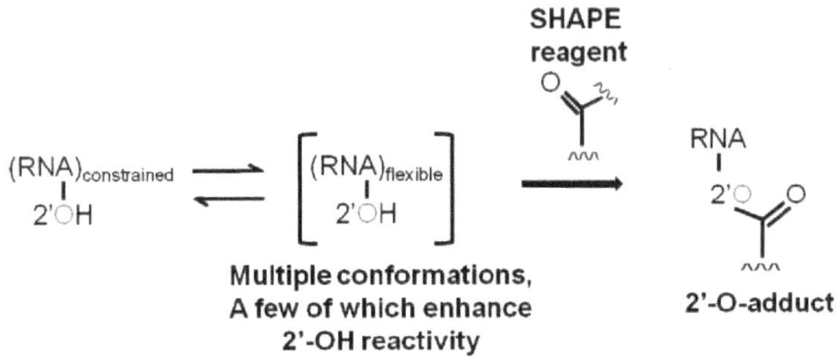

Figure 6.6 Scheme for SHAPE chemistry [30]. Nucleotides adopt a conformation that enhances the reactivity of the 2′-hydroxyl group.

chemical reagents, NMIA and 1M7 have the advantage of being largely unbiased toward identifying bases and very sensitive to conformational dynamics.

For nucleotides constrained by base-pairing interactions, close proximity of the 3′ phosphodiester anion minimizes reactivity of the 2′-hydroxyl resulting in lower rates of reactivity to an appropriate electrophile, for the formation of a 2′-*O*-adduct. However, flexible linkages allow conformations that will exhibit increased reactivity toward the electrophile. In other words, nucleotides that are constrained by base pairing (stem region) show less adduct formation than nucleotides that are unpaired (single-stranded region). Sites of 2′-*O*-adduct formation can be detected by reverse transcription, where the disruption of reverse transcriptase–aided complementary DNA (cDNA) synthesis occurred. Importantly, NMIA reactions are self-quenching by virtue of the fact that NMIA-mediated reactions undergo a parallel self-inactivating reaction. Due to these reasons, SHAPE is proved to be experimentally easier and is a rapid method for the prediction of paired and single-stranded regions on the interested target RNA. Adduct formation can be quantified for each nucleotide in a given RNA by reverse transcription using a cDNA primer, and this resulting fragment is compared with those from the control reaction (unmodified sample). This kind of analysis can be the method of choice to generate accurate secondary structure models [49].

In the past, SHAPE chemistry has been applied for the analyses of function inter-relationships in RNA [53], RNA–protein interactions [11, 28], RNA–small molecule interactions [9, 43, 51] and RNA-folding dynamics [24, 32]. McGinnis et al. [30] analyzed the constrained nucleotides poised by the efficient reactions at the 2′-hydroxyl group with SHAPE chemistry. To identify these nucleotides, they have carried out SHAPE on the crystal structure of the *Escherichia coli* ribosome and found the hyperreaction of 1,490 nucleotides in 16S rRNA. Similar to an in-line probing assay, SHAPE is considered as a choice of unbiased method to distinguish sequence-dependent, base-paired and single-stranded regions of RNA. Additionally, SHAPE is also applied for the analyses of riboswitch RNAs in the presence and absence of ligands [49]. SHAPE can also be applicable for in vitro transcripts and the RNAs from cellular and viral environments [23, 53].

6.5 Conclusions

There are different RNA structure-probing biochemical analyses reported with enzymatic or chemically modified RNA and subjected to spontaneous cleavages. Each of these methods has its own strengths and weaknesses. All the above-discussed electrophoretic-based biochemical RNA structure-probing analyses are considered as powerful methods to identify the single base-interactive sites. Among these methods NAIM is considered as one of the powerful methods to substitute the specific nucleotide functional group(s). However, to find a sequence-dependent local structural change, there is a need for an alternate method for RNA secondary structure prediction and conformational change investigation. To fulfill these, two methods in-line probing and SHAPE chemistry, were also discussed in this chapter. These methods can be used to identify the sequence-dependent paired and unpaired regions and are an ideal strategy to evaluate the software-based predictions of the RNA structure. These mappings will make way toward mutating the RNA bases to control the interactive molecules. For example, one can create receptor molecules with the identified interactive points and also stabilize the desired regions of the RNA molecule. Moreover, based on mapping analyses, RNA can be modified and shortened to be applied in a wide variety of applications such as RNA-based transfection.

References

1. Arabshahi, A., and Frey, P.A. (1994). A simplified procedure for synthesizing nucleoside 1-thiotriphosphates—dATPS, dGTPS, UTPS, and dTTPS. *Biochem. Biophys. Res. Commun.*, **204**, 150–155.
2. Basu, S., et al. (1998). A specific monovalent metal ion integral to the AA platform of the RNA tetraloop receptor. *Nat. Struct. Biol.*, **5**, 986–992.
3. Batey, R.T., et al. (2000). Crystal structure of the ribonucleoprotein core of the signal recognition particle. *Science*, **287**, 1232–1239.
4. Blount, K.F., and Breaker, R.R. (2006). Riboswitches as antibacterial drug targets. *Nat. Biotech.*, **24**, 1558–1564.

5. Boudvillain, M., and Pyle, A.M. (1998). Defining functional groups, core structural features and inter-domain tertiary contacts essential for group II intron self-splicing: a NAIM analysis. *EMBO J.*, **17**, 7091–7104.
6. Buck, A.H., Kazantsev, A.V., Dalby, A.B., and Pace, N.R. (2005). Structural perspective on the activation of RNAse P RNA by protein. *Nat. Struct. Mol. Biol.*, **12**, 958–964.
7. Cochrane, C., Batey, R.T., and Strobel, S.A. (2003). Quantitation of free energy profiles in RNA-ligand interactions by nucleotide analog interference mapping. *RNA*, **9**, 1282–1289.
8. Conrad, F., Hanne, A., Gaur, R.K., and Krupp, G. (1995). Enzymatic synthesis of 2′-modified nucleic acids—identification of important phosphate and ribose moieties in RNAse-P substrates. *Nucleic Acids Res.*, **23**, 1845–1853.
9. Dann, C.E., III et al. (2007). Structure and mechanism of a metal-sensing regulatory RNA. *Cell*, **130**, 878–892.
10. Das, S.R., Fong, R., and. Piccirilli, J.A. (2005). Nucleotide analogues to investigate RNA structure and function. *Curr. Opin. Chem. Biol.*, **9**, 585–593.
11. Duncan, C.D., and Weeks, K.M. (2008). SHAPE analysis of long-range interactions reveals extensive and thermodynamically preferred misfolding in a fragile group I intron RNA. *Biochemistry*, **47**, 8504–8513.
12. Gaur, R.K., and Krupp, G. (1993). Modification interference approach to detect ribose moieties important for the optimal activity of a ribozyme. *Nucleic Acids Res.*, **21**, 21–26.
13. Gopinath, S.C.B. (2009). Mapping of RNA and protein interactions. *Anal. Chim. Acta.*, **636**, 117–128.
14. Gopinath, S.C.B., et al. (2008). Insights into anti-termination regulation of the *hut* operon in *Bacillus subtilis*: importance of the dual RNA-binding surfaces of HutP. *Nucleic Acids Res.*, **36**, 3463–3473.
15. Gopinath, S.C.B., Hayashi, K., and Kumar, P.K.R. (2012). Aptamer that binds to the gD protein of herpes simplex virus-1 and efficiently inhibits viral entry. *J Virol.*, **86**, 6732–6744.
16. Gopinath, S.C.B., Matsugami, A., Katahira, K., and Kumar, P.K.R. (2005). Human vault-associated non-coding RNAs bind to mitoxantrone, a chemotherapeutic compound. *Nucleic Acids Res.*, **33**, 4874–4881.
17. Gopinath, S.C.B., et al. (2006). An RNA aptamer that distinguishes between closely related human influenza viruses and inhibits hemagglutinin-mediated membrane fusion. *J. Gen. Virol.*, **87**, 479–487.

18. Heidi, C., Pfeiffer, T., Nolan, J.M., and Hartmann, R.K. (1999). Guanosine 2-NH2 groups of Escherichia coli RNase P RNA involved in intramolecular tertiary contacts and direct interactions with tRNA. *RNA*, **5**, 102–116.
19. Jansen, J.A., McCarthy, T.J., Soukup, G.A., and Soukup, J.K. (2006). Backbone and nucleobase contacts to glucosamine-6-phosphate in the glmS ribozyme. *Nat. Struct. Mol. Biol.*, **13**, 517–523.
20. Kazantsev, A.V., and Pace, N.R. (1998). Identification by modification-interference of purine N-7 and ribose 2′-OH groups critical for catalysis by bacterial ribonuclease P. *RNA*, **4**, 937–947.
21. Kolev, N.G., and Steitz, J.A. (2006). In vivo assembly of functional U7 snRNP requires RNAbackbone flexibility within the Sm-binding site. *Nat. Struct. Mol. Biol.*, **13**, 347–353.
22. Kwon, M.Y., and Strobel, S.A. (2008). Chemical basis of glycine riboswitch cooperativity. *RNA*, **14**, 25–34.
23. Low, J.T., and Weeks, K.M. (2010). SHAPE-directed RNA secondary structure prediction. *Methods*, **52**, 150–158.
24. Lu, C., et al. (2011). Variable sequences outside the SAM-binding core critically influence the conformational dynamics of the SAM-III/S_{MK} box riboswitch. *J. Mol. Biol.*, **409**. 786–799.
25. Mandal, M., et al. (2003). Riboswitches control fundamental biochemical pathways in Bacillus subtilis and other bacteria. *Cell*, **113**, 577–586.
26. Mandal, M., and Breaker, R.R. (2004). Adenine riboswitches and gene activation by disruption of a transcription terminator. *Nat. Struct. Mol. Biol.*, **11**, 29–35.
27. Mandal, M., et al. (2004) A glycine-dependent riboswitch that uses cooperative binding to control gene expression. *Science*, **306**, 275–279.
28. Mayerle, M. Bellur, D.L., and Woodson, S.A. (2011). Slow formation of stable complexes during coincubation of minimal rRNA and ribosomal protein S4. *J. Mol. Biol.*, **412**, 453–465.
29. McConnell, T.S., Lokken, R.P., and Steitz, J.A. (2003). Assembly of the U1 snRNP involves interactions with the backbone of the terminal stem of U1 snRNA. *RNA*, **9**, 193–201.
30. McGinnis, J.L., Dunkel, J.A., Cate, J.H.D., and Weeks, K.M. (2012). The mechanisms of RNA SHAPE chemistry. *J. Am. Chem. Soc.*, **134**, 6617–6624.
31. Merino, E.J., Wilkinson, K.A., Coughlan, J.L., and Weeks, K.M. (2005). RNA structure analysis at single nucleotide resolution by selective 2′-hydroxyl acylation and primer extension (SHAPE). *J. Am. Chem. Soc.*, **127**, 4223–4231.

32. Mortimer, S.A., and Weeks, K.M. (2009). C2′-endo nucleotides as molecular timers suggested by the folding of an RNA domain. *Proc. Natl. Acad. Sci. USA*, **106**, 15622–15627.

33. Nahvi, A., et al. (2002). Genetic control by a metabolite binding mRNA. *Chem. Biol.*, **9**, 1043–1049.

34. Oyelere, A.K., Kardon, J.R., and Strobel, S.A. (2002). pKa perturbation in genomic hepatitis delta virus ribozyme catalysis evidenced by nucleotide analogue interference mapping. *Biochemistry*, **41**, 3667–3675.

35. Rox, C., Feltens, R., Pfeiffer, T., and Hartmann, R.K. (2002). Potential contact sites between the protein and RNA subunit in the Bacillus subtilis RNase P holoenzyme. *J. Mol. Biol.*, **315**, 551–560.

36. Ruffner, D.E, and Uhlenbeck, O.C. (1990). Thiophosphate interference experiments locate phosphates important for the hammerhead RNA self-cleavage reaction. *Nucleic Acids Res.*, **18**, 6025–6029.

37. Ryder, S.P, and Strobel, S.A. (1999). Nucleotide analog interference mapping of the hairpin ribozyme: Implications for secondary and tertiary structure formation. *J. Mol. Biol.*, **291**, 295–311.

38. Ryder, S.P., and Strobel, S.A. (1999). Nucleotide analog interference mapping. *Methods*, **18**, 38–50.

39. Schwans, J.P., Cortez, C.N., Olvera, J.M., and Piccirilli, J.A. (2003). 2′-mercaptonucleotide interference reveals regions of close packing within folded RNA molecules. *J. Am. Chem. Soc.*, **125**, 10012–10018.

40. Sekiya, S., Nishikawa, F., Fukuda, K., and Nishikawa, S. (2003). Structure/functional analysis of an RNA aptamer for Hepatitis C virus NS3 protease. *J. Biochem.*, **133**, 351–359.

41. Sood, V.D., Yekta, S., and Collins, R.A. (2002). The contribution of 2′-hydroxyls to the cleavage activity of the Neurospora VS ribozyme. *Nucleic Acids Res.*, **30**, 1132–1138.

42. Soukup, G.A., and Breaker, R.R. (1999). Relationship between internucleotide linkage geometry and the stability of RNA. *RNA*, **5**, 1308–1325.

43. Stoddard, C.D., Gilbert, S.D., and Batey, R.T. (2008). Ligand-dependent folding of the three-way junction in the purine riboswitch. *RNA*, **14**, 675–684.

44. Strobel, S.A., and Shetty, K. (1997). Defining the chemical groups essential for Tetrahymena group I intron function by nucleotide analog interference mapping. *Proc. Natl. Acad. Sci. USA.*, **94**, 2903–2908.

45. Sudarsan, N., et al. (2003). An mRNA structure in bacteria that controls gene expression by binding lysine. *Genes Dev.*, **17**, 2688–2697.

46. Suydam, I.T., and Strobel, S.A. (2009). Nucleotide analog interference mapping. *Methods Enzymol.*, **468**, 3–30.

47. Szewczak, L.B.W., DeGregorio, S.J, Strobel, S.A., and Steitz, J.A. (2002). Exclusive interaction of the 15.5 kD protein with the terminal box C/D motif of a methylation guide snoRNP. *Chem. Biol.*, **9**, 1095–1107.

48. Vaughan, R., Runnging, W., Qi, R., and Kao, C.C. (2012). Mapping protein-RNA interactions. *Virus Adapt. Treat.*, **4**, 29–41.

49. Wakeman, C.A., and Winkler, W.C. (2009). Analysis of the RNA backbone: structural analysis of riboswitches by in-line probing and selective 2′-hydroxyl acylation and primer extension. *Methods Mol. Biol.*, **540**, 173–191.

50. Waldsich, C., and Pyle, A.M. (2007). A folding control element for tertiary collapse of a group II intron ribozyme. *Nat. Struct. Mol. Biol.*, **14**, 37–44.

51. Wang, B. Wilkinson, KA., and Weeks, K.M. (2008). Complex ligand-induced conformational changes in $tRNA^{Asp}$ revealed by single nucleotide resolution SHAPE chemistry. *Biochemistry*, **47**, 3454–3461.

52. Weeks, K.M. (2010). Advances in RNA structure analysis by chemical probing. *Curr. Opin. Struct. Biol.*, **20**, 295–304.

53. Weeks, K.M., and Mauger, D.M. (2011). Exploring RNA structural codes with SHAPE chemistry. *Acc. Chem. Res.* **44**, 1280–1291.

54. Wilkinson, K.A., Merino, E.J., and Weeks, K.M. (2006). Selective 2′-hydroxyl acylation analyzed by primer extension (SHAPE): quantitative RNA structure analysis at single nucleotide resolution. *Nat. Prot.* **1**, 1610–1616.

55. Williamson, J.R. (2000). Induced fit in RNA-protein recognition. *Nat. Struct. Biol.*, **10**, 834–837.

Chapter 7

RNA Structural Analysis Based on Thermal Denaturation

Carla A. Theimer, Nakesha L. Smith, and Khadijah Onanuga-Islam

Department of Chemistry, University at Albany, SUNY, 1400 Washington Avenue, Albany, NY 12222, USA

ctheimer@albany.edu

Thermal denaturation methods have been used historically to investigate the thermodynamic stability and secondary and tertiary structures of nucleic acids, providing valuable insights into the folding behavior and conformations of RNA sequences. This chapter outlines the use of thermal denaturation methods for RNA structural analysis, specifically ultraviolet (UV) absorbance-monitored thermal denaturation and differential scanning calorimetry (DSC). In particular, the chapter describes the general theory and application of these techniques for RNA structural analysis, as well as methodologies that expand the scope of the information that can be determined.

RNA Nanotechnology
Edited by Bin Wang

ISBN 978-981-4411-64-6 (Hardcover), 978-981-4411-65-3 (eBook)
www.panstanford.com

7.1 Introduction

Helices are formed when complementary regions of primary nucleotide sequence are aligned such that each nucleic acid base in the sequence can form hydrogen bonds with an appropriate partner in the opposing sequence, forming hydrogen-bonded base pairs. While hydrogen bonding is an important factor in determining the specificity of secondary structure formation, hydrogen bonding does not drive the formation of nucleic acid helices. The major stabilizing force in ribonucleic acid (RNA) helical structure formation comes from base-stacking interactions [8]. Within a double helix, the planar nonpolar surfaces of the nitrogenous bases are sequestered from water when the hydrogen-bonded base pairs are stacked parallel to each other within the helical structure [54]. This arrangement allows noncovalent π stacking and hydrophobic interactions between the aromatic rings, resulting in an overall increase in the thermodynamic stability of the structure. In addition, the planar base pairs in a double helix are separated by a distance that is nearly optimal for van der Waals interactions. Therefore, it is the van der Waals interactions and the hydrophobic effect, not hydrogen bonding, that are the driving forces for double-helix formation by nucleic acid sequences.

This chapter describes how thermal denaturation experiments can and have been used to investigate the unfolding and refolding characteristics of nucleic acids [24, 31, 38, 49] and to determine the thermodynamic stability of nucleic acid structures (in terms of melting temperature (t_m, °C), enthalpy (ΔH_{vH} [van't Hoff], kcal mol^{-1}), and free energy under standard conditions, at a specific temperature (ΔG° [37°C], kcal mol^{-1}) [19]. The overall goal is to provide a perspective on the capabilities of these methods in determining the properties of RNA sequences of interest and to use that information for rational design of modified RNA sequences with required properties for RNA nanotechnology applications.

7.2 UV-Visible Absorption Spectroscopy

7.2.1 *The Hyperchromic Effect*

Denaturation of a nucleic acid structure, or unfolding, may result from a variety of sample conditions, including high temperature, high concentration of denaturants, and/or extreme pH conditions [54]. Ultraviolet (UV) absorption is a convenient way to monitor the folding and unfolding of nucleic acids by any of these methods, depending on the changes in their optical properties between the folded and unfolded states. It has been well established that there is a change in the absorbance of UV light (at 260 nm) by nucleic acids upon folding or unfolding [20, 36], (Fig. 1.1). The increase in UV absorption upon unfolding is called the hyperchromic effect (the decrease in absorbance upon folding, the hypochromic effect) [54]. Simply put, the hyperchromic effect is the observation that "double-stranded" (base-paired, folded) nucleic acids absorb less UV light than "single-stranded" (unfolded) nucleic acids [13].

UV absorption by RNA (and deoxyribonucleic acid (DNA)) results from the $n \rightarrow \pi$ and $\pi \rightarrow \pi^*$ transitions of the conjugated ring

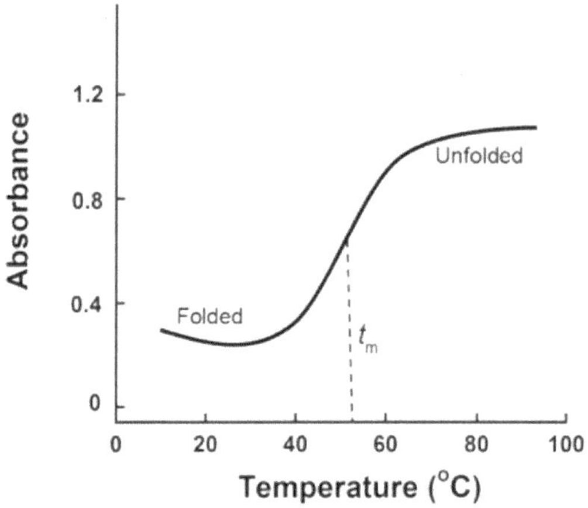

Figure 7.1 Typical thermal melting curve for a single nucleic acid double helix (folded) to unfolded transition. The vertical line indicates the melting temperature, t_m, where half the strands are in helical (folded) and half the strands in random coil (unfolded) conformations.

systems of the nucleobases. The change in UV light absorption by nucleic acid structures upon folding or unfolding is directly proportional to changes in the induced dipole moments of the bases [20, 36]. When a helix is formed, the bases of the helix have symmetric, roughly parallel, and stacked orientations. As a result, the dipole moments from the bases are stacked along the helix, causing the dipole moments in the two antiparallel strands to partially cancel each other out. Since absorption is directly proportional to the square of the dipole moment, a decrease in the dipole moment results in a decrease in absorption upon folding. The unfolding of a helical duplex to single strands eliminates the base alignment, and the magnitude of the dipole moments increase due to the random orientations of the transition moments, leading to hyperchromicity [54]. Therefore, since the hyperchromic effect is related to the extent to which a particular RNA (or DNA) adopts a folded secondary structure, the hyperchromic effect can be used to examine the folded state of an RNA and the pathway (series of unfolding steps) by which complicated RNA sequences unfold.

7.2.2 *Base Pair Content*

Based on Beer's law and a constant molar nucleic acid concentration, it was determined that changes in the absorbance associated with hyperchromicity (and hypochromicity) are the result of changes in the molar extinction coefficient (ε, M^{-1} cm^{-1}). Interestingly, the magnitude of the change in extinction coefficient ($\Delta\varepsilon$) for nucleic acids is different for A-U(T) and G-C base pairs at different wavelengths of absorbed light [20]. Equimolar mixtures of polyriboadenylic acid with polyribouridylic acid and polyriboguanylic acid with polyribocytidylic acid were used to generate polynucleotide helices containing all A-U or all G-C base pairs. Then, the difference spectra between the molar extinction coefficients of the double-stranded and single-stranded polynucleotides ($\Delta\varepsilon$) as a function of wavelength (λ) were examined [20].

The maximal increase in the extinction coefficient (hyperchromic effect) for A-U base pairs upon unfolding occurs around 260 nm, with little to no change in the extinction coefficient observed at 280 nm. For G-C base pairs, the maximal increase in the extinction coefficient upon unfolding is observed at ~230 nm and 275 nm, with a reduced, but nonzero, hyperchromic signal at 260 nm [20]. While the intensity of the hyperchromic effect for G-C base pairs is less than half that of A-U base pairs at 260 nm, it does contribute to the total hyperchromic effect observed at 260 nm. However, A-U base pairs contribute only negligibly to the hyperchromic effect at 275–280 nm. Therefore, the unfolding of G-C pairs can be optimally observed at a wavelength of 280 nm, and the optimal signal for A-U pairs can be observed at 260 nm. These observations can be utilized in thermal denaturation studies to identify the base pair content of individual transitions.

Investigating the base pair content is done by collecting unfolding data simultaneously at two wavelengths, 260 nm and 280 nm [18, 20, 43]. In a thermal denaturation experiment, absorption is graphed as a function of time (the "melting curve"). By plotting the derivative of the melting curve, $\delta A/\delta T$ versus temperature, the "melting profile" for the RNA can be obtained and melting transitions can be identified (Fig. 7.2). When the change in UV absorbance of a nucleic acid is measured at both wavelengths simultaneously, the ratio of these values (i.e., the $A260/A280$ ratio or hyperchromic signature) will give a qualitative measure of the base composition of the structural motif involved in a particular

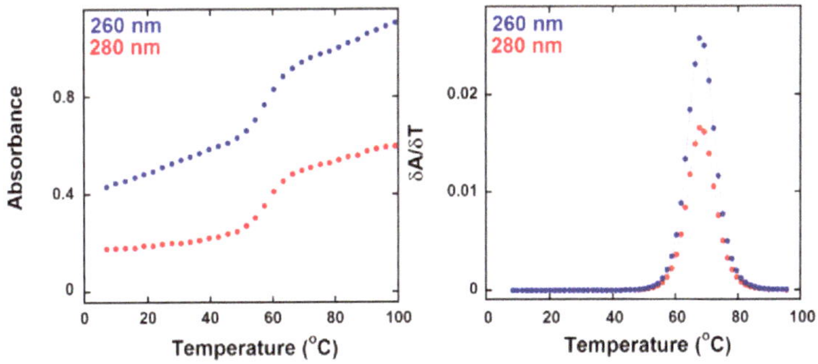

Figure 7.2 Thermal denaturation data for a simple hairpin collected at 260 nm (blue) and 280 nm (red). Data is shown as the raw absorbance (A) vs. temperature (T) data (left) as well as in the form of a first derivative "melting profile" (plotted as $\partial A/\partial T$ at 260 nm and 280 nm vs. temperature) (right).

unfolding transition [20], allowing identification of the structural element. A higher hyperchromic signature at 280 nm compared to the hyperchromic signature at 260 nm therefore indicates that the unfolding transition occurs in a region with a higher G-C base pair content than A-U base pair content.

This type of hyperchromic signature-based information, in combination with calculated thermodynamic parameters and mutational studies, was used to identify the unfolding transitions of individual helical stems (secondary structure) and the overall unfolding pathways of H-type RNA pseudoknots involved in programmed ribosomal frame-shifting in two different mouse retroviruses [49, 50], as well as the T2 and T4 gene 32 autoregulation [38, 48], and other systems. Similar hyperchromic information was also used, in combination with one-dimensional (1D) and two-dimensional (2D) nuclear magnetic resonance (NMR) spectroscopy, to detect the presence of significant tertiary structural contacts in the human telomerase RNA pseudoknot [52], identify the thermodynamic properties of human telomerase RNA hairpins [51–53], and investigate the unfolding pathway of the *Saccharomyces cerevisiae TLC1* telomerase RNA pseudoknot domain [33].

Since the magnitude of the hyperchromic signature for a nucleic acid is related to the amount of secondary structure present in the molecule, and the varying hyperchromic signatures at 260 nm and 280 nm report on base pair composition, even simple UV-monitored thermal denaturation experiments and absorbance versus temperature plots can be used to investigate and partially validate secondary structural predictions. For example, given two RNA sequences with the same number of nucleotides, the RNA with the larger hyperchromic signatures at 260 nm and 280 nm will have a more stacked helical structure. In addition, since the thermodynamic properties of an RNA structure are not dictated by the observation wavelength used to monitor the unfolding process, collecting thermal denaturation data at two wavelengths doubles the amount of data available for fitting. Fitting of this data to an unfolding model consisting of a set of sequential, independent two-state unfolding transitions (depending on the complexity of the RNA of interest) returns the thermodynamic parameters t_m and ΔH_{vH}, as well as the amplitudes of the hyperchromicity at each wavelength ($A260$ and $A280$, respectively) for each unfolding transition. Then, utilizing the experimentally determined t_m and ΔH_{vH} values, the free energy of a particular RNA can be calculated, (ΔG° (37°C)).

7.2.3 *The Two-State Assumption*

The two-state assumption states that an RNA duplex will exist in either a completely folded state or a completely unfolded state [57]. This assumption suggests that since no intermediate state is significantly populated, a discrete RNA helix can be considered to fold or unfold in a single step. Helices thus do not unzip one base pair at a time; all of the base pairs in a specific helix are disrupted simultaneously. Typically, short duplexes or helices do unfold in an all or nothing "two state" manner in which the nucleic acid is either completely base paired (F, folded) or completely single stranded (U, unfolded) and the two-state assumption is valid.

Deviations from the two-state assumption can be observed for long single helices, as well as some short ones, depending on their sequence, solution conditions, and other factors [57]. In the case of simple single helices, non-two-state behavior may be observable in melting profiles where the unfolding transition has a low-amplitude pretransition hyperchromic signal that cannot be fit accurately by a single Gaussian peak shape. It is possible to fit non-two-state melting behavior to a broad, low-amplitude unfolding transition, but the numerical values obtained for this type of transition are not always interpretable (since non-two-state behavior is being fit using a model that includes the two-state assumption).

When an RNA structure is made up of discrete structural elements or helices, these individual helices can fold or unfold independently of one another. If this is the case, each individual helix will generate an unfolding transition as it goes from the fully folded to fully unfolded state in a sequential manner. In addition, since the tertiary structure is formed after the secondary structure, there can be a discrete unfolding step (usually the first unfolding event) for tertiary structure unfolding. Therefore, the unfolding of an entire RNA structure that consists of multiple independent folding elements (typically helices) can be characterized by a series of discrete two-state unfolding steps (the "unfolding pathway") where each individual step reports on the two-state unfolding of a single helix or tertiary structural element.

7.2.4 *RNA Folding Studies Based on Thermal Denaturation*

UV-monitored thermal denaturation experiments are an excellent method to study the thermodynamics of nucleic acids. Due to the relatively low sample requirements (~1–5 nmoles per sample, depending on the specific RNA of interest), compared to other biophysical techniques, RNA can be studied under a wide variety of solution conditions and concentrations relatively quickly.

Of particular interest are the interactions of RNA with monovalent and divalent cations. RNA molecules have a large negative formal charge due to the negative charges on the phosphate groups in the phosphodiester backbone. As a result, RNAs in solution require the presence of positive counterions to form an ionic atmosphere around the molecule and minimize electrostatic repulsion. The presence of these counterions can significantly affect the folding and unfolding states of the RNA [1, 15, 45]. RNA secondary and tertiary structures can form under a variety of salt conditions. For small, less complicated RNAs, the presence of monovalent cations is sufficient to promote RNA folding; however, in larger, more complicated systems, like the large group I and group II introns and ribonuclease P (RNase P) RNA ribozymes, monovalent cations are not sufficient to activate these RNAs and divalent cations are required [11, 14, 17]. Divalent cations, such as magnesium, are often used to stabilize tertiary interactions because fewer ions are necessary and they can be utilized inside small folding pockets [6, 7, 25].

7.2.5 *Extending the Information Content of Thermal Denaturation: Fluorescent Base Analog Labeling*

The use of site-specific internal nucleotide labeling has become a fairly common practice to study specific local folding and unfolding events associated with nucleic acids [2]. The most common label used in UV-monitored thermal denaturation studies is the fluorescent base analog to adenine, 2-aminopurine (2AP). Performing UV-monitored thermal melting with internally labeled adenines allows the discrimination and identification of specific A-U base pairs present in an RNA structure. Typical UV-monitored thermal melting experiments monitor absorbance at 260 nm and 280 nm; however,

2AP has a red-shifted absorbance spectrum that can be monitored at a maximum wavelength of 370 nm [44, 65]. The basis of 2AP detection by UV-monitored thermal denaturation is the same as that for normal nucleobases, which results in a hyperchromic shift from stacking interactions with neighboring bases. This tool is especially useful in elucidating more complicated melting profiles. Recently, 2AP-based thermal denaturation experiments have been used to investigate the rearrangement of the human immunodeficiency virus (HIV) SL1 stem loop from a kissing stem–loop dimer to an extended dimer [34] and to study trinucleotide repeat sequences [10].

Another base analog that maintains Watson–Crick base pairing, 3-[beta-D-2-ribofuranosyl]-6-methylpyrrolo[2,3-d]pyrimidin-2 (3H)-one (pyrroloC or pyrolocytidine), a fluorescent analog of the nucleoside cytidine, has also been used for monitoring the RNA secondary structure [55]. Similar to the 2AP nucleoside, the fluorescence of pyrrolocytidine becomes reversibly quenched upon base pairing, most likely due to stacking interactions with neighboring bases. Further permutations of fluorescent properties can be obtained based on the nature of the position 6 functional group on pyrollocytidine, in addition to the sequence position of the modified nucleoside [61, 62]. In addition to thermal denaturation studies, pyrrolocytidine derivatives have been used for other types of RNA investigations, including investigations of small interfering RNA (siRNA) trafficking in real-time imaging [63], and HIV-1 RT RNase H activity [64]. Thus, strategic use of fluorescent base analogs can make it possible to investigate multiple aspects of RNA function based on a single substitution in the RNA sequence.

7.3 Differential Scanning Calorimetry

7.3.1 *Calorimetry Measurements*

Calorimetry is a complementary technique to UV-monitored thermal denaturation experiments for characterization of RNA unfolding. Two experimental calorimetric techniques that have been widely used for this purpose are differential scanning calorimetry (DSC) and isothermal titration calorimetry (ITC). The primary advantage of calorimetry over UV-monitored thermal denaturation studies is the ability of calorimetric methods to directly measure the thermodynamics of biological systems, rather than using indirect methods based on changes in other observable physical properties. Additionally, since calorimetry is a direct measure of the heat absorbed or released by the system, thermodynamic information can be obtained for RNA structural transitions that have little to no optical signature, and the total calorimetric enthalpy measured is model independent.

Calorimetry measures the excess differential heat capacity (Cp) relative to a reference sample as a function of temperature, and the enthalpy of the system can then be determined using the difference in heat capacity. Heat capacity is defined as the amount of energy needed to raise the sample temperature by one degree. Typically, if an oligonucleotide unfolds consistent with a true two-state model-unfolding process, even if there are multiple unfolding transitions, the model-independent measured calorimetric enthalpy, and the two-state model-dependent calculated enthalpy from the UV-monitored thermal denaturation data will agree. However, if intermediate non-two-state unfolding states are significantly populated, the optically observed ΔH will be smaller than the true ΔH determined by calorimetry. Despite providing more accurate thermodynamic data than UV-monitored thermal denaturation studies, the major limitations of this technique are the availability of RNase-contamination-free instrumentation, the relatively large amount of material required for the experiment, and its destructiveness. Instruments for RNA calorimetry typically have to be specially ordered since RNase is often used as a testing standard for new instruments. In addition, rigorous use and cleaning protocols must be maintained to prevent RNase contamination if protein-containing samples are also tested on the instrument. DSC samples also require typically ~25–100 times more material than UV-melting samples, and the RNA sample is significantly degraded after a single DSC scan that runs up to 120–140°C, depending on instrument configuration and solution conditions.

In addition to the model independent total enthalpy, ΔH_{cal}, the model dependent enthalpy, $\Delta H_{\mathrm{vH,cal}}$ (van't Hoff), can also be

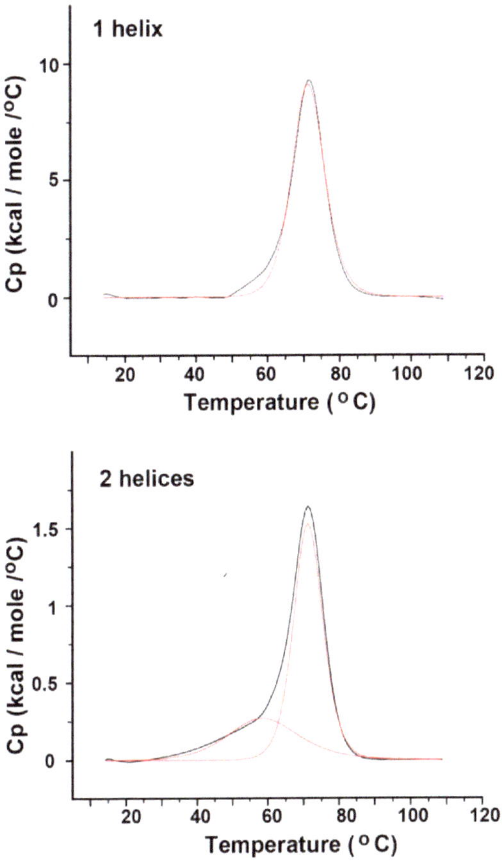

Figure 7.3 DSC data for two different RNAs, one containing a single helix (top) and a second containing two helical segments (bottom). Data is shown in black, with the calculated van't Hoff transitions indicated in red. Note the small deviation between experimental and fit data in the 1 helix plot immediately before the large transition, consistent with a small amount of non-two-state unfolding behavior.

determined by fitting the calorimetric enthalpy using the two-state model (Fig. 7.3) after setting the pre- and post-transition baselines. The model-dependent enthalpy determined from the calorimetry data should be in good agreement with the UV-monitored thermal denaturation data.

One of the primary concerns with UV-monitored thermal denaturation experiments is the assignment of the correct number of transitions, since addition of extra transitions will always lower the chi-squared value, even if they are physically meaningless. However, since the combined enthalpy of the model-dependent transitions from the UV-monitored experiment should not exceed the total calorimetric enthalpy determined by DSC, this combination of techniques can prevent the incorporation of additional unnecessary transitions. UV-monitored experiments can also miss transitions that have small optical signatures, but as long as there is sufficient enthalpy associated with the transition, low hyperchromicity transitions can be detected in the DSC experiment based on their thermal properties. Thus, the combination of the DSC and UV-monitored thermal denaturation experiments makes it possible to more accurately determine the total energy of the system, and the accurate number of transitions in more complicated RNA systems.

7.3.2 *Application of DSC to Biochemical Investigations*

DSC is used in many studies involving molecular recognition and interactions between biomolecules [9, 27]. DSC measurements have been used to investigate the thermodynamic parameters of samples, including, but not limited to, nucleic acid conformations under various conditions [39], protein–protein or protein–lipid interactions, protein–ligand interactions, *trp*-RNA-binding attenuation protein activation by tryptophan [35], and the thermodynamics and molecular mechanisms of drug–DNA interactions [3, 4]. Research involving the inhibition of the expression of mutant genes that cause diseases in humans and other immunology studies has also benefited from DSC techniques, including studies of Huntington's disease [21], the role of stem–loop structures in coronavirus replication [32], and the role of RNA rearrangement in memory T-cell differentiation [66].

Calorimetric techniques have been widely used for RNA-related investigations, including analyzing the effect of various conditions on protein–RNA complexes in viruses [16, 60], the biophysical characterization of RNA from viruses [28, 30], the comparison of synthesized 4′-thioRNAs in human serum to natural RNA [26], and the folding pathways of higher-order RNA structures [58]. DSC has also been applied to studies targeted at improving siRNAs in conjunction with X-ray diffraction techniques to compare the efficacy of modified RNAs in vitro and in vivo with unmodified siRNAs [40]. Other in vitro studies examined finding effecting delivery vectors for siRNA using DSC and other techniques [47].

In addition, DSC has been widely applied in the pharmaceutical industry for determining drug purity and examining different drug delivery methods [42]. Some investigations include the use of nanoparticles in cytosolic drug delivery, the efficacy of transfection reagents for siRNA, and the effect of nucleoside modifications on siRNA activity [12, 29, 46]. The interactions of nucleic acids with lipid membranes are essential to applications of biotechnology in gene delivery and drug carriers as well as in biological mechanisms [22]. Researchers used DSC to examine the conditions under which transfer RNA (tRNA) interacts with phospholipid bilayers, and monitored how the lipid bilayers were affected by this interaction [37].

7.3.3 *Instrumentation Advances*

Many advances have been made in DSC instrumentation. Traditional DSC involves the measurement of enthalpy change in a sample as a function of temperature. The heat absorbed or released in the sample is measured against a reference material as the temperature of both is increased in a controlled manner. In addition to traditional DSC, DSC instruments have been designed to be heated by infrared radiation, perform parallel nanocalorimetry, and use microelectromechanical systems. Additional types are the gas-flow-modulated, pressure perturbation, modulated-temperature, self-referenced, and high-performance DSC instruments [23].

As technology advances, more sensitive commercial instruments have been developed including a sensitive DSC instrument connected to a robotic system that handles samples loading, cleaning between experiments, and data collection. After the initial setup, the system can work automatically performing about 10 to 50 experiments per day [41]. Two miniature calorimeters have been made through microscale technology that could facilitate studies of enzymatic reactions and determination of heats of binding. One machine is equipped with an enthalpy array, which permits the mixing of two 250 nanoliter drops for analysis [56]. Another was designed to enable parallel data collection, making it possible to use a sample volume of 5 microliters [59]. Some scientists, however, prefer the commercial microcalorimeters to high-efficiency calorimeters because the high efficiency falls short in terms of sensitivity [5]. Nevertheless, calorimetry instruments are moving toward having smaller sample requirements, multiple-sample-handling capacity, and automation, which will make them more and more powerful tools in the future.

7.4 Outlook

Overall, UV-monitored thermal denaturation and DSC experiments are typically most useful at the point where a secondary structural or tertiary structural model has been proposed and the model is being evaluated for validation purposes. In addition, these techniques can also be used to facilitate the rational design of new RNA sequences and structures. RNA systems and structures that are rationally designed are far more likely to yield the properties required for downstream experimentation. Thus, a clear understanding of the underlying thermodynamic principles of folding and structure makes the design of RNA platforms and nanosystems more attainable and paves the way for advances in RNA nanotechnology.

References

1. Bai, Y., et al. (2007). Quantitative and comprehensive decomposition of the ion atmosphere around nucleic acids. *JACS*, **129**(48), 14981–14988.
2. Ballin, J.D., et al. (2007). Site-specific variations in RNA folding thermodynamics visualized by 2-aminopurine fluorescence. *Biochemistry*, **46**(49), 13948–13960.
3. Barone, G., Del Vecchio, P., Giancola, C., and Graziano, G. (1995). The liquid amide transfer model and the unfolding thermodynamics of small globular proteins. *Int. J. Biol. Macromol.*, **17**(5), 251–257.

4. Bruylants, G., Wouters, J., and Michaux, C. (2005). Differential scanning calorimetry in life science: thermodynamics, stability, molecular recognition and application in drug design. *Curr. Med. Chem.*, **12**(17), 2011–2020.
5. Chaires, J.B. (2008). Calorimetry and thermodynamics in drug design. *Annu. Rev. Biophys.*, **37**, 135–151.
6. Chin, K., Sharp, K.A., Honig, B., and Pyle, A.M. (1999). Calculating the electrostatic properties of RNA provides new insights into molecular interactions and function. *Nat. Struct. Biol.*, **6**(11), 1055–1061.
7. Conn, G.L., et al. (2002). A compact RNA tertiary structure contains a buried backbone-K^+ complex. *JMB*, **318**(4), 963–973.
8. Crothers, D.M., and Zimm, B.H. (1964). Theory of the melting transition of synthetic polynucleotides: evaluation of the stacking free energy. *JMB*, **9**(1), 1–9.
9. Das, A., Bhadra, K., and Suresh Kumar, G. (2011). Targeting RNA by small molecules: comparative structural and thermodynamic aspects of aristololactam-ß-D-glucoside and daunomycin binding to $tRNA^{phe}$. *PLoS ONE*, **6**(8), e23186.
10. Degtyareva, N.N., Reddish, M.J., Sengupta, B., and Petty, J.T. (2009). Structural studies of a trinucleotide repeat sequence using 2-aminopurine. *Biochemistry*, **48**(11), 2340–2346.
11. Derose, V.J. (2003). Metal ion binding to catalytic RNA molecules. *Curr. Opin. Struct. Biol.*, **13**(3), 317–324.
12. Dobbs, W., et al. (2009). Mesomorphic imidazolium salts: new vectors for efficient siRNA transfection. *JACS*, **131**(37), 13338–13346.
13. Doty, P., et al. (1959). Secondary structure in ribonucleic acids. *PNAS*, **45**, 482–499.
14. Doudna, J.A., and Cech, T.R. (2002). The chemical repertoire of natural ribozymes. *Nature*, **418**(6894), 222–228.
15. Draper, D.E. (2004). A guide to ions and RNA structure. *RNA*, **10**(3), 335–343.
16. Fang, H.-J., et al. (2009). Thermostability of the N-terminal RNA-binding domain of the SARS-CoV nucleocapsid protein: experiments and numerical simulations. *Biophys. J.*, **96**(5), 1892–1901.
17. Fedor, M.J., and Westhof, E. (2002). Ribozymes: the first 20 years. *Mol. Cell*, **10**(4), 703–704.
18. Felsenfeld, G., and Hirschman, S.Z. (1965). A neighbor-interaction analysis of the hypochromism and spectra of DNA. *JMB*, **13**(2), 407–427.

19. Freier, S.M., Petersheim, M., Hickey, D.R., and Turner, D.H. (1984). Thermodynamic studies of RNA stability. *J. Biomol. Struct. Dyn.*, **1**(5), 1229–1242.
20. Fresco, J.R., Klotz, L.C., and Richards, E.G. (1963). A new spectroscopic approach to the determination of helical secondary structure in ribonucleic acids. *Cold Spring Harb. Symp. Quant. Biol.*, **28**, 83–90.
21. Gagnon, K.T., et al. (2010). Allele-selective inhibition of mutant huntingtin expression with antisense oligonucleotides targeting the expanded CAG repeat. *Biochemistry*, **49**(47), 10166–10178.
22. Giatrellis, S., and Nounesis, G. (2011). Nucleic acid-lipid membrane interactions studied by DSC. *J. Pharm. Bioallied Sci.*, **3**(1), 70–76.
23. Gill, P., Moghadam, T., and Ranjbar, B. (2010). Differential scanning calorimetry techniques: applications in biology and nanoscience. *J. Biomol. Tech.*, **21**, 167–193.
24. Gluick, T.C., Wills, N.M., Gesteland, R.F., and Draper, D.E. (1997). Folding of an mRNA pseudoknot required for stop codon readthrough: effects of mono- and divalent ions on stability. *Biochemistry*, **36**(51), 16173–16186.
25. Grilley, D., Soto, A.M., and Draper, D.E. (2006). Mg^{2+}–RNA interaction free energies and their relationship to the folding of RNA tertiary structures. *PNAS*, **103**(38), 14003–14008.
26. Hoshika, S., Minakawa, N., and Matsuda, A. (2004). Synthesis and physical and physiological properties of 4'-thioRNA: application to post-modification of RNA aptamer toward NF-kappaB. *NAR*, **32**(13), 3815–3825.
27. Islam, M.M., Sinha, R., and Kumar, G.S. (2007). RNA binding small molecules: studies on t-RNA binding by cytotoxic plant alkaloids berberine, palmatine and the comparison to ethidium. *Biophys. Chem.*, **125**(2–3), 508.
28. Jin, E., et al. (2000). Aminoglycoside binding in the major groove of duplex RNA: the thermodynamic and electrostatic forces that govern recognition. *JMB*, **298**(1), 95–110.
29. Kenski, D.M., et al. (2010). Analysis of acyclic nucleoside modifications. *NAR*, **38**(2), 660–671.
30. Kuznetsov, Y.G., et al. (2010). Biophysical and atomic force microscopy characterization of the RNA from satellite tobacco mosaic virus. *NAR*, **38**(22), 8284–8294.
31. Laing, L.G., Gluick, T.C., and Draper, D.E. (1994). Stabilization of RNA structure by Mg ions: specific and non-specific effects. *JMB*, **237**(5), 577–587.

32. Li, L., et al. (2008). Structural lability in stem-loop 1 drives a 5' UTR-3' UTR interaction. *JMB*, **377**(3), 790–803.
33. Liu, F., Kim, Y., Cruickshank, C., and Theimer, C.A. (2012). Thermodynamic characterization of the *Saccharomyces cerevisiae* telomerase RNA pseudoknot domain in vitro. *RNA*, **18**(5), 973–991.
34. Livshits, M.A., Kaluzhny, D.N., and Minyat, E.E. (2011). Melting of model HIV-1 stem-loop 1 RNA dimers monitored by 2-aminopurine fluorescence. *J. Biomol. Struct. Dyn.*, **28**(6), 939–947.
35. Mcelroy, C.A. (2006). Thermodynamics of tryptophan-mediated activation of the trp. *Biochemistry*, **45**(25), 7844–7853.
36. Mergny, J.-L., et al. (2005). Thermal difference spectra: a specific signature for nucleic acid structure. *NAR*, **33**(16), e138-.
37. Michanek, A., et al. (2009). RNA and DNA interactions with zwitterionic and charged lipid membranes: a DSC and QCM-D study. *Biochim. Biophys. Acta*, **1798**(4), 829–838.
38. Nixon, P.L., Theimer, C.A., and Giedroc, D.P. (1999). Thermodynamics of stabilization of RNA pseudoknots by cobalt(III) hexaammine. *Biopolymers*, **50**(4), 443–458.
39. O' Brien, R., and Haq, I. (2005). Applications of biocalorimetry: binding, stability and enzyme kinetics, in *Biocalorimetry 2,* Ladbury, J.E., and Doyle, M.L. (eds.), John Wiley & Sons, Chichester, UK, 1–34.
40. Pallan, P.S., et al. (2011). Unexpected origins of the enhanced pairing affinity of 2'-fluoro-modified RNA. *NAR*, **39**(8), 3482–3495.
41. Plotnikov V, R.A. (2002). An autosampling differential scanning calorimeter. *Assay Drug Dev. Tech.*, **1**, 83–90.
42. Prenner, E., and Chiu, M. (2011). Differential scanning calorimetry: an invaluable tool for a detailed thermodynamic characterization of macromolecules and their interactions. *J. Pharm. Bioallied Sci.*, **3**(1), 39–59.
43. Puglisi, J.D., and Tinoco, I.J. (1989). Absorbance melting curves of RNA. *Methods Enzymol.*, **180**, 304–325.
44. Rist, M.J., and Marino, J.P. (2002). Fluorescent nucleotide base analogs as probes of nucleic acid structure, dynamics and interactions. *Curr. Org. Chem.*, **6**(9), 775–793.
45. Sarah A, W. (2005). Metal ions and RNA folding: a highly charged topic with a dynamic future. *Curr. Opin. Chem. Biol.*, **9**(2), 104–109.
46. Sedlák, M., Konák, C., and Dybal, J. (2010). Nanoparticles for cytosolic delivery of important biomolecular drugs. *Open Conf. Proc. J.*, **1**, 87–90.

47. Techaarpornkul, S.W., et al. (2010). Chitosan-mediated siRNA delivery in vitro: effect of polymer molecular weight, concentration and salt forms. *AAPS PharmSciTech*, **11**(1), 64–72.
48. Theimer, C.A., et al. (1998). Non-nearest neighbor effects on the thermodynamics of unfolding of a model mRNA pseudoknot. *JMB*, **279**(3), 545–564.
49. Theimer, C.A., and Giedroc, D.P. (1999). Equilibrium unfolding pathway of an H-type RNA pseudoknot which promotes programmed −1 ribosomal frameshifting. *JMB*, **289**(5), 1283–1299.
50. Theimer, C.A., and Giedroc, D.P. (2000). Contribution of the intercalated adenosine at the helical junction to the stability of the *gag-pro* frameshifting pseudoknot from mouse mammary tumor virus. *RNA*, **6**(3), 409–421.
51. Theimer, C.A., Finger, L.D., and Feigon, J. (2003). YNMG tetraloop formation by a dyskeratosis congenita mutation in human telomerase RNA. *RNA*, **9**(12), 1446–1455.
52. Theimer, C.A., Finger, L.D., Trantirek, L., and Feigon, J. (2003). Mutations linked to dyskeratosis congenita cause changes in the structural equilibrium in telomerase RNA. *PNAS*, **100**(2), 449–454.
53. Theimer, C.A., et al. (2007). Structural and functional characterization of human telomerase RNA processing and cajal body localization signals. *Mol. Cell*, **27**(6), 869–881.
54. Tinoco, I. (1960). Hypochromism in polynucleotides. *JACS*, **82**(18), 4785–4790.
55. Tinsley, R.A., and Walter, N.G. (2006). Pyrrolo-C as a fluorescent probe for monitoring RNA secondary structure formation. *RNA*, **12**(3), 522–529.
56. Torres, F.E., et al. (2004). Enthalpy arrays. *PNAS*, **101**(26), 9517–9522.
57. Turner, D.H., Sugimoto, N., and Freier, S.M. (1988). RNA structure prediction. *Annu. Rev. Biophys. Biophys. Chem.*, **17**(1), 167–192.
58. Vander Meulen, K.A., et al. (2008). Thermodynamics and folding pathway of tetraloop receptor-mediated RNA helical packing. *JMB*, **384**(3), 702–717.
59. Verhaegen, K., Baert, K., Simaels, J., and Van Driessche, W. (1998). Design of a high-throughput microphysiometer, in *Biocalorimetry: Applications of Calorimetry in the Biological Sciences*, Ladbury, J., and Chowdhry, B. (eds.), John Wiley & Sons, New York, 227–231.
60. Virudachalam, R., Low, P.S., Argos, P., and Markley, J.L. (1985). Turnip yellow mosaic virus and its capsid have thermal stabilities with opposite

ph dependence: studies by differential scanning calorimetry and 31P nuclear magnetic resonance spectroscopy. *Virology*, **146**(2), 213–220.

61. Wahba, A.S., Damha, M.J., and Hudson, R.H. (2008). RNA containing pyrrolocytidine base analogs: good binding affinity and fluorescence that responds to hybridization. *Nucleic Acids Symp. Ser. (Oxf.)*, (52), 399–400.
62. Wahba, A.S., Damha, M.J., and Hudson, R.H. (2008). RNA containing pyrrolocytidine base analogs: increased binding affinity and fluorescence that responds to hybridization. *Nucleic Acids Symp. Ser. (Oxf.)*, (52), 397–398.
63. Wahba, A.S., Esmaeili, A., Damha, M.J., and Hudson, R.H. (2010). A single-label phenylpyrrolocytidine provides a molecular beacon-like response reporting HIV-1 RT RNase H activity. *NAR*, **38**(3), 1048–1056.
64. Wahba, A.S., et al. (2011). Phenylpyrrolocytosine as an unobtrusive base modification for monitoring activity and cellular trafficking of siRNA. *ACS Chem. Biol.*, **6**(9), 912–919.
65. Ward, D.C., Reich, E., and Stryer, L. (1969). Fluorescence studies of nucleotides and polynucleotides. *JBC*, **244**(5), 1228–1237.
66. Wu, Z., et al. (2008). Memory T cell RNA rearrangement programmed by heterogeneous nuclear ribonucleoprotein hnRNPLL. *Immunity*, **29**(6), 863–875.

Chapter 8

Probing the Structural Basis of Retroviral RNA Functions via NMR Spectroscopy

Blanton S. Tolbert, Carrie Rollins, Jeffrey D. Levengood, Le Luo, and Prashant Rajan

Case Western Reserve University, 10900 Euclid Avenue,
Cleveland, OH 44106-7708, USA
bst18@case.edu

8.1 Introduction

Ribonucleic acid (RNA) molecules fold into elaborate structures that underlie their molecular functions. Knowledge of the structural and dynamical properties of RNA thus informs on mechanistic principles. Nuclear magnetic resonance (NMR) spectroscopy is the only analytical tool that can report both on the high-resolution structure and the conformational dynamics of RNA under solution conditions. Recent advances in preparation of isotopically labeled ribonucleosidetriphosphates (rNTPs) [1, 2], improvements in RNA segmental ligation [3, 4], and the continued development of NMR methods [5, 6] have allowed for larger and more complex RNA molecules to be studied by NMR. Unlike proteins, the secondary

RNA Nanotechnology
Edited by Bin Wang

ISBN 978-981-4411-64-6 (Hardcover), 978-981-4411-65-3 (eBook)
www.panstanford.com

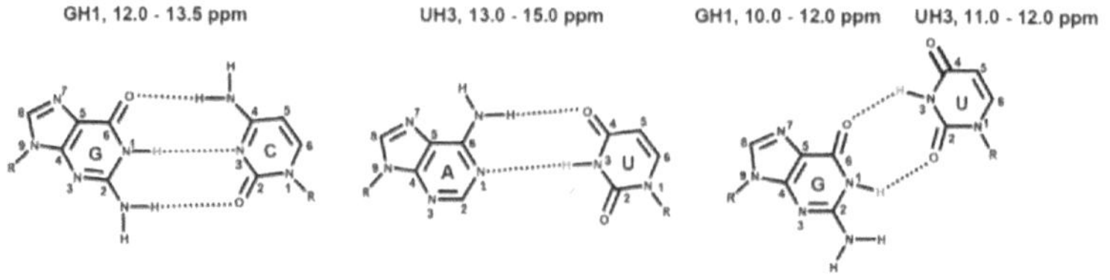

Figure 8.1 Schematic representation of Watson–Crick G:C, A:U, and G:U base pairs. Imino protons are colored red, and their approximate ^{1}H NMR chemical shift ranges are shown above.

structure of a given RNA molecule tightly constrains its tertiary folding space [7]. Hence, an accurate description of the RNA two-dimensional (2D) conformation provides a robust framework to probe function with nucleotide resolution. The imino protons of guanosines and uracils are sensitive reporters of the RNA secondary structure since their NMR signals depend on the base pair type (Fig. 8.1). Furthermore, the H2 proton of adenosines in Watson–Crick A–U pairs gives rise to the readily identifiable nuclear Overhauser effect (nOe) patterns that are sensitive to the identity of the nearest neighbors. The sensitivity of the imino and H2 chemical shifts to RNA conformation thus sheds light on molecular function, even in the absence of high-resolution structures. In this chapter, we discuss how recent advancements in NMR methods and labeling technologies have provided insight into retroviral RNA functions. Retroviral genomes contain highly structured RNA elements that self-assemble to carry out essential viral mechanisms; as such, they represent evolutionarily optimized biological "nanomachines" that are excellent model systems for probing structure-property-function relationships. In many cases, solution NMR spectroscopy has been the go-to method for elucidating structural-based mechanisms of retroviral RNA functions. Here, we briefly provide an account of more recent NMR studies of retroviral RNA genomes. Our approach is to place greater emphasis on how NMR has shed light on the biological functions of these RNA nanomachines, as opposed to offering a technical overview of the NMR methodology. Several excellent reviews on the latter topic have already been written [8–10].

8.2 Methods for Preparing Segmentally Labeled RNA

NMR spectroscopic analyses of functional RNA molecules require rNTPs enriched with ^{13}C, ^{15}N, and, in some cases, ^{2}H isotopes. Isotopic enrichment and segmentally labeled constructs provide the advantage of distinguishing structural characteristics of the RNA otherwise difficult to resolve. Since isotopic-enriched rNTPs with diverse site-specific labeling patterns are commercially available, we will not discuss their (bio)synthetic schemes here. Recent review articles written by the Dayie and Summers' groups cover these topics in detail [1, 2], and new approaches for preparing novel site-specific probes continue to be described [11–15]. Here, we briefly discuss methods for preparing segmentally labeled RNA constructs, with examples taken from the primary literature and our own attempts to collect high-quality NMR data on an ~175-nucleotide splicing regulatory element from human immunodeficiency virus (HIV)-1.

The Crothers' group first reported a procedure for preparing segmentally labeled RNA constructs used for NMR studies of the 5′ half of *Caenorhabditis elegans* (*C. elegans*) spliced leader (5CESL) RNA 1 [16]. Their method employs three enzymatic reactions: (i) T7 RNA polymerase (RNAP)-dependent in vitro transcription, (ii) ribonuclease (RNase) H site-specific cleavage, and (iii) T4 deoxyribonucleic acid (DNA)-mediated ligation. Each reaction step is followed by urea gel purification. As is commonly known, T7 RNAP produces transcripts with heterogeneous ends due to its tendency to abort initiations and to add nontemplate nucleotides at the 3′ termini of nascent transcripts [17, 18]. End heterogeneity leads to a distribution of transcripts that might complicate the ligation reaction. To alleviate this complication, the Crothers' group used an RNase H digestion step that allows for site-specific cleavage and formation of ligation constructs with well-defined termini, namely, a donor substrate with a 5′ monophosphate and an acceptor substrate with a 3′ hydroxyl. The 5CESL RNA was thus in vitro–transcribed using unlabeled as well as ^{15}N-labeled rNTPs, digested with RNase H to yield appropriate fragments, and cross-ligated with T4 DNA ligase. The T4 DNA ligase step requires the use of an annealed DNA splint, which creates a double-strand junction at the donor and acceptor termini. The complex formed between the RNA donor, acceptor, and DNA splint is referred to as a ligation competent complex (LCC) [19]. Incubation of the LCC with T4 DNA ligase yielded two ^{15}N segmentally labeled 5CESL constructs (18–54% yield) that were used to differentiate between two competing RNA secondary structural models. Thus, segmental ligation provided the necessary NMR chemical shift resolution and signatures to unambiguously determine the 5CESL secondary structure.

T4 DNA ligation yields are sensitive to several physical parameters, including RNA structure and DNA splint length. Using size exclusion chromatography, Kurschat et al. demonstrated that splint length affects the extent of LCC formed and the structural integrity of the resultant complex [19]. In their work, tRNA^{lys} (73 nts) ligation reactions were carried out with DNA splints that ranged in size from 30 to 73 nts. Splints of short (30 nts) to intermediate lengths (40–50 nts) led to either no LCC formed or LCCs with different hydrodynamic properties (structural isomers that reduced ligation yields), respectively. Complete hybridization with a 73 nt splint resulted in a single, well-behaved LCC that produced maximum ligation yield, greater than 95% after size exclusion chromatography. Presumably, the longer splint was necessary to disrupt the competing RNA structure.

In our lab, we have attempted to prepare segmentally labeled constructs of an RNA fragment derived from the HIV-1 splice site A7 (ssA7) locus. The ssA7 fragment is ~60 kDa, contains three stem loops connected by flexible linkers, and comprises *cis* regulatory elements that modulate splicing efficiency (Fig. 8.2) [20, 21]. As a first step toward collecting high-quality NMR spectra on an intact ssA7 construct, we screened conditions for segmentally ligating SL12 to SL3. Both SL12 and SL3 fragments were prepared by runoff transcription; however, SL3 was primed with excess guanosine 5′-monophosphate (GMP) to produce the appropriate 5′ donor terminus. Using DNA splints of different lengths (33–81 nts), LCC formation was monitored by native gel electrophoresis (Fig. 8.2). The 33 nt DNA splint produced low levels of stable LCC under our native gel conditions. Incubations of SL12 and SL3 with longer splints led to more stable LCCs; however, only the 81 nt splint produced a robust and well-defined LCC (Fig. 8.2).

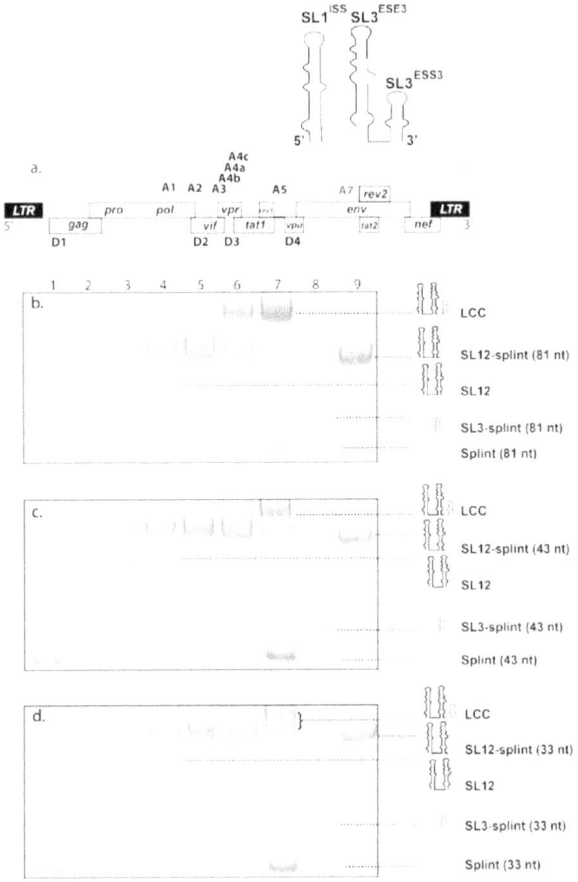

Figure 8.2 Formation of a ssA7 ligation competent complex. (a) Overview of the HIV-1 genome, highlighting the nine open reading frames along with the 5′ splice donor (D1–D4) and 3′ splice acceptor (A1–A7) sites. The experimentally determined RNA secondary structure at ssA7 is drawn schematically. Incubations of SL12 and SL3 fragments with increasing amounts of (b) 81 nt splint, (c) 43 nt splint, and (d) 33 nt splint. Lane numbers are as follows: (1) free splint, (2) free SL12, (3–7) fixed SL12/SL3 with increasing splint amount, (8) SL3+splint, and (9) SL12+splint.

Several alternatives to the Crothers' segmental ligation protocol have been described, where major differences include the use of either *cis*- or *trans*-cleaving ribozymes to produce appropriate termini [3], ligations with T4 RNA rather than T4 DNA ligase [4, 22, 23], or non-polyacrylamide gel electrophoresis (PAGE) purifications of the RNA fragments [3]. More recently, the Allain group described a sequence-independent method for preparing multiple segmental ligations. Their approach combines *cis*-acting ribozymes for end preparation along with RNase H digestion steps to prepare sequence-independent ligation-ready termini. In sum, segmental ligation of RNA with fragments that contain NMR-active nuclei allows detection of RNA structural features that would be difficult or impossible to observe using unlabeled constructs. The commercial availability of isotopically enriched rNTPs, along with improved ligation protocols could potentially make segmental ligation a routine platform for those interested in characterizing complex RNA nanostructures by NMR spectroscopy. A few examples of how these methods have provided structural insight into retroviral RNA functions are highlighted below.

8.3 NMR Detection of RNA Structures Involved in Retroviral Packaging Mechanisms

8.3.1 *The Murine Leukemia Virus*

All retroviruses selectively package a dimeric copy of their unspliced RNA genomes. Genome dimerization is promoted through RNA structures located mostly within the 5′-untranslated region (UTR), and these structures overlap with *cis*-packaging signals (psi or Ψ) that confer specific recognition of viral genomic RNA by the Gag polyprotein. There is now a general consensus that genome dimerization and packaging are mechanistically linked. However, the molecular details as to how dimerization manifests into selective recognition of Ψ signals by Gag is just starting to be revealed [24]. Traditional RNA modification techniques and new selective 2′-hydroxyl acylation analyzed by primer extension (SHAPE) methods have shown differential reactivity patterns between monomeric and dimeric genomes, leading to the suggestion that RNA conformational changes promote selective packaging of the dimer [25–34]. The first structure-based mechanism was contributed by NMR studies of the Ψ signal from the murine leukemia virus (MLV) [35–38]. The MLV Ψ region encompasses four conserved stem loops: DIS-1 (or SLA), DIS-2 (or SLB), SL-C, and SL-D, where DIS refers to the dimer initiation site. DIS-1 and DIS-2 contain palindromic segments that upon genome dimerization form intermolecular duplexes, whereas stem loops C and D contain highly conserved GACG tetraloop motifs that form intermolecular kissing interactions [39–41]. Isolated fragments of DIS-2 (and DIS-1) undergo a concentration-dependent stem-loop-to-duplex transition that is easily detected by native gel electrophoresis or one-dimensional (1D) ^{1}H NMR spectroscopy [36, 38]. In particular, the H2 chemical shifts of *A*289 and *A*293 move upfield to around 6.45 and 6.72 ppm, respectively, in the duplex conformation. These protons resonate within a well-resolved region of the spectrum and thus are diagnostic of an MLV Ψ dimer. Early chemical modification experiments showed that DIS-2 undergoes a register shift where nucleotides U304–U309 are exposed in the dimer but remain base-paired in the monomer [30]. More recent calorimetric titrations of isolated DIS-2 fragments, engineered to retain secondary structural features of either the stem loop or the duplex, with the nucleocapsid (NC) domain of Gag showed that only the duplex-like structure binds NC with high affinity ($K_d = 173 \pm 32$ nM for DIS-2_{duplex}) [36]. Structural insight into the determinants of high-affinity binding were provided by solving the NMR structure of the NC-($\Psi^{\text{DIS-2 / SLC / SLD}}$) complex where the apical loops were replaced with GNRA tetraloops, which forced the monomeric RNA into a dimer-like conformation [36]. The structure provided a rationale for why MLV genome dimers are selectively packaged over monomers (Fig. 8.3). Dimerization induces an RNA structural switch in DIS-2 (and DIS-1) that exposes high-affinity NC recognition sites (UCUG elements). These sites are sequestered by base pairing in monomeric genomes and are thus not recognized by the NC domain of Gag.

The GNRA substitution at DIS-2 forced it into a dimer-like conformation; however, analogous substitutions at SL-C and SL-D prevented complete characterization of their tertiary interactions.

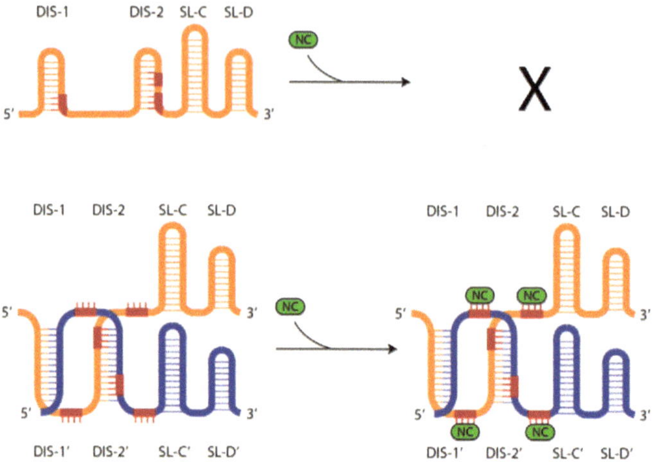

Figure 8.3 NMR-derived mechanism of retroviral RNA genome dimerization and packaging for the MLV retrovirus. (Top) Monomeric genomes are not recognized by NC because the high-affinity UCUG recognition sites (shown in red) are sequestered by base pairing. (Bottom) Genome dimerization induces a register shift in DIS-1 and DIS-2 that exposes UCUG elements, which are recognized by NC with high affinity. This figure has been adapted from reference [36].

What structures do SL-C and SL-D adopt in the context of a dimeric genome? Phylogenetic and biochemical studies confirm the importance of these stem loops in MLV dimerization/packaging mechanisms [39–42], and ^{15}N/^{13}C-filtered 2D NMR analysis of an isolated SL-D fragment showed that it forms a homodimeric kissing complex stabilized via C–G Watson–Crick base pairs located within the conserved GACG apical loop [43]. This observation suggest that at least two types of quaternary interactions are possible for the tandem SL-C/SL-D stem-loop motif within the context of the Ψ dimer, C:C′/D:D′ or C:D′/C′:D, where primes denote the second RNA molecule. Weeks et al. probed the quaternary structure using site-directed hydroxyl radical cleavage experiments, where their data best fit a C:D′/C′:D kissing complex [32, 33]. This method requires chemical modification of the RNA substrate along with semiquantitative interpretation of the various conformers that might satisfy the observed cleavage patterns. Definitive evidence

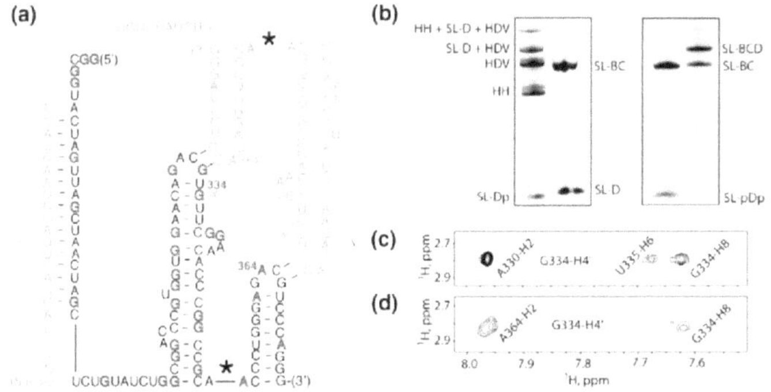

Figure 8.4 Segmental ligation construct of the MLV Ψ dimer provides direct evidence for the cross-kissing topology (a) Secondary structure of the MLV Ψ dimer used to determine kissing topology. (b) Denaturing PAGE results of the HH and HDV ribozyme cleavage (left) and T4 RNA ligase ligation products (right). (c) NMR evidence obtained for an isolated SLC_2 fragment, showing key intermolecular nOe cross-peaks that define the kissing interaction. (d) NMR evidence obtained for the (G^H)SL-BC/(A^H)SL-D ligation construct that proves the cross-kissing topology. This figure has been reprinted with permission from reference [44].

of the quaternary structure was provided using a hybrid NMR and CryoET approach, however [44]. Segmental ligation constructs of the Ψ dimer were prepared, where a (G^H)SL-BC fragment was ligated to an (A^H)SL-D fragment (Fig. 8.4). The G^H/A^H nomenclature denotes that only those nucleotides are NMR active within the respective stem loops; all others are perdueterated. Analysis of 2D nuclear Overhauser enhancement spectroscopy (nOeSY) spectra of the (G^H)SL-BC/(A^H)SL-D construct readily identified a key nOe interaction between G334 H4′ and A364 H2, which is only possible in the C:D′/C′:D quaternary arrangement (Fig. 8.4).

To further characterize the C:D′/C′:D structure, Summers et al. prepared multiple Ψ^{CD} (SL-B deletion) constructs incorporating differential ^{2}H labeling schemes. Deuterium labeling was necessary to unambiguously assign the 2D NOESY spectra, given the inherent poor spectral properties of the large RNA construct (132 residues). Moreover, this approach revealed critical nOe interactions at the SLC/SLD junction, SLC internal bulge loop (G338-A341), and other noncanonical regions, that were poorly resolved in spectra recorded

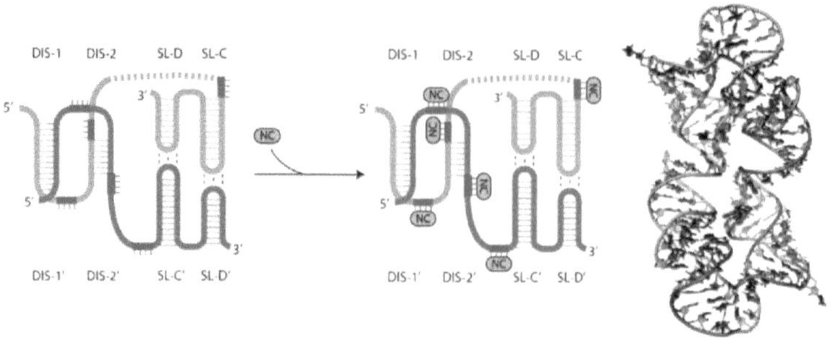

Figure 8.5 (Left) A refined model of the MLV genome dimerization and packaging mechanism based on NMR and CryoET data. Notice that in this model for dimerization/packaging, the cross-kissing interactions between stem loops C and D promote exposure of a second UCUG element at the base of DIS-2, allowing for a higher NC occupancy. (Right) Solution NMR structure of MLV Ψ_2^{CD} determined by a hybrid NMR and CryoET method, PDB code 2L1F [44]. *Abbreviation*: PDB, Protein Data Bank.

on the fully protonated construct. MLV Ψ^{CD} forms an extended homodimer where stem loops C and D from one monomer coaxially stack and engage in cross-kissing interactions with C′ and D′ from the second monomer, respectively (Fig. 8.5). The overall extended shape of the Ψ^{CD} homodimer was confirmed by CryoET projections, and disagree with a tightly compact model derived from chemical cross-linking [32–34]. A potential biological implication of the C:D′/C′:D nanostructure would be to constrain the overall geometrical topology of the Ψ dimer, thus maintaining full UCUG exposure and promoting cooperative Gag–Gag interactions and viral genome assembly (Fig. 8.5).

8.3.2 *The Human Immunodeficiency Virus-1*

Unlike MLV, there is less agreement on the RNA structures that mediate dimerization/packaging in HIV-1 [24]. This is due in part to the apparent requirement for a much larger region of the viral RNA genome to promote efficient packaging [24]. In addition, multiple secondary structural models have been proposed for the 5′-UTR, which further complicates defining a clear mechanism [24]. The 5′-UTR is the most conserved region of the HIV-1 genome, and it

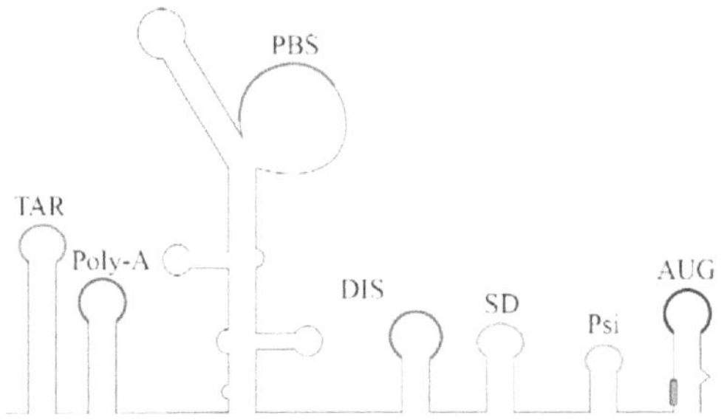

Figure 8.6 Schematic representation of the HIV-1 5′-UTR showing functional RNA stem loops: tar, transactivating response; poly-A; PBS, primer-binding site; DIS; SD, major splice donor; psi; and AUG, Gag start codon.

contains RNA structural elements that function in several steps of the viral replication cycle, including dimerization and packaging (Fig. 8.6). Efforts to determine the series of molecular events that link RNA dimerization and packaging in HIV-1 have primarily been carried out using RNA modification experiments complemented by functional assays (reviewed in [24, 45, 46]), although several NMR and crystal structures have been determined for isolated stem loops with and without NC bound [47–54].

Berkhout et al. first proposed that the HIV-1 5′-UTR exists in a conformational equilibrium between two alternating secondary structures, long-distance interaction (LDI) and branched multiple hairpin (BMH) [55–58]. In the LDI conformation, the dimer initiation stem loop (DIS) is engaged in a long-range base-pairing interaction with the poly-A stem loop (Fig. 8.7), thereby preventing genome dimerization and favoring translation. The GCGCGC palindrome (Lai isolate) of DIS is exposed in the BMH conformer where it can form a kissing-loop interaction with DIS from a second RNA molecule. The BMH conformer promotes genome dimerization and is also believed to repress translation via sequestering the *gag* initiation codon in a long-range interaction with the U5 region (Fig. 8.7). Hence, the HIV-1 5′-UTR is a dynamic RNA nanomachine where different structures act as checkpoints to regulate viral RNA function [59].

A recent NMR study of the 356 nucleotide 5′-UTR derived from the NL4 isolate has provided additional insight into the mechanisms of dimerization and packaging [60]. To probe conformational

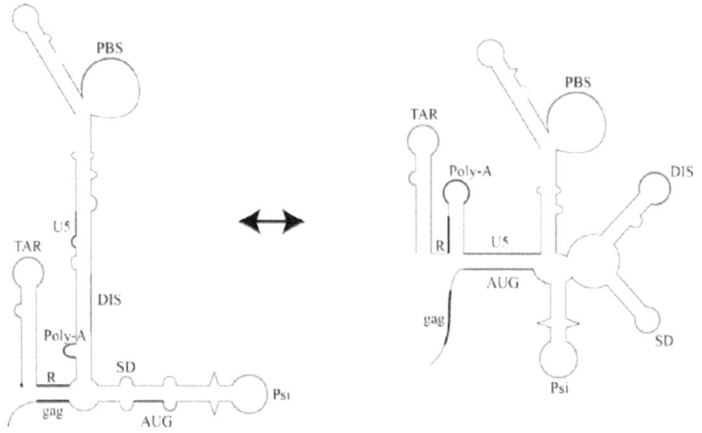

Figure 8.7 Equilibrium structures of the HIV-1 5'-UTR as determined by biochemical methods; (right) long-distance interaction and (left) branched multiple hairpin conformers. Color coding of functional stem loops is the same as Fig. 8.6.

properties of the *gag* AUG region, Summers et al. prepared a RNA construct where a ^{13}C enriched *gag* AUG fragment (nucleotides 328–356) was segmentally ligated to unlabeled 5′-UTR-spanning nucleotides 1–327. Previous NMR experiments of an isolated *gag* AUG fragment (nucleotides 338–353) showed that this RNA adopts a stem-loop structure capped by a GAGA tetraloop [52]. Under conditions that favor monomers, the ^{1}H-^{13}C heteronuclear multiple quantum coherence (HMQC) chemical shifts and peak intensities for the ligated AUG fragment matched those of isolated AUG, implying that this region adopts the stem-loop structure within the context of a monomeric genome [60]. Upon genome dimerization, the ^{1}H-^{13}C HMQC signals either shift to new positions or broaden beyond detection (Fig. 8.8). The detectable ^{1}H and ^{13}C chemical shifts matched those of a "control" U5:AUG duplex, suggesting that the long-range U5:AUG interaction forms within the context of a dimeric genome [60].

To detect nOe evidence of the U5:AUG duplex within the 712 nucleotide dimer, Summers et al. developed an innovative approach—long-range probing by adenosine interaction detection (lr-AID). The experimental premise of lr-AID is that adenosine H2

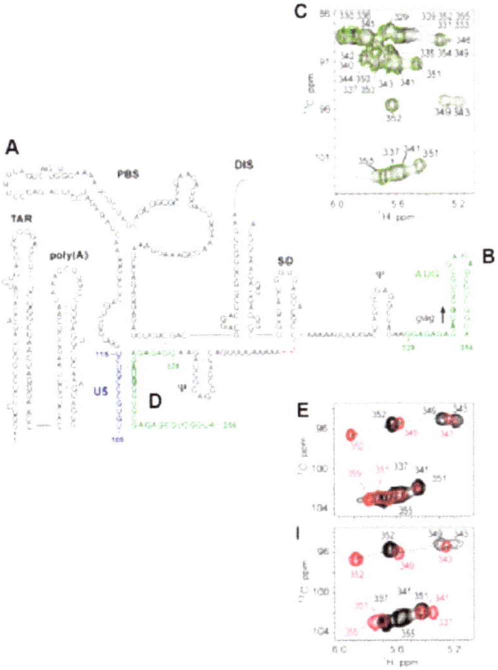

Figure 8.8 NMR evidence of a dimerization-induced structural switch at the AUG stem loop. (A) Sequence and secondary structure of the HIV-1 5′-UTR NL4 isolate. (B) Conformation of AUG in the context of a monomeric genome. (C) Overlay of ^{1}H-^{13}C HMQC spectra recorded for an AUG ligation construct (green) and an isolated AUG hairpin (black). (D) Conformation of AUG in the context of a dimeric genome. (E) Overlay of ^{1}H-^{13}C HMQC spectra recorded for an AUG ligation construct where the 5′-UTR is (black) mostly monomeric and (red) dimeric. (F) Overlay of ^{1}H-^{13}C HMQC spectra recorded for an (black) isolated AUG stem loop and (red) for an isolated AUG:U5 duplex. This figure has been reprinted with permission from reference [60].

protons give rise to a signature chemical shift and nOe patterns that depend on the base pair type and nearest neighbors. For example, in a stretch of adjacent AU base pairs of the motif [5′-$U_iU_jA_k$-3′]:[3′-$A_nA_mU_l$-5′], the H2 chemical shift of A_m resonates in a well-resolved region of the spectrum (~6.5 ppm) and gives rise to signature nOe patterns: (i) A_m H2 to A_n H1′,H2 and (ii) A_m H2 to A_kH1′,H2 (Fig. 8.9). Hence, detection of an upfield A_m H2 chemical shift along with the signature nOe patterns (i) and (ii) is strong evidence that the [5′-$U_iU_jA_k$-3′]:[3′-$A_nA_mU_l$-5′] helical stretch forms. To implement this NMR method in practice, the AU base pair motif either naturally occurs or should be engineered into the RNA sequence [60]. Natural sites that are not being probed are substituted with compensatory base pairs (such as C_j:G_m or some other combination) to eliminate background nOe cross-peaks. A reasonable assumption is that the substituting element does not augment the helical properties or function of the native RNA, which should be verified experimentally. Lastly, a major advantage of the lr-AID approach is that structural information can be obtained using conventional 2D nOe experiments only.

The U5:AUG duplex was directly detected by preparing an HIV-1 5′-UTR construct (nucleotides 1–356) where the [5′-$U_iU_jA_k$-3′]:[3′-$A_nA_mU_l$-5′]-substituting element replaced native sequences in U5 (C110–G112) and AUG (G338–G339), respectively (Fig. 8.9). NMR analysis was further simplified by in vitro–transcribing the RNA with fully ^{2}H-labeled rCTP, rUTP, rGTP, and rATP (only C8 position deuterated), which affords 2D nOe spectra of all adenosine H2 and ribose interactions that are ≤5 Å apart. Under conditions that favor the dimer, signature A338 H2 chemical shifts and nOe cross-peak patterns where easily detected and agreed with NMR signals of a "control" lr-AID-modified U5:AUG duplex. Further mutational studies confirmed the U5:AUG interaction is characteristic of the dimer, whereas the AUG stem-loop structure forms in the monomer. The question remained, what is the structural link between genome dimerization, packaging, and translational control? Given the partial sequence complimentary between the DIS palindrome and the U5 region, Summers et al. postulated that DIS base pairs with U5 in the monomer (AUG stem-loop conformer) but forms loop–loop kissing interactions in the dimer (U5:AUG conformer).

8.4 Mechanistic Insights into tRNA Priming of Reverse Transcription in HIV Monitored by NMR Spectroscopy

HIV replication is dependent on reverse transcription of the viral genomic RNA into DNA, which is catalyzed by the virally encoded reverse transcriptase (RT). RT initiates reverse transcription from

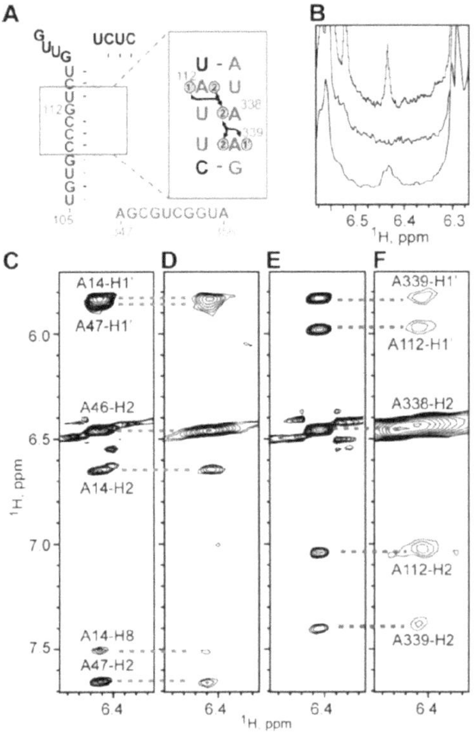

Figure 8.9 nOe evidence of the long-range base-pairing interaction between AUG and U5 within the context of a dimeric genome. (A) Implementation of the [5′-$U_iU_jA_k$-3′]:[3′-$A_nA_mU_l$-5′] substituting element to directly probe for the AUG:U5 base-pairing interaction by lr-AID. (B) 1D NMR signal of (top) TAR A46 H2 in wild-type 5′-UTR, (middle) TAR A46G substitution, and (bottom) A338 H2 for the lr-AID construct. (C-D) Signature nOe cross-peaks involving TAR A46 H2 in spectra recorded for an (C) isolated TAR stem loop and (D) native dimeric 5′-UTR. (E-F) Signature nOe cross-peaks involving AUG-substituted A338 H2 for (E) isolated lr-AID U5:AUG and (F) lr-AID substitution within the context of a dimeric 5′-UTR. This figure has been reprinted with permission from reference [60].

a specific complex formed between the 3′ end of a host $tRNA^{Lys}{}_3$ and a structural element located within the HIV 5′-UTR known as the PBS [61–63]. The initiation complex is primarily stabilized by an 18-base-pair intermolecular duplex; however, ancillary interactions involving the $tRNA^{Lys}{}_3$ anticodon loop and an A-rich segment located

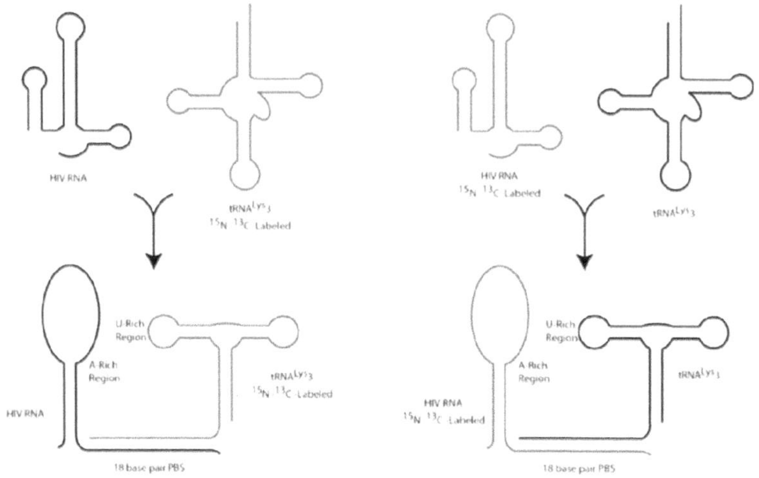

Figure 8.10 NMR-detected secondary structures of a minimal HIV-1 PBS:tRNA$_3^{Lys}$ initiation complex. Figure adapted from a recent NMR study of the initiation complex by the Puglisi group [65].

upstream of the PBS increases reverse transcription efficiency [63, 64]. Efforts to determine the RNA structures and conformational changes of the initiation complex have been studied mostly by chemical modification, although the Puglisi group recently used NMR methods to define the secondary structure of the initiation complex [65].

A minimal construct (69 mer), which retained the PBS and A-rich regions, was used to map the NMR signals of the initiation complex. Isolated tRNA$^{Lys}{}_3$ and the 69-mer fragment gave rise to imino chemical shifts and nOe cross-peak patterns consistent with the predicted secondary structures, whereas spectra of the complex were markedly distinct [65]. The NMR spectral perturbation supports a model where large-scale conformational changes in the PBS and tRNA$^{Lys}{}_3$ occur upon forming the initiation complex. To probe the conformational changes with nucleotide resolution, Puglisi et al. developed a clever hybridization scheme that consisted of preparing differential mixtures of uniformly $^{13}C/^{15}N$-labeled and $^{13}C/^{15}N$-unlabeled tRNA$^{Lys}{}_3$ and a 69-mer PBS (Fig. 8.10) [65]. Standard 2D nOe and transverse relaxation optimized spectroscopy (TROSY) heteronuclear single quantum coherence (HSQC) spectra were recorded to assign the imino chemical shifts, and 2D HNN-COSY (correlated spectroscopy that measures ^{15}N–^{15}N scalar couplings) was used to filter out intramolecular base pairing ($tRNA^{Lys}{}_3$:$tRNA^{Lys}{}_3$ and 69 mer:69 mer) from intermolecular base pairing ($tRNA^{Lys}{}_3$:69 mer). Using this NMR scheme, the 18-base-pair intermolecular duplex was directly confirmed and new intramolecular secondary structural features were observed in both the PBS and $tRNA^{Lys}{}_3$: an 8-base-pair duplex in the PBS and pairing interactions between the 5′ acceptor and T stems of $tRNA^{Lys}{}_3$. Their NMR-derived model of the initiation complex supports previous biochemical and mapping data and provides additional evidence that RNA structures, within the initiation complex, regulate RT by forming a kinetic block.

8.5 Conclusion

Retroviral genomes are RNA nanomachines that have been evolutionarily optimized to sustain viral fitness; thus, they represent excellent model systems to determine principles of RNA molecular assembly, conformational adaptation, and function. The 5′-UTR of MLV and HIV-1 genomes direct essential viral functions, each predicated on the dynamic folding and refolding of the viral RNA structure as a monomer, homodimer, or heterodimer. Knowledge of the physicochemical bases of the conformational polymorphisms is necessary to deduce molecular mechanisms. As reviewed in this chapter, modern NMR experiments coupled with isotopic labeling schemes are a strategic way for observing structural features of retroviral RNA genomes and will play an important role in characterizing other natural and synthetic RNA nanostructures.

References

1. Lu, K., Miyazaki, Y., and Summers, M.F. (2010). Isotope labeling strategies for NMR studies of RNA. *J. Biomol. NMR*, **46**(1), 113–125.
2. Dayie, K.T. (2008). Key labeling technologies to tackle sizeable problems in RNA structural biology. *Int. J. Mol. Sci.*, **9**(7), 1214–1240.

3. Duss, O., et al., (2010). A fast, efficient and sequence-independent method for flexible multiple segmental isotope labeling of RNA using ribozyme and RNase H cleavage. *Nucleic Acids Res.*, **38**(20), e188.
4. Nelissen, F.H.T., et al., (2008). Multiple segmental and selective isotope labeling of large RNA for NMR structural studies. *Nucleic Acids Res.*, **36**(14), e89.
5. Tolbert, B.S., et al., (2010). Major groove width variations in RNA structures determined by NMR and impact of 13C residual chemical shift anisotropy and 1H-13C residual dipolar coupling on refinement. *J. Biomol. NMR*, **47**(3), 205–219.
6. Grishaev, A., Ying, J., and Bax, A. (2006). Pseudo-CSA restraints for NMR refinement of nucleic acid structures. *J. Am. Chem. Soc.*, **128**, 10010–10011.
7. Bailor, M.H., Sun, X., and Al-Hashimi, H.M. (2010). Topology links RNA secondary structure with global conformation, dynamics, and adaptation. *Science*, **327**(5962), 202–206.
8. Scott, L.G., and Hennig, M. (2008). RNA structure determination by NMR. *Methods Mol. Biol.*, **452**, 29–61.
9. Latham, M.P., et al., (2005). NMR methods for studying the structure and dynamics of RNA. *ChemBioChem: Eur. J. Chem. Biol.*, **6**(9), 1492–1505.
10. Furtig, B., et al., (2003). NMR spectroscopy of RNA. *ChemBioChem: Eur. J. Chem. Biol.*, **4**(10), 936–962.
11. Thakur, C.S., et al., (2012). Biomass production of site selective 13C/15N nucleotides using wild type and a transketolase E. coli mutant for labeling RNA for high resolution NMR. *J. Biomol. NMR*, **52**(2), 103–114.
12. Thakur, C.S., and Dayie, T.K. (2012). Asymmetry of 13C labeled 3-pyruvate affords improved site specific labeling of RNA for NMR spectroscopy. *J. Biomol. NMR*, **52**(1), 65–77.
13. Thakur, C.S., and Dayie, T.K. (2011). Asymmetry of (13)C labeled 3-pyruvate affords improved site specific labeling of RNA for NMR spectroscopy. *J. Biomol. NMR*, **51**(4), 505–517.
14. Thakur, C.S., et al., (2010). Selective 13C labeling of nucleotides for large RNA NMR spectroscopy using an E. coli strain disabled in the TCA cycle. *J. Biomol. NMR*, **48**(4), 179–192.
15. Dayie, T.K., and Thakur, C.S. (2010). Site-specific labeling of nucleotides for making RNA for high resolution NMR studies using an E. coli strain disabled in the oxidative pentose phosphate pathway. *J. Biomol. NMR*, **47**(1), 19–31.

16. Xu, J., Lapham, J., and Crothers, D.M. (1996). Determining RNA solution structure by segmental isotopic labeling and NMR: application to Caenorhabditis elegans spliced leader RNA 1. *Proc. Natl. Acad. Sci. U S A*, **93**(1), 44–48.
17. Milligan, J.F., and Uhlenbeck, O.C. (1989). Synthesis of small RNAs using T7 RNA polymerase. *Methods Enzymol.*, **180**, 51–62.
18. Milligan, J.F., et al., (1987). Oligoribonucleotide synthesis using T7 RNA polymerase and synthetic DNA templates. *Nucleic Acids Res.*, **15**(21), 8783–9798.
19. Kurschat, W.C., et al., (2005). Optimizing splinted ligation of highly structured small RNAs. *RNA*, **11**(12), 1909–1914.
20. Damgaard, C.K., Tange, T.O., and Kjems, J. (2002). hnRNP A1 controls HIV-1 mRNA splicing through cooperative binding to intron and exon splicing silencers in the context of a conserved secondary structure. *RNA*, **8**(11), 1401–1415.
21. Marchand, V., et al., (2002). A Janus splicing regulatory element modulates HIV-1 tat and rev mRNA production by coordination of hnRNP A1 cooperative binding. *J. Mol. Biol.*, **323**(4), 629–652.
22. Tzakos, A.G., Easton, L.E., and Lukavsky, P.J. (2007). Preparation of large RNA oligonucleotides with complementary isotope-labeled segments for NMR structural studies. *Nat. Prot.*, **2**(9), 2139–2147.
23. Tzakos, A.G., Easton, L.E., and Lukavsky, P.J. (2006). Complementary segmental labeling of large RNAs: economic preparation and simplified NMR spectra for measurement of more RDCs. *J. Am. Chem. Soc.*, **128**(41), 13344–13345.
24. Lu, K., Heng, X., and Summers, M.F. (2011). Structural determinants and mechanism of HIV-1 genome packaging. *J. Mol. Biol.*, **410**(4), 609–633.
25. Ly, H., et al., (2000). Functional characterization of the dimer linkage structure RNA of Moloney murine sarcoma virus. *J. Virol.*, **74**, 9937–9945.
26. Fisher, J., and Goff, S.P. (1998). Mutational analysis of stem-loops in the RNA packaging signal of the moloney murine leukemia virus. *Virology*, **244**, 133–145.
27. De Tapia, M., et al., (1998). Dimerization of MoMuLV genomic RNA: redefinition of the role of the palindromic stem-loop H1 (278-303) and new roles for stem-loops H2 (310-352) and H3 (355-374). *Biochemistry*, **37**, 6077–6085.

28. Mougel, M., Zhang, Y. and Barklis, E. (1996). cis-active structural motifs involved in specific encapsidation of Moloney Murine leukemia virus RNA. *J. Virol.*, **70**, 5043–5050.
29. Mougel, M., et al., (1993). Conformational analysis of the 5′ leader and the gag initiation site of Mo-MuLV RNA and allosteric transitions induced by dimerization. *Nucleic Acids Res.*, **21**(20), 4677–4684.
30. Tounekti, N., et al., (1992). Effect of dimerization on the conformation of the encapsidation psi domain of Moloney murine leukemia virus RNA. *J. Mol. Biol.*, **223**, 205–220.
31. Gherghe, C., et al., (2010). Secondary structure of the mature ex virio Moloney murine leukemia virus genomic RNA dimerization domain. *J. Virol.*, **84**(2), 898–906.
32. Badorrek, C.S., and Weeks, K.M. (2006). Architecture of a gamma retroviral genomic RNA dimer. *Biochemistry*, **45**, 12664–12672.
33. Badorrek, C.S., Gherghe, C.M., and Weeks, K.M. (2006). Structure of an RNA switch that enforces stringent retroviral genomic RNA dimerization. *Proc. Natl. Acad. Sci. U S A*, **103**, 13640–13645.
34. Gherghe, C., et al., (2010). Definition of a high-affinity Gag recognition structure mediating packaging of a retroviral RNA genome. *Proc. Natl. Acad. Sci. U S A*, **107**(45), 19248–19253.
35. D'Souza, V., and Summers, M.F. (2005). How retroviruses select their genomes. *Nat. Rev. Microbiol.*, **3**, 643–655.
36. D'Souza, V., and Summers, M.F. (2004). Structural basis for packaging the dimeric genome of Moloney murine leukaemia virus. *Nature*, **431**, 586–590.
37. D'Souza, V., et al., (2004). NMR structure of the 101 nucleotide core encapsidation signal of the Moloney murine leukemia virus. *J. Mol. Biol.*, **337**, 427–442.
38. D'Souza, V., et al., (2001). Identification of a high-affinity nucleocapsid protein binding site within the Moloney Murine Leukemia Virus Ψ-RNA packaging signal. Implications for genome recognition. *J. Mol. Biol.*, **314**, 217–232.
39. Konings, D.A.M., et al., (1992). Novel GACG-hairpin pair motif in the 5′ untranslated region of type C retrovirus related to murine leukemia virus. *J. Virol.*, **66**(2), 632–640.
40. Ly, H., and Parslow, T.G. (2002). Bipartite signal for genomic RNA dimerization in the Moloney Murine Leukemia Virus. *J. Virol.*, **76**, 3135–3144.

41. Mougel, M., and Barklis, E. (1997). A role for two hairpin structures as a core RNA encapsidation signal in murine leukemia virus virions. *J. Virol.*, **71**, 8061–8065.
42. Gherghe, C., and Weeks, K.M. (2006). The SL1-SL2 (stem loop) domain is the primary determinant for stability of the Gamma retroviral genomic RNA dimer. *J. Biol. Chem.*, **281**, 37925–37961.
43. Kim, C.-H., and Tinoco Jr., I.A. (2000). retroviral RNA kissing complex containing only two G-C base pairs. *Proc. Natl. Acad. Sci. U S A*, **97**, 9396–9401.
44. Miyazaki, Y., et al., (2010). Structure of a conserved retroviral RNA packaging element by NMR spectroscopy and cryo-electron tomography. *J. Mol. Biol.*, **404**(5), 751–772.
45. Mujeeb, A., et al., (1998). Structure of the dimer initiation complex of HIV-1 genomic RNA. *Nat. Struct. Biol.*, **5**, 432–436.
46. Mujeeb, A., et al., (1999). NMR structure of the mature dimer initiation complex of HIV-1 genomic RNA. *FEBS Lett.*, **458**, 387–392.
47. De Guzman, R.N., et al., (1998). Structure of the HIV-1 nucleocapsid protein bound to the SL3 Ψ-RNA recognition element. *Science*, **279**, 384–388.
48. Amarasinghe, G.K., et al., (2000). NMR structure of the HIV-1 nucleocapsid protein bound to stem-loop SL2 of the Ψ-RNA packaging signal. *J. Mol. Biol.*, **301**, 491–511.
49. Amarasinghe, G.K., et al., (2000). NMR structure of stem loop SL2 of the HIV-1 Ψ-RNA packaging signal reveals a novel A-U-A base triple platform. *J. Mol. Biol.*, **299**, 145–156.
50. Amarasinghe, G.K., et al., (2001). Stem-loop SL4 of the HIV-1 Ψ-RNA packaging signal exhibits weak affinity for the nucleocapsid protein. Structural studies and implications for genome recognition. *J. Mol. Biol.*, **314**, 961–969.
51. Lawrence, D.C., et al., (2003). Structure of the intact stem adn bulge of HIV-1 Ψ-RNA stem loop SL1. *J. Mol. Biol.*, **326**, 529–542.
52. Borer, P.N., et al., (1995). Proton NMR and structural features of a 24-nucleotide RNA hairpin. *Biochemistry*, **34**, 6488–6503.
53. Ennifar, E., et al., (1999). The crystal structure of the dimerization initiation site of genomic HIV-1 RNA reveals an extended duplex with two adenine bulges. *Structure*, **7**, 1439–1449.
54. Ennifar, E., et al., (2001). Crystal structures of coaxially stacked kissing complexes of the HIV-1 RNA dimerization initiation site. *Nat. Struct. Biol.*, **8**, 1064–1068.

55. Huthoff, H., and Berkhout, B. (2001). Two alternating structures of the HIV-1 leader RNA. *RNA*, **7**, 143–157.
56. Huthoff, H., and Berkhout, B. (2002). Multiple secondary structure rearrangements during HIV-1 RNA dimerization. *Biochemistry*, **41**, 10439–10445.
57. Abbink, T.E.M., and Berkhout, B. (2003). A novel long distance base-pairing interaction in human immunodeficiency virus type 1 RNA occludes the Gag start codon. *J. Biol. Chem.*, **278**, 11601–11611.
58. Ooms, M., et al., (2004). A riboswitch regulates RNA dimerization and packaging in human immunodeficiency virus type 1 virions. *J. Virol.*, **78**, 10814–10819.
59. Berkhout, B., et al., (2002). In vitro evidence that the untranslated leader of the HIV-1 genome is an RNA checkpoint that regulates multiple functions through conformational changes. *J. Biol. Chem.*, **277**(22), 19967–19975.
60. Lu, K., et al., (2011). NMR detection of structures in the HIV-1 5′-leader RNA that regulate genome packaging. *Science*, **334**(6053), 242–245.
61. Barraud, P., et al., (2007). New insights into the formation of HIV-1 reverse transcription initiation complex. *Biochimie*, **89**(10), 1204–1210.
62. Mougel, M., Houzet, L., and Darlix, J.L. (2009). When is it time for reverse transcription to start and go? *Retrovirology*, **6**, 24.
63. Isel, C., et al., (1993). Modified nucleotides of tRNA(3Lys) modulate primer/template loop-loop interaction in the initiation complex of HIV-1 reverse transcription. *J. Biol. Chem.*, **268**(34), 25269–25272.
64. Isel, C., et al., (1998). Mutational analysis of the tRNA3Lys/HIV-1 RNA (primer/template) complex. *Nucleic Acids Res.*, **26**(5), 1198–1204.
65. Puglisi, E.V., and Puglisi, J.D. (2011). Secondary structure of the HIV reverse transcription initiation complex by NMR. *J. Mol. Biol.*, **410**(5), 863–874.

Chapter 9

Probing the Folding and Structural Dynamics in RNAs by Single-Molecule Approaches

Guangtao Song[a] **and Maumita Mandal**[a,b,c]

[a] *Department of Chemistry, Carnegie Mellon University, Pittsburgh, PA 15213, USA*
[b] *Department of Physics, Carnegie Mellon University, Pittsburgh, PA 15213, USA*
[c] *Department of Biomedical Engineering, Carnegie Mellon University, Pittsburgh, PA 15213, USA*
mandal@cmu.edu

9.1 Introduction

Over the last two decades, advances in deoxyribonucleic acid (DNA) nanotechnology have inherently provided the impetus toward the development of ribonucleic acid (RNA) nanotechnology [1–6]. Of all the techniques, single-molecule spectroscopy has emerged as a powerful tool and contributed largely toward this progression. Various single-molecule approaches have empowered researchers to perform novel experiments on individual biomolecules, opening up new frontiers in molecular and cellular biology. The fundamental processes that were once studied by bulk biochemical methods are

RNA Nanotechnology
Edited by Bin Wang

ISBN 978-981-4411-64-6 (Hardcover), 978-981-4411-65-3 (eBook)
www.panstanford.com

being revisited, leading to new theoretical and mechanistic insights. In particular, the force-measuring optical tweezers and the single-molecule fluorescence spectroscopy have paved the way for a new field often referred to as "single-molecule biophysics," wherein the dynamics of single RNA molecules (or any biomolecule) are studied in real time—from binding the effectors to performing complex molecular motor functions as in ribosomes.

Natural RNA-based systems such as ribozymes [7], riboswitches [8, 9], small interfering RNAs (siRNAs) and micro RNAs (miRNAs) [10, 11] offer a wide array of structural, chemical, and functional diversity. This is mainly because single-stranded RNA (ssRNA) can fold upon itself by using canonical and noncanonical base-pairing elements, which are further stabilized by tertiary interactions, such as pseudoknots, kissing loops, stem stacking, etc, [12, 13]. Understanding the structural and folding dynamics of these secondary and tertiary structures is of paramount importance to understand the conformational rearrangements in RNA that are directly or indirectly responsible for genetic regulations. Most of our current knowledge about RNA structures comes from X-ray crystallography or nuclear magnetic resonance (NMR) [14]. Several biochemical and biophysical methods such as circular dichriosm (CD), isothermal titration calorimetry (ITC), ultraviolet (UV) melting, and chemical footprinting have been used traditionally to monitor the conformational changes undergone by RNA during its folding and assembly [15–19]. However, all these techniques provide ensemble averaged signals from a large number of molecules, which restricts the information due to rare conformers in the solution. The increasingly sophisticated single-molecule approaches such as optical tweezers and single-molecule fluorescence resonance energy transfer (FRET) are beginning to aptly fill this gap [20–25].

Single-molecule methods are especially attractive and powerful tools for the study of biological systems and processes in real time [26, 27]. This is due mainly because the molecular properties are measured one molecule at a time, yielding a direct measurement of the distributions along the reaction coordinate. Such a distribution is always considered a near-complete description of the system as it reflects not only the significantly large behavior but also the rare occurrences that would be averaged in an ensemble method.

Second, the dynamics of the system can be measured under equilibrium conditions in single-molecule studies. Dynamic heterogeneity occurs when a sample contains subpopulations of molecules that can interconvert over the timescale of the observation. Additionally, the single-molecule manipulation allows insights into the energy landscape of reactions and thus allows direct probing into elusive structural features such as transition states and energy barriers.

Since the first single-molecule RNA study in 1999 [28, 29], single-molecule approaches have been widely used as a powerful tool to probe the structural folding and dynamics of several RNA-based systems. Here, we will summarize some of the recent developments in the field of RNA science using single-molecule techniques. The discussions are mainly based on optical tweezers and single-molecule FRET methods.

9.2 Single-Molecule Approaches: An Overview

Various single-molecule experiments are roughly categorized into two groups, single-molecule force spectroscopy [30] and single-molecule fluorescence spectroscopy [31]. Each group has its own features that make it more appropriate for certain types of studies. Recently, a third approach that combines both the force and fluorescence capabilities is also emerging [32–35].

9.2.1 *Single-Molecule Force Spectroscopy*

The importance of forces in biology has been recognized for quite some time, but only in the past 20 years, we have acquired theoretical and experimental breakthroughs to make major advances in instrumentation and methodology to directly measure interactive forces at subnanometer precision at the level of single biological macromolecules and/or their complexes [30]. Among the most popular single-molecule force techniques are atomic force microscopy (AFM), magnetic tweezers, and optical tweezers.

9.2.1.1 Atomic force spectroscopy

AFM was first developed as a part of scanning probe microscopy in order to overcome the limitations of the scanning tunneling microscope in imaging nonconductive samples [36]. However, the possibility of modifying the surface and manipulating individual molecules made AFM an ideal tool for biological applications [37–39]. Recent improvements in the temporal and spatial resolution of AFM allows measurement of inter- and intramolecular interaction forces at piconewton (pN) resolution [40, 41]. A typical AFM instrument generally contains a moveable piezoelectric sample stage; a small and highly flexible cantilever, which ends in a narrow tip; and a quadrant photodiode detector A sample is connected between the surface and the tip using various chemistries. A laser beam is reflected off the back surface of the cantilever and its reflection is detected with high precision using the quadrant detector. Since a change in the deflection of the cantilever will change the reflection angle of the laser beam, the detector signal provides a high-resolution measurement of the cantilever deflection. The deflection of the cantilever (within approximately a harmonic potential) provides readout of the force being exerted on the molecule, while the distance between the surface and the tip is simultaneously controlled. This can provide a force–extension curve (FEC) for the molecule.

9.2.1.2 Magnetic tweezers

Magnetic tweezers use magnetic fields to apply forces on the molecules. Similar to AFM, the molecule of interest is tethered between two surfaces, one to the magnetic bead which is manipulated by a strong magnetic field, and the second to a coverslip. The fields generated by sharp electromagnetic tips [42] or small permanent magnets [43] have been used to apply forces in excess of 200 pN on micrometer-sized magnetic particles. This technique is especially useful in some cases because, in addition to the application of linear forces, it also allows torque to be easily applied to individual molecules [44].

9.2.1.3 Optical tweezers

Optical tweezers, also known as optical traps, are arguably the most versatile single-molecule manipulation technique [45]. An optical trap is created by focusing a laser to a diffraction-limited spot with a high numerical aperture (NA $>$ 1) microscope objective. Particles with high refractive indexes are attracted to the most intense part of the laser beam and held at the focus. The focused beam creates a nearly harmonic potential so that the force acting on a bead is proportional to the displacement of the bead from the center of the laser beam. For small displacements ($\sim$150 nm) of the trapped object from its equilibrium position the force is linearly proportional to the displacement and the optical trap can be well approximated as a linear spring. The spring constant, or stiffness, depends on the steepness of the optical gradient (how tightly the laser is focused), the laser power and the polarizability of the trapped object. For the optical tweezers experiment, a molecule of interest is tethered between two surfaces: typically a bead is trapped by the laser which is manipulated by an optical trap, and the second surface is offered by a coverslip or a second bead which is held in place by suction using a micropipette or another optical trap [22, 25, 127]. Different RNA systems that have been studied using optical tweezers are discussed in the later sections.

9.2.2 *Single-Molecule Fluorescence Spectroscopy*

Fluorescence spectroscopy and microscopy uses several properties of fluorescent biomolecules to monitor their location, structure and dynamics. Single-molecule fluorescence detection is achieved by repeatedly exciting the fluorophore of interest, followed by detection and analysis of the emitted fluorescence photons [46, 47]. Since the rate and number of photons emitted by a single fluorophore are limited, fluorescence detection at single-molecule resolution has several key requirements, including the use of high-efficiency and low-background photon detection and bright fluorophores [48]. To achieve these goals, two main formats of design are used confocal [49] and total internal reflection fluorescence (TIRF) microscopy [50].

9.2.2.1 Confocal microscopy

Here continuous or pulsed laser light is focused through the objective lens, exciting only a very small (femtoliter-size) volume. The emitted fluorescence can be split into multiple channels for acquisition. Confocal detection is achieved by introducing a small aperture (pinhole) at an image focal point in the detection pathway such that outoffocus light is rejected by the pinhole. The benefit of confocal microscopy is that a microsecond time resolution can be achieved accurately.

9.2.2.2 TIRF microscopy

In TIRF microscopy, the surface-bound molecules are excited by an evanescent wave generated by total internal reflection of a laser beam. This approach reduces the excitation volume to a thin layer at the interface between the glass slide and the buffer and decreases the background signal and excessive photobleaching of molecules in a wide observation field with the use of a charge-coupled device (CCD) camera. Two types of TIRF microscopes are commonly used prism type and objective type, which differ only by the method of how the excitation light is brought to the sample. The benefit of TIRF over confocal microscopy is that a larger area of the slide can be imaged simultaneously, and this provides more data quickly. However, the time resolution of current CCD cameras is still lower than that of avalanche photodiodes (APDs) and photomultipliers that are used in confocal fluorescence microscopy.

9.2.2.3 smFRET

FRET is a widely used technique for studying the intra- and intermolecular motions of biomolecules in real time [51, 52]. In this assay, a pair of fluorescent donor and acceptor molecules is attached to the host biomolecule(s) of interest. Upon exciting the donor fluorophore, the energy is transferred to the acceptor, leading to a decrease in the donor emission and an increase in the acceptor emission. The energy transfer efficiency E is given by $E = 1/(1 + (R/R)^6)$, where R is the distance between the donor and the acceptor. As the Forster radius R is typically 3 nm to 8 nm, it enables FRET sensitive to changes of a few nanometers in the donor–acceptor distance, which can reflect either the conformation changes of the host molecule or the relative motions between two molecules. The power of single-molecule detection combined with the sensitivity of FRET has made single-molecule FRET (smFRET) one of the most widely used real-time detection techniques to monitor the structural dynamics, kinetics and thermodynamics of biomolecules.

9.2.3 *Combined Approaches*

Although each single-molecule method has its own merits, no individual singlemolecule technique allows us to simultaneously study all the features of complex biological systems. Hence, multiple single-molecule techniques have often been applied to study different aspects of many important biological systems and processes. The force- and fluorescence-based approaches are highly complementary: force spectroscopy reports on more global structural and mechanical rearrangements in biomolecules, while fluorescence measures local conformational changes [35]. Hybrid instruments combining fluorescent capabilities and mechanical manipulation provide a promising direction for single-molecule spectroscopy, and recently some instruments have been successfully constructed [32–35].

9.3 Single-Molecule Studies on RNA

9.3.1 *Folding of Simple Secondary and Tertiary Structures*

Traditional investigation of RNA secondary structures is mainly based on the thermal melting data or biochemical methods used for structural probing [53]. However, RNA can adopt alternative conformations with similar stability, and misfoldings are common in in vitro experiments, leading to a rugged folding energy landscape [54]. The use of mechanical force to induce the unfolding and refolding of single RNA molecules provides a promising method to study the RNA conformers in real time [22, 23, 55]. Liphardt

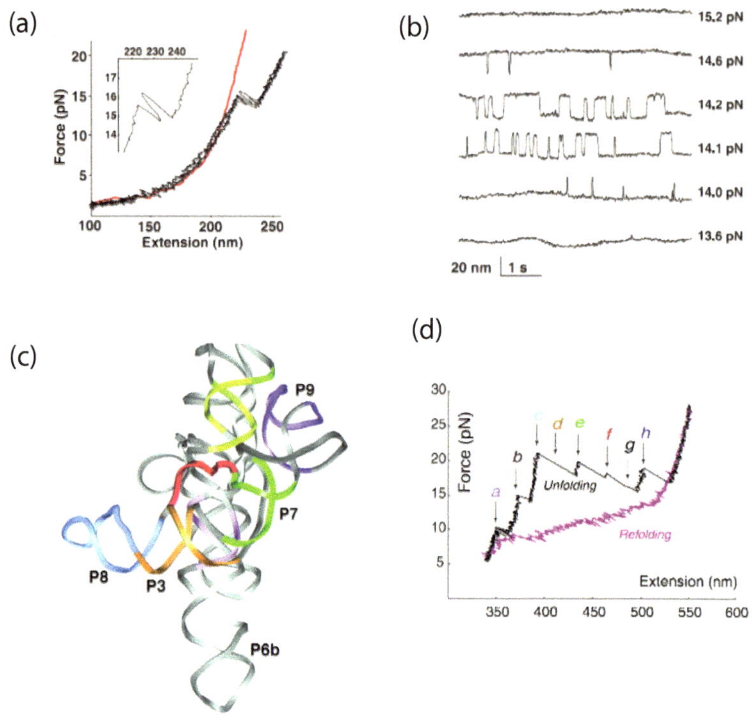

Figure 9.1 Representative single-molecule force measurements. (a) FECs for the P5ab RNA hairpin. The red curve shows the FEC of handles alone. In a typical single-molecule experiment, long DNA–RNA hybrid handles are attached at the two ends of the RNA, which are, in turn, attached to the beads. Inset shows the hopping of the molecule. (b) Extension vs. time traces of the P5ab hairpin at various constant forces [56]. (c) Crystal structure of the *Tetrahymena* ribozyme. (d) Representative unfolding (black) and refolding (pink) FECs of the L-21 RNA, displaying six unfolding events (rips). Letters and arrows correspond to the positions assigned to the kinetic barriers.

et al. first characterized the folding of a single P5ab group I intronic hairpin RNA using optical tweezers (Fig. 9.1) [56]. With slow loading rates, the folding and unfolding of the hairpin RNA can be achieved in equilibrium. By varying the refolding speed and refolding time at the end of a pulling-relax cycle, different folding pathways are observed along with their partially folded intermediates [57]. Constant force trajectories, as in Fig. 9.1B, directly reveal properties such as the net extension changes between the folded and unfolded states and the kinetics. Moreover, the transition states along the folding pathway, which are difficult to measure, are calculated for the force-dependent kinetics data [58, 59]. The free-energy difference between the folded and unfolded states is readily calculated from the FECs being folded or unfolded at equilibrium and compares well with the theoretical predictions. Remarkably, the similar information could be even obtained from FECs by fitting results to kinetic models or by applying theorems relating the nonequilibrium work to the equilibrium free energy [60–63]. In contrast to optical tweezers, the smFRETbased methods are constrained in studying the RNA secondary structure because of its inherent high stability at room temperature. Recently, Gell et al. used a temperature-controlled smFRET setup and successfully observed the heterogeneous folding of a simple RNA stem loop [64]. The results showed that a single-base mutation in the loop could dramatically alter the folding behavior of the hairpin.

Tertiary contacts in RNA are responsible for forming compact three-dimensional structures and are important for functions such as catalysis and regulation [13]. Unlike the highly cooperative, all-or-none folding process observed for most protein domains, RNAs generally fold in a hierarchical manner. The structural dynamics and diversity of these tertiary interactions are now a subject of intense investigation. Ha et al. first demonstrated the utility of smFRET for real-time detection of conformational changes in a three-way junction upon binding of a ligand [28]. After that, a series of specific tertiary motifs of RNA were studied with smFRET and/or optical tweezers.

Helical junctions, a common architectural feature in RNA, are classified according to the number of helical arms [65], for example, four-way junctions. Typically, the RNA helices adopt a compact fold by pairwise coaxial stacking of helices. Formation of a stacked structure lowers the symmetry of the junction from fourfold to twofold and thus creates two distinguishable kinds of strands, termed "antiparallel" and "parallel" conformers. Using smFRET methods, Hohng et al. demonstrated that unlike DNA, both antiparallel and parallel conformations can be stably populated in RNA and there is continual interconversion between these forms

[66]. The rate of interconversion becomes slower at higher Mg^{2+} concentrations. They further suggested that the interconversion proceeds via an open intermediate with reduced cation binding in which coaxial stacking between helices is disrupted.

The kink turns (or K-turns) is another structural motif found in RNA molecules that introduces a tight kink into the helical axis [67]. K-turns play an important architectural role in RNA structures and serve as binding sites for a number of proteins. Wozniak et al. used smFRET methods to study the folding of the spliceosomal U4 small nuclear RNA (snRNA) K-turn [68]. They demonstrated that the U4 snRNA K-turn structure could stay folded only upon binding of the 15.5K protein, which indicated that the K-turns do not provide a thermodynamic driving force for RNA tertiary folding but rather require cooperation from surrounding proteins in order to stabilize them.

The tetraloop–receptor motif is a type of long-range RNA tertiary interaction in which a 4 nt terminal hairpin loop docks with an asymmetric internal loop elsewhere in the molecule [69]. This motif has been observed in almost every large RNA crystal structure and been employed as a building block for the construction of RNA nanostructures [70]. An smFRET study by Hodak et al. revealed that the RNA folding due to the ubiquitous GAAA tetraloop–receptor docking interaction is highly dynamic [71]. The rate constants for docking and undocking are measured as a function of Mg^{2+}, revealing a complex dependence on metal ion concentration: increasing Mg^{2+} induces the equilibrium to the docking event; however, no cooperative of Mg^{2+} binding and docking was observed. A further study explored the effect of temperature on the docking/undocking events. Kinetic heterogeneity in the tetraloop–receptor construct is unaffected over the temperature range investigated, indicating a large energy barrier for interconversion between the actively docking and nondocking subpopulations [72].

Another two typical long-range tertiary interactions in RNA are the kissing loop and the pseudoknot. Long-range base pairings between hairpin stem loops are known as "kissing" loops. These interactions are mediated by loop nucleotides that interact through complementary Watson–Crick base pairs. Li et al. used optical tweezers to study the folding behavior of RNA kissing complexes [73, 74]. The results revealed that in contrast to the RNA hairpin, the kissing interactions show distinct folding behaviors. The loop–loop interaction is exceptionally stable, leading to a high unfolding force. The unfolding rate was found to be relatively insensitive to force, although Mg^{2+} dependent. Increasing Mg^{2+} leads to decreasing of the unfolding rate of the kissing complexes. Pseudoknots involve long-range Watson–Crick base pairing between the loop nucleotides of a hairpin and a complementary region of the RNA. The long-range base pairs form a second helix which coaxially stacks upon the first one. Pseudoknots are critical in programmed ribosomal frame-shifting, telomere structure, and ribozyme activity [75]. Using optical tweezers, Hansen et al. first proved that the efficiency of infectious bronchitis virus (IBV) frame-shifting is related to the mechanical strength of the RNA pseudoknot [76]. The one with a higher unfolding force gives a higher degree of frame-shifting. A further study by Green et al. revealed that similar to the kissing complex, the mechanical stability of pseudoknots is affected by the sequence and length of their stems, the rate of unfolding changes little with increasing force, and the presence of Mg^{2+} increases the kinetic barriers to the unfolding of the IBV pseudoknot [77]. The combined biochemical and single-molecule studies by Chen et al. further proved that the enhancement of pseudoknot stem stability and torsional resistance improved the frame-shifting efficiency [78]. Unlike the relative simple one step unfolding behavior, the refolding of pseudoknots seems much more complicated By stepwise force-drop experiments, Chen et al. showed that the folding of a pseudoknot in human telomerase RNA takes complex pathways, and they also observed collapsed nonnative structures due to the misfolding [79]. Using smFRET, Mihalusova et al. studied the importance of the pseudoknot structure for the function of telomerase [80]. They showed that the isolated pseudoknot sequence stably folds into a pseudoknot. However, in the context of the full-length telomerase RNA, interference by other parts of the RNA prevents the formation of the pseudoknot. The protein subunits of the telomerase holoenzyme counteract RNA-induced misfolding and allow a significant fraction of the ribonucleoproteins (RNPs) to form the pseudoknot structure. Only

RNP complexes containing a properly folded pseudoknot are catalytically active. These results from single-molecule experiments not only demonstrate the functional importance of the pseudoknot but also reveal the critical role played by telomerase proteins in pseudoknot folding.

Recently, using a P4–P6 domain derived from the *Tetrahymena* ribozyme as a model system, and based on an smFRET assay, the Herschlag lab established a new assay to study the cooperativity in directing the assembly of the entire catalytically active ribozyme [81, 82]. With at least two more domains and various interdomain and active site tertiary contacts, it is likely that cooperativity plays a pivotal role not only in orchestrating the global architecture of the intron, but also in organizing the critical tertiary interactions at the active site that are responsible for splice sit selection and catalysis.

9.3.2 *Folding and Structural Dynamics of Large Functional RNA Structures*

9.3.2.1 Ribozyme

Ribozymes are RNA molecules that perform catalytic activity [7]. The ribozymes have been found for about 30 years, and significant progress has been made in understanding the chemistry involved in their catalytic functions. There are complete or partial crystal structures for most of the known ribozymes highlighting the atomic details of the secondary and tertiary interactions. Several single-molecule experiments have seen considerable success in studying the folding pathways and kinetics of ribozymes [83].

Despite differences in size, structure, and residues involved in catalysis, almost all ribozymes catalyze by site-specific and reversible internal-backbone cleavage reactions, which generates 2′,3′-cyclic phosphate and 5′-hydroxyle termini in the products. The hairpin ribozyme is the smallest ribozyme known and has been extensively studied by smFRET [20, 84]. Zhuang et al. first performed smFRET experiments on the minimal hairpin ribozyme with the two-way junction [85]. FRET trajectories of individual molecules directly revealed the undocked, docked, and substrate-free states of the ribozyme. Despite its structural simplicity, the hairpin ribozyme showed highly complex folding dynamics. By fitting the lifetime distributions to multiple exponential functions, smFRET revealed that there were multiple populations of molecules with distinct undocking kinetics. The heterogeneous kinetics of the hairpin ribozyme was further explored by characterizing the effect of site-specific mutations and metal ion titrations on the transition state of the two-state folding reaction [86–89]. The persistence of molecular heterogeneity with memory effect for all variants under different buffer conditions confirmed the previously identified unique nature of the hairpin ribozyme. Surprisingly, most modifications not only affect the docking and undocking but also significantly impact the internal chemistry rate constants over a substantial distance from the site of catalysis, revealing the presence of a network of coupled molecular motions that connects distant parts of the RNA with its reaction site.

Large ribozymes typically show more complicated folding dynamics due to the many stable interactions that can form between these interactions. Onoa et al. used optical tweezers to study the unfolding of the *Tetrahymena* ribozyme from the native state to a fully extended ssRNA (Fig. 9.2) [92]. The FECs of a single molecule show eight discrete unfolding steps, indicating the presence of eight intermediate states. They successfully mapped the location of kinetic barrier in each step by using mutations or antisense oligonucleotides. Unlike thermal unfolding, the intermediates contain both secondary and tertiary contacts. Moreover, by pulling the molecule many times, they were able to map the possible unfolding pathways and their relative frequency. The ribozyme was also studied extensively by smFRET methods (Fig. 9.2) [90], wherein the folding transition state was probed by comparing the effects of mutations and denaturants on the docking rate (k_{dock}) and equilibrium constant (K_{dock}). Modifications of eight groups in the P1 duplex that formed the tertiary interactions decreased the docking equilibrium to a greater extent than the docking rate constant [91]. More recently, by using smFRET, Solomatin et al. demonstrated that the *Tetrahymena* ribozyme folds into multiple distinct native states that interconvert on a timescale much longer than that of catalysis [93]. Their results provided strong evidence for multiple conformations of the native states that interconvert but are separated by large energetic barriers.

In both these earlier examples, the power of single-molecule approaches is obvious. Dynamic properties that are difficult to analyze with ensemble experiments that include noncumulative intermediate states, parallel folding pathways, and equilibrium conformational fluctuations are directly observed in single-molecule trajectories.

Group II intron ribozymes rank among the largest protein-free, multidomain ribozymes found in nature [94, 95]. The secondary structure consists of six domains that radiate from a central core, and only nucleotides involved in tertiary interactions are highly conserved. By using the minimal active form consisting of the necessary components for catalysis, Steiner et al. studied the folding pathway of the group II intron ribozyme [96]. The smFRET data revealed three distinct structural conformations with a previously unidentified on-pathway folding intermediate. The observed fast folding rates between different states indicated that these conformations are connected by small activation barriers. Interestingly, it was found that increasing Mg^{2+} concentration leads to an increase in structural dynamics. Further analysis revealed that the native conformation occurs only transiently at 20 mM or higher Mg^{2+} concentrations. A subsequent study by Karunatilaka et al. monitored the effect of Mss116 on the folding dynamics of this minimal active construct [97]. The authors showed that Mss116 stimulates dynamic sampling between states along the folding pathway, an effect previously observed only with high Mg^{2+} concentrations. Furthermore, the data indicate that Mss116 promotes folding through discrete adenosine triphosphate (ATP)-independent and ATP-dependent steps. They proposed that Mss116 stimulates group II intron folding through a multistep process that involves electrostatic stabilization of early intermediates and ATP hydrolysis during the final stages of native state assembly.

9.3.2.2 Riboswitch

Riboswitches are recently discovered genetic regulatory elements found in certain 5′ untranslated regions of certain messenger RNAs (mRNAs). These *cis*-acting genetic elements regulate adjacent gene expression by binding to small molecules [8, 9]. Riboswitches are composed of two overlapping domains, an aptamer domain, which binds the metabolite, and a downstream expression platform. Switching from a repressed gene expression to an overexpression state or vice versa is accomplished by the expression platform, which adopts one of two mutually exclusive secondary structures (in the case of transcriptional regulation, either an antiterminator or a terminator element), depending upon whether the aptamer domain is bound to the appropriate ligand. A terminator hairpin structure will interrupt the RNA polymerase to halt prematurely thus resulting in a repressed state, while an antiterminator will cause the transcription to continue normally. Riboswitch-dependent regulatory processes depend significantly on the native folding of the aptamer. While the kinetics of RNA polymerase has been shown to play a vital role in FMN riboswitch, the underlying thermodynamics of ligand-induced RNA folding cannot be ruled out either.

Purine riboswitches are among the simplest riboswitches that have been extensively studied by several single-molecule studies. The purine aptamers with "tuning fork" structures bind either guanine or adenine with high specificity imparted by a cytosine residue in the binding pocket (Fig. 9.2a) [98, 99]. By using an smFRET assay, Lemay et al. found that under low Mg^{2+} conditions (<50 mM) the *add* riboswitch experiences three folding states: an unfolded state (in which the RNA has all of its secondary structural elements formed but with no tertiary structure), a folded state (a globally organized state), and a new intermediate state (Fig. 9.2b) [100]. While the authors do not speculate as to the nature of this conformation, combined with other studies, Gilbert et al. proposed a model for this process (Fig. 9.2c) [101]: in the absence of a ligand, the three-way junction is not completely disordered; rather, one strand of the three-way junction is preorganized to present a pyrimidine residue that forms the Watson–Crick pair with exogenous adenine or guanine. Only upon productive pairing between the ligand and this pyrimidine, the surrounding junctions close around the ligand, forming the extensive set of interactions that make the nucleobase almost completely solvent and inaccessible. And the intermediate

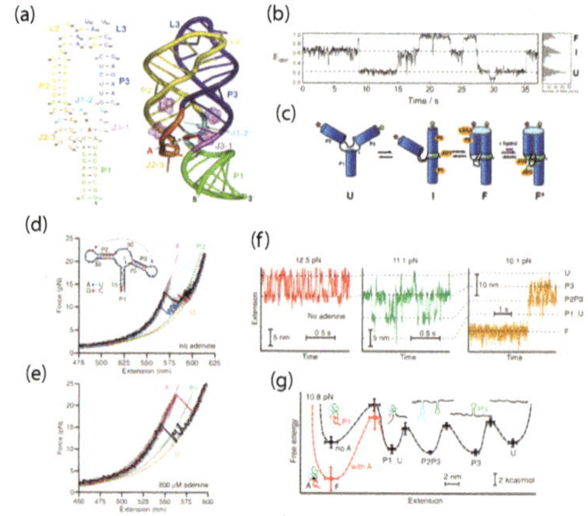

Figure 9.2 Single-molecule studies of the folding and structural dynamics of the adenine riboswitch. (a) Secondary (left) and crystal (right) of the *add* riboswitch. (b) An intermediate folding state in the adenine aptamer. FRET trace and histogram for a single molecule (50 ms integration time) in the presence of 0.02 mM Mg^{2+} clearly showing interconversion between three states: U (FRET = 0.25), I (FRET = 0.5), and F (FRET = 0.9). (c) A proposed mechanism by which folding and ligand binding are coupled in the adenine riboswitch. The U, I, and F/F* states were observed directly in smFRET studies. The elements of the structure highlighted by orange circles are those that are organized in each step of the folding/binding mechanism. (d) FECs in the absence of adenine reveal two intermediate states corresponding to the unfolding of everything except hairpins P2 and P3, followed by hairpin P2. Three FECs (black, blue, red) are plotted above the aggregated data from 700 FECs (gray dots). WLC fits are shown for the four states: F, fully folded (purple); U, fully unfolded (yellow); P3P3, folded (green); and P2P3, both P2 and P3 folded (cyan). (e) Adenine binding resulted in similar behavior but higher unfolding forces. The state P2P3 was observed less frequently, if at all. (f) The extension as a function of time at different levels of constant force in the absence of adenine reveals five distinct states corresponding to the major structural features: fully unfolded at the largest extension ('U'), then P3 folded ('P3'), then both P2 and P3 folded but no loop-loop interaction ('P2P3'), then the P2P3 kissing loop complex (labeled 'P1_U', P1 unfolded), and finally fully folded ('F'). Only short segments of the full records are displayed. (g) Folding energy landscapes. The key features of the energy landscapes for the five-state native folding pathway were reconstructed from piecewise two-state analyses of each transition. Energies and positions are plotted with reference to the P2P3 state. Error bars show SEM. Dotted lines indicate notional landscape shapes in the presence (red) and absence (black) of adenine. Cartoons indicate the structural features associated with each state.

state might represent the coaxial stacking of the P1 and P3 helices along with the bases in J3/1 in the absence of the L2–L3 interaction. In this state, L2 and L3 would be brought closer together, leading to the detected intermediate FRET signal.

During transcription of the riboswitch, the aptamer domain is first synthesized by RNA polymerase and folds independently of the expression platform. A stretch of uridines following the aptamer domain causes the polymerase to stall temporarily, enabling the temporal separation of the two folding events. By using optical tweezers, Greenleaf et al. developed a new assay that mimics cotranscriptional folding of adenine riboswitches [102]. For the *add* riboswitch [103], as seen in Fig. 9.2d–g, the authors were able to see the folding of each structural element in the riboswitch and taking the length measurements of the extended RNA molecule subjected to a stepwise decreasing force. The order of helix formation was P3 followed by P2. Although the P2 and P3 differed by only 2 nt, Greenleaf et al. were able to identify with precision which helix was formed in each case by using antisense oligonucleotides directed against each of the helices, blocking their folding separately. Folding of helices P2 and P3 was shown to be independent of the presence of adenine. However, the formation of the next intermediate was shown to be adenine dependent that organized the adenine pocket; however helix P1 is not formed. Finally, the state showing the shortest length represents the completely folded, adenine-bound state, in which P1, P2 and P3 helices are present.

Finally two other riboswitches have also been studied using single-molecule approaches *S*-adenosylmethionine (SAM) and thymine pyrophosphate (TPP)-responsive riboswitches. Both SAM and TPP riboswitches are widespread in bacteria [104, 107]. The SAM riboswitch is reported to exist in five different bacterial families, highlighting the relevance of SAM regulation [104]. Although the SAM-I and SAM-II riboswitches share the same cellular function, that is, both switch off gene expression in response to SAM, they have distinct structural folds in their ligand-binding aptamer domains, different phylogenetic distributions, and different mechanisms of regulation. The SAM-I secondary structure consists of four base-paired stems that meet in a four-way junction, whereas the SAM-II secondary structure consists of a single stem loop that forms a pseudoknot with the 3′ sequence. SAM-I riboswitches typically operate via ligand-dependent formation of an intrinsic transcription terminator, which results in premature termination of the transcript when levels of SAM are high. In contrast, SAM-II riboswitches typically operate via ligand-dependent sequestration of the ribosome-binding sequence, which results in the inhibition of translation when levels of SAM are high. Using smFRET and other biophysical techniques, two separate groups studied the conformational switching behavior of the SAM-I and SAM-II riboswitches, providing insights into gene regulation by these RNAs [105, 106]. Their studies revealed another common thread between the two riboswitches: both show some structural preorganization in the presence of Mg^{2+} ions that is not present without Mg^{2+}. Furthermore, both seem to transiently adopt a conformation in the absence of the ligand SAM that at least partially resembles the ligand-bound conformation, which is further stabilized upon SAM binding. For these two riboswitches, it is possible that ligand binding does not change the RNA fold but instead drives a preexisting equilibrium containing a minor population of native-like states toward the ligand-bound conformation.

More recently, Anthony et al. extended the single-molecule approaches to investigate the folding and energetics of the TPP aptamer using optical tweezers. The TPP aptamer is significantly larger than the adenine aptamer [108]. By measuring the energy landscape of the TPP aptamer and its sensitivity to ligand binding and mutations, the authors explored the hierarchy of folding events in this aptamer. In the absence of ligand, the aptamer unfolded through a series of transitions observed in both nonequilibrium and equilibrium measurements. Similar transitions were observed in RNA constructs containing only portions of the aptamer sequence. Constant-force measurements carried out at equilibrium were used to calculate the distance and kinetics associated with each folding transition. The probability density distribution is further used to reconstruct the energy landscape for secondary structure formation [108]. To study aspects of tertiary structure formation, they compared aptamer folding for wild-type and mutant sequences in the presence and absence of ligands. The data suggested that ligand binding cannot occur until the entire aptamer has been transcribed and the final elements of the secondary structure have formed, which is very different from that of the adenine riboswitch.

9.3.2.3 Ribosome

The ribosome is a structurally and functionally conserved molecular assembly composed of RNA and protein molecules to form a supramolecular machine responsible for protein biosynthesis [109, 110]. This RNA-rich enzyme works collaboratively with other cellular RNAs and proteins to convert mRNA-encoded genetic information into a specific polypeptide according to the messenger RNA sequence. Studying the relationship between the conformational dynamics of a ribosome, its transfer RNA (tRNA) substrates, and its translation cofactors, as well as the mechanical, catalytic, and regulatory events that drive protein synthesis, is greatly advancing our understanding of translation [111]. X-ray crystallography and cryo–electron microscopy have obtained static structural snapshots of ribosomes in a variety of functionally relevant conformations. However, neither of these two methods could provide information on the evolving dynamic translation process itself. Single-molecule spectroscopic and force-based techniques have provided fresh, new perspectives to the protein biosynthesis [24, 112].

In 1999, Sytnik et al. first showed that the surface-immobilized ribosome complexes could retain their activity and can be detected at the level of single-molecule assembly [29]. This work opened up a new dimension of ribosome study: studying the translation at the single-molecule level. Using fluorescence-labeled tRNAs and smFRET, Blanchard et al. resolved a multistep movement of the tRNA into the ribosome and identified several intermediate states in the aa-tRNA selection process [113]. Subsequent studies investigated the rapid tRNA thermal fluctuations in the aminoacyl (A) site during tRNA selection [114]. The results showed that the cognate tRNA has a reduced dissociation rate over near-cognate tRNAs from the A site both before and after guanosine triphosphate (GTP) hydrolysis by EF-Tu. Accordingly, rapid conformational sampling by tRNAs ensured that near-cognate tRNAs are quickly shuttled into unstable binding conformations, thereby only selecting the cognate tRNA for accommodation. These results provided the first glimpse into the structural basis for the initial cognate tRNA selection, kinetic proofreading, and thus explain the fidelity of translation by the ribosome.

During the elongation phase of translation, tRNAs rapidly and directionally translocate in ~30 Å steps through structurally distinct aminoacyl (A), peptidyl (P), and exit (E) sites at the interface of the two ribosomal subunits (30S and 50S in bacteria). The speed and accuracy of these translocation processes are fueled by elongation factor–dependent GTP hydrolysis. There is a general agreement that before translocation the ribosome shifts into an unlocked state characterized by intersubunit rotation and rapid tRNA fluctuations between the classical and hybrid tRNA configurations. Using fluorescently labeled tRNAs, Munro et al. showed that tRNA molecules bound at the A site and P site and that the pretranslocation ribosome dynamically samples three kinetically and structurally distinct states: one classical state and two hybrid states [115]. Other labeling strategies also demonstrated that tRNA fluctuations and subunit ratcheting are intrinsically dynamics processes that can occur spontaneously [116–119]. The translational machinery modulates the equilibrium of these fluctuations to guide the ribosome and tRNAs into a desirable configuration for translocation.

Using specifically fluorescently labeled ribosome subunits, Aiken et al. employed an intersubunit ratcheting signal to monitor global ribosomal dynamics throughout initiation followed by multiple elongation cycles [120]. They successfully monitored the global conformational dynamics of an elongationg ribosome codon by codon in real time. The results showed that the early steps of ribosomal translocation are particularly slow and could represent a point of potential regulation. More recently, by introducing the new zero-mode waveguide (ZMW) technology, Uemura et al. accomplished real-time observation of translation at physiologically relevant micromolar ligand concentrations (Fig. 9.3) [121]. Translation at each codon is monitored by stable binding of fluorescence-labeled tRNA to the translating ribosomes, which allows direct detection of the identity of tRNA molecules bound to the ribosome and therefore the underlying mRNA sequence. Further analysis demonstrated that ribosomes are only briefly occupied by the two tRNA molecules and

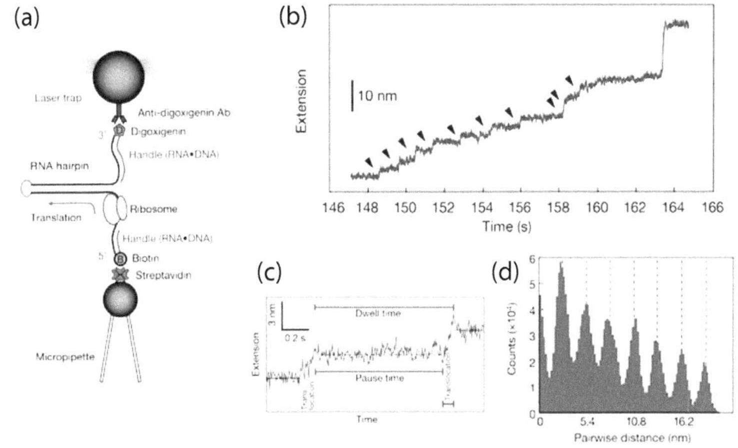

Figure 9.3 Single-molecule translation using optical tweezers. (a) Experimental design. The ribosome was stalled at the 5′-end of the mRNA hairpin construct, which was then held between two polystyrene beads coated with antidigoxigenin antibody (top; approximately 3 um in diameter) and streptavidin (bottom; approximately 2 um). The micropipette was substituted by a second laser trap when VE274hp was used for translation. Drawings are schematic and not to scale. (b) Extension and force trajectories during translation. Discrete steps are indicated by arrowheads. The 18 nm rip at 163 seconds corresponds to spontaneous opening of the remaining approximately 18 base-pair hairpin ahead of the translating ribosome. (c) The pattern of the translation process illustrated by a close-up view of the 148–150 seconds region. (d) Pairwise distance analysis of the extension trajectory after correction for drift.

that the release of tRNA from the E site is uncoupled with the next (Fig. 9.3) tRNA arrival to the A site tRNA.

In addition to the smFRET study of tRNA dynamics in translation, optical tweezers have been used to explore the structural basis for ribosome movement along the mRNA. Vanzi et al. used optical tweezers to study the stability of ribosome complexes on a polyuridine template, which first demonstrated the utility of force measurements in monitoring ribosome–mRNA interactions [122]. Uemura et al. further studied the rupture forces exerted between the ribosome and mRNA at different stages of the translation cycle [123]. They demonstrated that the presence of the Shine–

Dalgarno (SD) sequence could significantly enhance the stability of the ribosomal complex on the mRNA. In contrast, the presence of a peptidyl tRNA analog, *N*-acetyl-Phe-$tRNA^{Phe}$, in the A site, which mimicked the postpeptidyl transfer state, weakened the rupture force, which suggested that formation of the first peptide bond destabilizes the SD interaction, resulting in the weakening of the force with which the ribosome grips an mRNA. This might be an important requirement to facilitate movement of the ribosome along mRNA during the first translocation step.

Wen et al. used high-resolution optical tweezers to monitor the motion of a single *Escherichia coli* (*E. coli*) ribosome on the hairpin mRNA templates during translation [124]. As shown in Fig. 9.3, the ends of the mRNA are held at a force below that which is necessary to unwind the hairpin [124]. As the ribosome translates from the 5′-end of the hairpin, the double helix is opened, leading to an increase in the end-to-end distance of the RNA. Further analysis revealed that translation occurs in a series of translocation-pause-translocation events. The pause duration that controls the overall translation rate, in turn, depends on the secondary structure of the mRNA. By contrast, the translocation steps, lasting for less than 0.1 seconds, are independent of the experimental variables. Follow-up studies reported that the ribosome uses two distinct mechanism to unwind mRNA structures [125]: It destabilizes the helical junction at the mRNA entry site by biasing its thermal fluctuations toward the open state and thus increases the probability of the ribosome translocation that progresses unhindered, and it mechanically pulls apart the mRNA single strands of the closed junction during the conformational changes that accompany ribosome translocation. More recently, by using optical tweezers, Kaiser et al. proved that the ribosome could also affect the folding process of nascent protein [126]. Using the T4 lysozyme as the model system, they showed that incomplete T4 lysozyme polypeptides misfold and prefer to aggregate when free in solution, but they remain folding competent near the ribosome surface. The ribosome slows the formation of stable tertiary interactions and keeps the nascent peptide at native state relative to the free protein.

9.4 Summary and Prospects

During the past 20 years single-molecule experiments have highly progressed, and been widely used in the field of RNA science. Single-molecule experiments have already yielded important insights into RNA folding by detecting new folding intermediate states and pathways and by revealing the complexity in the conformational dynamics of both large and small RNAs. In addition, single-molecule approaches allowed us to better dissect the complex catalytic reactions by RNA enzymes by resolving individual reaction steps and discovering new reaction intermediates that are critical for understanding the reaction mechanisms. While the progress up to now has been extraordinary, there are still many challenges and limitations yet to overcome in the future. New methods such as the combination of different single-molecule approaches will surely help solve the outgrowing troublesome problems.

References

1. Afonin, K.A., et al. (2010). In vitro assembly of cubic RNA-based scaffolds designed in silico. *Nat. Nanotechnol.*, **5**, 676–682.
2. Famulok, M., and Ackermann, D. (2010). RNA nanotechnology: inspired by DNA. *Nat. Nanotechnol.*, **5**, 634–635.
3. Guo, P. (2010). The emerging field of RNA nanotechnology. *Nat. Nanotechnol.*, **5**, 833–842.
4. Delebecque, C.J., Lindner, A.B., Silver, P.A., and Aldaye, F.A. (2011). Organization of intracellular reactions with rationally designed RNA assemblies. *Science*, **333**, 470–474.
5. Shu, D., et al. (2011). Thermodynamically stable RNA three-way junction for constructing multifunctional nanoparticles for delivery of therapeutics. *Nat. Nanotechnol.*, **6**, 658–667.
6. Shukla, G.C., et al. (2011). A boost for the emerging field of RNA nanotechnology. *ACS Nano*, **5**, 3405–3418.
7. Doudna, J.A., and Cech, T.R. (2002). The chemical repertoire of natural ribozymes. *Nature*, **418**, 222–228.
8. Mandal, M., and Breaker, R.R. (2004). Gene regulation by riboswitches. *Nat. Rev. Mol. Cell Biol.*, **5**, 451–463.

9. Roth, A., and Breaker, R.R. (2009). The structural and functional diversity of metabolite-binding riboswitches. *Annu. Rev. Biochem.*, **78**, 305–334.
10. Sontheimer, E.J., and Carthew, R.W. (2005). Silence from within: endogenous siRNAs and miRNAs. *Cell*, **122**, 9–12.
11. Carthew, R.W., and Sontheimer, E.J. (2009). Origins and mechanisms of miRNAs and siRNAs. *Cell*, **136**, 642–655.
12. Hendrix, D.K., Brenner, S.E., and Holbrook, S.R. (2005). RNA structural motifs: building blocks of a modular biomolecule. *Q. Rev. Biophys.*, **38**, 221–243.
13. Butcher, S.E., and Pyle, A.M. (2011). The molecular interactions that stabilize RNA tertiary structure: RNA motifs, patterns, and networks. *Acc. Chem. Res.*, **44**, 1302–1311.
14. Holbrook, S.R. (2005). RNA structure: the long and the short of it. *Curr. Opin. Struct. Biol.*, **15**, 302–308.
15. Pan, T., and Sosnick, T.R. (1997). Intermediates and kinetic traps in the folding of a large ribozyme revealed by circular dichroism and UV absorbance spectroscopies and catalytic activity. *Nat. Struct. Biol.*, **4**, 931–938.
16. Sclavi, B., et al. (1998). RNA folding at millisecond intervals by synchrotron hydroxyl radical footprinting. *Science*, **279**, 1940–1943.
17. Treiber, D.K., Rook, M.S., Zarrinkar, P.P., and Williamson, J.R. (1998). Kinetic intermediates trapped by native interactions in RNA folding. *Science*, **279**, 1943–1946.
18. Gilbert, S.D., Stoddard, C.D., Wise, S.J., and Batey, R.T. (2006). Thermodynamic and kinetic characterization of ligand binding to the purine riboswitch aptamer domain. *J. Mol. Biol.*, **359**, 754–768.
19. Feig, A.L. (2007). Applications of isothermal titration calorimetry in RNA biochemistry and biophysics. *Biopolymers*, **87**, 293–301.
20. Zhuang, X. (2005). Single-molecule RNA science. *Annu. Rev. Biophys. Biomol. Struct.*, **34**, 399–414.
21. Aleman, E.A., Lamichhane, R., and Rueda, D. (2008). Exploring RNA folding one molecule at a time. *Curr. Opin. Chem. Biol.*, **12**, 647–654.
22. Li, P.T.X., Vieregg, J., and Tinoco, I. (2008). How RNA unfolds and refolds. *Annu. Rev. Biochem.*, **77**, 77–100.
23. Woodside, M.T., Garcia-Garcia, C., and Block, S.M. (2008). Folding and unfolding single RNA molecules under tension. *Curr. Opin. Chem. Biol.*, **12**, 640–646.

24. Aitken, C.E., Petrov, A., and Puglisi, J.D. (2010). Single ribosome dynamics and the mechanism of translation. *Annu. Rev. Biophys.*, **39**, 491–513.
25. Tinoco, I., Chen, G., and Qu, X.H. (2010). RNA reactions one molecule at a time. Cold Spring Harb. *Perspect. Biol.*, 2.
26. Deniz, A.A., Mukhopadhyay, S., and Lemke, E.A. (2008). Single-molecule biophysics: at the interface of biology, physics and chemistry. *J. R. Soc. Interface*, **5**, 15–45.
27. Kapanidis, A.N., and Strick, T. (2009). Biology, one molecule at a time. *Trends Biochem. Sci.*, **34**, 234–243.
28. Ha, T., et al. (1999). Ligand-induced conformational changes observed in single RNA molecules. *Proc. Natl. Acad. Sci. U S A*, **96**, 9077–9082.
29. Sytnik, A., et al. (1999). Peptidyl transferase center activity observed in single ribosomes. *J. Mol. Biol.*, **285**, 49–54.
30. Neuman, K.C., and Nagy, A. (2008). Single-molecule force spectroscopy: optical tweezers, magnetic tweezers and atomic force microscopy. *Nat. Methods*, **5**, 491–505.
31. Joo, C., et al. (2008). Advances in single-molecule fluorescence methods for molecular biology. *Annu. Rev. Biochem.*, **77**, 51–76.
32. Ishijima, A., et al. (1998). Simultaneous observation of individual ATPase and mechanical events by a single myosin molecule during interaction with actin. *Cell*, **92**, 161–171.
33. Lang, M.J., et al. (2004). Simultaneous, coincident optical trapping and single-molecule fluorescence. *Nat. Methods*, **1**, 133–139.
34. Gumpp, H., et al. (2009). Ultrastable combined atomic force and total internal reflection fluorescence microscope [corrected]. *Rev. Sci. Instrum.*, **80**, 063704.
35. Comstock, M.J., Ha, T., and Chemla, Y.R. (2011). Ultrahigh-resolution optical trap with single-fluorophore sensitivity. *Nat. Methods*, **8**, 335–340.
36. Binnig, G., Quate, C.F., and Gerber, C. (1986). Atomic force microscope. *Phys. Rev. Lett.*, **56**, 930–933.
37. Fritz, J., Anselmetti, D., Jarchow, J., and Fernandez-Busquets, X. (1997). Probing single biomolecules with atomic force microscopy. *J. Struct. Biol.*, **119**, 165–171.
38. Engel, A., and Muller, D.J. (2000). Observing single biomolecules at work with the atomic force microscope. *Nat. Struct. Biol.*, **7**, 715–718.

39. Muller, D.J., and Dufrene, Y.F. (2008). Atomic force microscopy as a multifunctional molecular toolbox in nanobiotechnology. *Nat. Nanotechnol.*, **3**, 261–269.

40. Crampton, N., et al. (2007). Fast-scan atomic force microscopy reveals that the type III restriction enzyme EcoP15I is capable of DNA translocation and looping. *Proc. Natl. Acad. Sci. U S A*, **104**, 12755–12760.

41. Heus, H.A., et al. (2011). Atomic force microscope-based single-molecule force spectroscopy of RNA unfolding. *Anal. Biochem.*, **414**, 1–6.

42. Fisher, J.K., et al. (2006). Thin-foil magnetic force system for high-numerical-aperture microscopy. *Rev. Sci. Instrum.*, **77**.

43. Yan, J., Skoko, D., and Marko, J.F. (2004). Near-field-magnetic-tweezer manipulation of single DNA molecules. *Phys. Rev. E*, **70**.

44. Strick, T.R., Croquette, V., and Bensimon, D. (2000). Single-molecule analysis of DNA uncoiling by a type II topoisomerase. *Nature*, **404**, 901–904.

45. Moffitt, J.R., Chemla, Y.R., Smith, S.B., and Bustamante, C. (2008). Recent advances in optical tweezers. *Annu. Rev. Biochem.*, **77**, 205–228.

46. Weiss, S. (1999). Fluorescence spectroscopy of single biomolecules. *Science*, **283**, 1676–1683.

47. Weiss, S. (2000). Measuring conformational dynamics of biomolecules by single molecule fluorescence spectroscopy. *Nat. Struct. Biol.*, **7**, 724–729.

48. Haustein, E., and Schwille, P. (2004). Single-molecule spectroscopic methods. *Curr. Opin. Struct. Biol.*, **14**, 531–540.

49. Nie, S., Chiu, D.T., and Zare, R.N. (1994). Probing individual molecules with confocal fluorescence microscopy. *Science*, **266**, 1018–1021.

50. Funatsu, T., et al. (1995). Imaging of single fluorescent molecules and individual ATP turnovers by single myosin molecules in aqueous solution. *Nature*, **374**, 555–559.

51. Stryer, L. (1978). Fluorescence energy transfer as a spectroscopic ruler. *Annu. Rev. Biochem.*, **47**, 819–846.

52. Selvin, P.R. (1995) Fluorescence resonance energy transfer. *Methods Enzymol.*, **246**, 300–334.

53. Lu, Z.J., Turner, D.H., and Mathews, D.H. (2006). A set of nearest neighbor parameters for predicting the enthalpy change of RNA secondary structure formation. *Nucleic Acids Res.*, **34**, 4912–4924.

54. Tinoco, I., and Bustamante, C. (1999). How RNA folds. *J. Mol. Biol.*, **293**, 271–281.

55. Tinoco, I., Jr., Li P.T., and Bustamante, C. (2006). Determination of thermodynamics and kinetics of RNA reactions by force. *Q. Rev. Biophys.*, **39**, 325–360.

56. Liphardt, J., et al. (2001). Reversible unfolding of single RNA molecules by mechanical force. *Science*, **292**, 733–737.

57. Li, P.T., Bustamante, C., and Tinoco, I., Jr. (2007). Real-time control of the energy landscape by force directs the folding of RNA molecules. *Proc. Natl. Acad. Sci. U S A*, **104**, 7039–7044.

58. Woodside, M.T., et al. (2006). Direct measurement of the full, sequence-dependent folding landscape of a nucleic acid. *Science*, **314**, 1001–1004.

59. Woodside, M.T., et al. (2006). Nanomechanical measurements of the sequence-dependent folding landscapes of single nucleic acid hairpins. *Proc. Natl. Acad. Sci. U S A*, **103**, 6190–6195.

60. Jarzynski, C. (1997). Equilibrium free-energy differences from non-equilibrium measurements: A master-equation approach. *Phys. Rev. E*, **56**, 5018–5035.

61. Liphardt, J., et al. (2002). Equilibrium information from nonequilibrium measurements in an experimental test of Jarzynski's equality. *Science*, **296**, 1832–1835.

62. Collin, D., et al. (2005). Verification of the Crooks fluctuation theorem and recovery of RNA folding free energies. *Nature*, **437**, 231–234.

63. Gupta, A.N., et al. (2011). Experimental validation of free-energy-landscape reconstruction from non-equilibrium single-molecule force spectroscopy measurements. *Nat. Phys.*, **7**, 631–634.

64. Gell, C., et al. (2008). Single-molecule fluorescence resonance energy transfer assays reveal heterogeneous folding ensembles in a simple RNA stem-loop. *J. Mol. Biol.*, **384**, 264–278.

65. Lilley, D.M., et al. (1995). A nomenclature of junctions and branchpoints in nucleic acids. *Nucleic Acids Res.*, **23**, 3363–3364.

66. Hohng, S., et al. (2004). Conformational flexibility of four-way junctions in RNA. *J. Mol. Biol.*, **336**, 69–79.

67. Schroeder, K.T., McPhee, S.A., Ouellet, J., and Lilley, D.M. (2010). A structural database for k-turn motifs in RNA. *RNA*, **16**, 1463–1468.

68. Wozniak, A.K., et al. (2005). Detecting protein-induced folding of the U4 snRNA kink-turn by single-molecule multiparameter FRET measurements. *RNA*, **11**, 1545–1554.

69. Costa, M., and Michel, F. (1997). Rules for RNA recognition of GNRA tetraloops deduced by in vitro selection: comparison with in vivo evolution. *EMBO J.*, **16**, 3289–3302.

70. Geary, C., Baudrey, S., and Jaeger, L. (2008). Comprehensive features of natural and in vitro selected GNRA tetraloop-binding receptors. *Nucleic Acids Res.*, **36**, 1138–1152.

71. Hodak, J.H., et al. (2005). Docking kinetics and equilibrium of a GAAA tetraloop-receptor motif probed by single-molecule FRET. *Proc. Natl. Acad. Sci. U S A*, **102**, 10505–10510.

72. Downey, C.D., et al. (2006). Metal ion dependence, thermodynamics, and kinetics for intramolecular docking of a GAAA tetraloop and receptor connected by a flexible linker. *Biochemistry*, **45**, 3664–3673.

73. Li, P.T., Bustamante, C., and Tinoco, I., Jr. (2006). Unusual mechanical stability of a minimal RNA kissing complex. *Proc. Natl. Acad. Sci. U S A*, **103**, 15847–15852.

74. Li, P.T., and Tinoco, I., Jr. (2009). Mechanical unfolding of two DIS RNA kissing complexes from HIV-1. *J. Mol. Biol.*, **386**, 1343–1356.

75. Staple, D.W., and Butcher, S.E. (2005). Pseudoknots: RNA structures with diverse functions. *PLoS Biol.*, **3**, e213.

76. Hansen, T.M., Reihani, S.N.S., Oddershede, L.B., and Sorensen, M.A. (2007). Correlation between mechanical strength of messenger RNA pseudoknots and ribosomal frame-shifting. *Proc. Natl. Acad. Sci. U S A*, **104**, 5830–5835.

77. Green, L., Kim, C.H., Bustamante, C., and Tinoco, I., Jr. (2008). Characterization of the mechanical unfolding of RNA pseudoknots. *J. Mol. Biol.*, **375**, 511–528.

78. Chen, G., et al. (2009). Triplex structures in an RNA pseudoknot enhance mechanical stability and increase efficiency of −1 ribosomal frame-shifting. *Proc. Natl. Acad. Sci. U S A*, **106**, 12706–12711.

79. Chen, G., Wen, J.D., and Tinoco, I., Jr. (2007). Single-molecule mechanical unfolding and folding of a pseudoknot in human telomerase RNA. *RNA*, **13**, 2175–2188.

80. Mihalusova, M., Wu, J.Y., and Zhuang, X. (2011). Functional importance of telomerase pseudoknot revealed by single-molecule analysis. *Proc. Natl. Acad. Sci. U S A*, **108**, 20339–20344.

81. Sattin, B.D., et al. (2008). Direct measurement of tertiary contact cooperativity in RNA folding. *J. Am. Chem. Soc.*, **130**, 6085–6087.

82. Solomatin, S.V., Greenfeld, M., and Herschlag, D. (2011). Implications of molecular heterogeneity for the cooperativity of biological macromolecules. *Nat. Struct. Mol. Biol.*, **18**, 732–734.
83. Lilley, D.M. (2005). Structure, folding and mechanisms of ribozymes. *Curr. Opin. Struct. Biol.*, **15**, 313–323.
84. Ha, T. (2004). Structural dynamics and processing of nucleic acids revealed by single-molecule spectroscopy. *Biochemistry*, **43**, 4055–4063.
85. Zhuang, X., et al. (2002). Correlating structural dynamics and function in single ribozyme molecules. *Science*, **296**, 1473–1476.
86. Bokinsky, G., et al. (2003). Single-molecule transition-state analysis of RNA folding. *Proc. Natl. Acad. Sci. U S A*, **100**, 9302–9307.
87. Rueda, D., et al. (2004). Single-molecule enzymology of RNA: essential functional groups impact catalysis from a distance. *Proc. Natl. Acad. Sci. U S A*, **101**, 10066–10071.
88. Liu, S., Bokinsky, G., Walter, N.G., and Zhuang, X. (2007). Dissecting the multistep reaction pathway of an RNA enzyme by single-molecule kinetic "fingerprinting." *Proc. Natl. Acad. Sci. U S A*, **104**, 12634–12639.
89. Ditzler, M.A., et al. (2008). A rugged free energy landscape separates multiple functional RNA folds throughout denaturation. *Nucleic Acids Res.*, **36**, 7088–7099.
90. Zhuang, X., et al. (2000). A single-molecule study of RNA catalysis and folding. *Science*, **288**, 2048–2051.
91. Bartley, L.E., et al. (2003). Exploration of the transition state for tertiary structure formation between an RNA helix and a large structured RNA. *J. Mol. Biol.*, **328**, 1011–1026.
92. Onoa, B., et al. (2003). Identifying kinetic barriers to mechanical unfolding of the T. thermophila ribozyme. *Science*, **299**, 1892–1895.
93. Solomatin, S.V., Greenfeld, M., Chu, S., and Herschlag, D. (2010). Multiple native states reveal persistent ruggedness of an RNA folding landscape. *Nature*, **463**, 681–684.
94. Lehmann, K., and Schmidt, U. (2003). Group II introns: structure and catalytic versatility of large natural ribozymes. *Crit. Rev. Biochem. Mol. Biol.*, **38**, 249–303.
95. Pyle, A.M., Fedorova, O., and Waldsich, C. (2007). Folding of group II introns: a model system for large, multidomain RNAs? *Trends Biochem. Sci.*, **32**, 138–145.

96. Steiner, M., Karunatilaka, K.S., Sigel, R.K., and Rueda, D. (2008). Single-molecule studies of group II intron ribozymes. *Proc. Natl. Acad. Sci. U S A*, **105**, 13853–13858.
97. Karunatilaka, K.S., Solem, A., Pyle, A.M., and Rueda, D. (2010). Single-molecule analysis of Mss116-mediated group II intron folding. *Nature*, **467**, 935–939.
98. Mandal, M., and Breaker, R.R. (2004). Adenine riboswitches and gene activation by disruption of a transcription terminator. *Nat. Struct. Mol. Biol.*, **11**, 29–35.
99. Kim, J.N., and Breaker, R.R. (2008). Purine sensing by riboswitches. *Biol. Cell*, **100**, 1–11.
100. Lemay, J.F., et al. (2006). Folding of the adenine riboswitch. *Chem. Biol.*, **13**, 857–868.
101. Gilbert, S.D., and Batey, R.T. (2006). Riboswitches: fold and function. *Chem. Biol.*, **13**, 805–807.
102. Greenleaf, W.J., et al. (2008). Direct observation of hierarchical folding in single riboswitch aptamers. *Science*, **319**, 630–633.
103. Neupane, K., et al. (2011). Single-molecule force spectroscopy of the add adenine riboswitch relates folding to regulatory mechanism. *Nucleic Acids Res.*, **39**, 7677–7687.
104. Poiata, E., Meyer, M.M., Ames, T.D., and Breaker, R.R. (2009). A variant riboswitch aptamer class for *S*-adenosylmethionine common in marine bacteria. *RNA*, **15**, 2046–2056.
105. Heppell, B., et al. (2011). Molecular insights into the ligand-controlled organization of the SAM-I riboswitch. *Nat. Chem. Biol.*, **7**, 384–392.
106. Haller, A., et al. (2011). Conformational capture of the SAM-II riboswitch. *Nat. Chem. Biol.*, **7**, 393–400.
107. Montange, R.K., and Batey, R.T. (2008). Riboswitches: emerging themes in RNA structure and function. *Annu. Rev. Biophys.*, **37**, 117–133.
108. Anthony, P.C., Perez, C.F., Garcia-Garcia, C., and Block, S.M. (2012). Folding energy landscape of the thiamine pyrophosphate riboswitch aptamer. *Proc. Natl. Acad. Sci. U S A*, **109**, 1485–1489.
109. Korostelev, A., Ermolenko, D.N., and Noller, H.F. (2008). Structural dynamics of the ribosome. *Curr. Opin. Chem. Biol.*, **12**, 674–683.
110. Schmeing, T.M., and Ramakrishnan, V. (2009). What recent ribosome structures have revealed about the mechanism of translation. *Nature*, **461**, 1234–1242.

111. Wilson, D.N., and Nierhaus, K.H. (2007). The weird and wonderful world of bacterial ribosome regulation. *Crit. Rev. Biochem. Mol. Biol.*, **42**, 187–219.

112. Petrov, A., et al. (2011). Dynamics of the translational machinery. *Curr. Opin. Struct. Biol.*, **21**, 137–145.

113. Blanchard, S.C., et al. (2004). tRNA dynamics on the ribosome during translation. *Proc. Natl. Acad. Sci. U S A*, **101**, 12893–12898.

114. Blanchard, S.C., et al. (2004). tRNA selection and kinetic proofreading in translation. *Nat. Struct. Mol. Biol.*, **11**, 1008–1014.

115. Munro, J.B., Altman, R.B., O'Connor, N., and Blanchard, S.C. (2007). Identification of two distinct hybrid state intermediates on the ribosome. *Mol. Cell*, **25**, 505–517.

116. Cornish, P.V., Ermolenko, D.N., Noller, H.F., and Ha, T. (2008). Spontaneous intersubunit rotation in single ribosomes. *Mol. Cell*, **30**, 578–588.

117. Fei, J., Kosuri, P., MacDougall, D.D., and Gonzalez, R.L., Jr. (2008). Coupling of ribosomal L1 stalk and tRNA dynamics during translation elongation. *Mol. Cell*, **30**, 348–359.

118. Cornish, P.V., et al. (2009). Following movement of the L1 stalk between three functional states in single ribosomes. *Proc. Natl. Acad. Sci. U S A*, **106**, 2571–2576.

119. Munro, J.B., et al. (2010). Spontaneous formation of the unlocked state of the ribosome is a multistep process. *Proc. Natl. Acad. Sci. U S A*, **107**, 709–714.

120. Aitken, C.E., and Puglisi, J.D. (2010). Following the intersubunit conformation of the ribosome during translation in real time. *Nat. Struct. Mol. Biol.*, **17**, 793–800.

121. Uemura, S., et al. (2010). Real-time tRNA transit on single translating ribosomes at codon resolution. *Nature*, **464**, 1012–1017.

122. Vanzi, F., et al. (2005). Mechanical studies of single ribosome/mRNA complexes. *Biophys. J.*, **89**, 1909–1919.

123. Uemura, S., et al. (2007). Peptide bond formation destabilizes Shine-Dalgarno interaction on the ribosome. *Nature*, **446**, 454–457.

124. Wen, J.D., et al. (2008). Following translation by single ribosomes one codon at a time. *Nature*, **452**, 598–603.

125. Qu, X., et al. (2011). The ribosome uses two active mechanisms to unwind messenger RNA during translation. *Nature*, **475**, 118–121.

126. Kaiser, C.M., et al. (2011). The ribosome modulates nascent protein folding. *Science*, **334**, 1723–1727.

127. Abbondanzieri, E.A., et al. (2005). Direct observation of base-pair stepping by RN Apolymerase. *Nature*, **438**, 460–465.

Chapter 10

Application of Single-Molecule Fluorescence in RNA Biology

Pan T.X. Li[a] and Lori Goldner[b]

[a] *Department of Biological Sciences and the RNA Institute University at Albany—SUNY, 1400 Washington Ave. Albany, NY 12222, USA*
[b] *Department of Physics, University of Massachusetts, Amherst, MA 01003, USA*
pli@albany.edu, lgoldner@physics.umass.edu

10.1 Introduction

Single-molecule-sensitive fluorescence microscopy is a relatively new technique that is still developing. In the past two decades, instruments capable of detecting single fluorophores have been rapidly developed and improved, allowing new types of applications in a variety of fields. In addition to the "conventional" in vitro studies, it is now also possible to image single ribonucleic acid (RNA) molecules inside cells using super-resolution fluorescence microscopy [1, 2]. Together, these techniques have revealed detailed molecular dynamics of RNAs in various biological contexts. To date, most studies have been performed on home-built microscopes [3–5], but the recent availability of commercial instruments capable

RNA Nanotechnology
Edited by Bin Wang

ISBN 978-981-4411-64-6 (Hardcover), 978-981-4411-65-3 (eBook)
www.panstanford.com

of single-molecule-sensitive fluorescence imaging has led to dramatically increased interest in the technique and enormous efforts at developing better, faster, and cheaper instruments. In this chapter, we focus our discussion on applications of the single-molecule fluorescence techniques in RNA biology. Comprehensive and detailed reviews on the techniques and instruments are covered in several recently published books [6–8].

10.1.1 *Single-Molecule Approach to the RNA Structure Problem*

Single-molecule experiments are still more costly and more time consuming to perform than bulk fluorescence measurements. Therefore, before embarking on single-molecule-sensitive measurements, the following question must be addressed: what can be learned from such an experiment that cannot be obtained using conventional methods? Single-molecule techniques can elucidate the details of an ensemble and can be used to track the conformational changes of individual molecules. They are therefore useful in situations where a sample contains heterogeneous molecules and/or where dynamics occur in the bandwidth of a fluorescence measurement, typically between 10 milliseconds and minutes.

In RNA biology, justification for the use of single-molecule techniques is rather straightforward, as RNA molecules are inherently heterogeneous and dynamic. A single RNA sequence can fold into multiple thermodynamically stable structures [9, 10]. In response to changes in the cellular environment, a folded RNA can change its structure in a timely fashion, as occurs in transcription attenuation [11] and in riboswitches [12, 13] in gene regulation. For their biological functions, many RNAs bind to various small molecules, other RNAs, and proteins. These selective interactions depend again on the cellular environment. Moreover, RNAs with enzymatic activity, such as ribozymes and ribosomes [14], show time-dependent structural dynamics. Let us imagine a scenario that half of the RNA molecules in a solution are folded and the other half unfolded. A bulk measurement would reveal an average structure that resembles neither folded nor unfolded RNA. While an increasing number of RNA structures have been determined by crystallography and nuclear magnetic resonance (NMR), these studies focus on a few classes of well-folded RNAs [15]. In contrast, single-molecule approaches are almost ideal to study structural and functional polymorphism of RNAs. This distinctive advantage of single-molecule techniques eliminates the need for model-dependent deconvolution that is often required for interpreting bulk measurements, such as ultraviolet (UV) melting. Furthermore, by monitoring many individual molecules over time, it is possible to reconstruct dynamic ranges of RNA structures and therefore better interpret the results observed from ensemble experiments, such as NMR and chemical footprinting [16, 17].

In this postgenomic era, an entire genome of an individual can be sequenced rapidly and relatively cheaply [18]. However, it remains a daunting challenge to interpret the biological meaning from large datasets of sequences, particularly those of noncoding RNAs [19]. A large fraction of each genome is transcribed into RNAs but not translated into proteins [20]. In addition to the classic ribosomal RNA (rRNA) and transfer RNA (tRNA) genes, most of these noncoding RNAs are not experimentally characterized, and their potential biological functions remain unknown [21]. It is still difficult to predict RNA structures from a sequence, and it is even more difficult to guess biological functions. For instance, how can potential regulatory RNAs be identified, particularly those that function by alternating their structures [22]? Further, when and where in living cells does a specific RNA function? All these unanswered questions make single-molecule approaches not only interesting to biophysicists but also attractive to biological and biomedical scientists in general.

10.1.2 *Fluorescence Probes for Single-Molecule Studies*

Intrinsic fluorescence of RNA is relatively low and not distinctive from many other biological molecules [23]. Therefore, fluorophores suitable for single-molecule fluorescence spectroscopy and microscopy have to be incorporated into RNA experiments. This is not a trivial task as chemical modifications are only commercially available for small RNAs. Various clever schemes have been designed to make single RNA molecules fluoresce.

In addition, only a limited selection of fluorophores, with high quantum yield and low photobleaching characteristics, are suitable for use in single-molecule studies. High quantum yield means relatively less light required to excite the dye, which minimizes photodamage to biological samples and cells. Photobleaching, which occurs suddenly and completely for single fluorophores, limits the time over which observations can be made. The solubility and size of the dyes are also important. Two dyes that are widely used to label RNA are Cy5 and Cy3. Their popularity largely arises from their commercial availability and the fact that they work well as a donor–acceptor pair in fluorescence resonance energy transfer (FRET). Other brilliant dyes and quenchers, such as tetramethyl rhodamine (TMR), can also be used to modify RNAs and their ligands [24]. In in vivo experiments, probes labeled with organic dyes can be used for in situ hybridization [25]. In addition, green fluorescence protein (GFP) or its derivatives can be fused to RNA-binding proteins, allowing fluorescent tagging at specific sites in RNA inside living cells [1, 26]. Also, RNA aptamers have been developed to bind the chromophore portion of GFP [27, 28]. When bound to their ligand, the aptamers show similar fluorescence properties as GFP. New innovations in labeling are still needed. For instance, the recent advancement of super-resolution microscopy necessitates development of new classes of fluorophores for imaging single RNAs inside cells [29, 30].

Organic dyes can be chemically attached to RNAs using solid-phase synthesis, but the size of fluorescently labeled RNAs is limited to about 50 nucleotides. On occasions, up to three fluorophores have been attached to single nucleic acid molecules [31, 32]. In addition to fluorophores, other chemical modifications, such as biotin, are also used to attach RNAs to surfaces or particles. However, RNAs with more than one label have relatively low yield and a stricter size limit than those with a single chemical modification. The most common approach to circumvent the size limit is to anneal small, modified oligonucleotides to the RNA of interest [33]. Alternatively, a dual-labeled RNA can be made by ligating two small oligonucleotides each labeled with a fluorophore. The ligated RNA can be purified from unreacted oligonucleotides by denatured polyacrylamide gel electrophoresis. Using this approach, a pair of donor and acceptor fluorophores can be added to a large RNA for use in FRET. A drawback of such multiple-step preparations is low yield. On the lucky side, single-molecule fluorescence spectroscopy is relatively tolerant to sample purity. For instance, a solution of dual-labeled samples inevitably contains molecules with none, one, or two fluorophores. In a single-molecule experiment, the three types of molecules can be distinguished by their fluorescence properties.

When interpreting single-molecule fluorescence data, it is important to tease out effects of the dyes from the properties of RNAs. Fluorophores are fragile and short lived, and photobleaching, triplet blinking, charge transfer, and isomerization can mimic the effects of molecular dynamics. Care needs to be taken to systematically isolate effects of dye photophysics and the attachment chemistry from fluorescence signals due to changes in the biomolecular structure [34, 35]. As demonstrated for Cy5 and Cy3 on double-stranded deoxyribonucleic acid (DNA), dye properties, including their positions and orientations, can change as a function of where and how they are attached to the DNA [36–38].

10.1.3 *Single-Molecule-Sensitive Microscopies*

In addition to sufficiently bright and stable dyes, single-molecule-sensitive measurements require minimizing the background and maximizing detection sensitivity. This is typically achieved using either confocal microscopy with high efficiency point detectors or a total internal reflection fluorescence (TIRF) microscope with full-field imaging on a charge-coupled device (CCD) detector. In both cases, innovations in detector technologies in the last 20 years have facilitated single-molecule-sensitive measurement. The remaining problem of reducing the background to the point where a single fluorophore can be observed is typically approached by minimizing the detection volume. For example, in a confocal microscope, the entire detection volume can be less than 1 femtoliter, limited by the size of the excitation light focus and the filtering pinhole used on the detector side. In a TIRF microscope, the detection volume is limited to within about 200 nm from the surface at which the light is reflected, typically a microscope coverslip. In this case, the sample to be observed must be confined within 200 nm of the surface. All the major microscope manufacturers now sell systems capable of single-molecule-sensitive TIRF microscopy. Super-resolution microscopes are often built onto existing TIRF instruments and require mostly additional lasers and software. Details of these instruments are covered in several recently published books [6–8].

10.2 Monitoring the Dynamics of Single RNA Molecules Using FRET

The function of RNA is often associated with structural changes. To monitor such processes, a technique known as FRET is employed to monitor conformational dynamics of molecules [39]. FRET is based on the phenomenon that light absorbed by a donor dye can be transferred to an acceptor fluorophore and be subsequently emitted. The efficiency of energy transfer between the dyes, E_{FRET}, critically depends on the distance between the two dyes, R, as described by the following equation:

$$E_{\mathrm{FRET}} = \frac{1}{1 + R^6/R_0^6} \tag{10.1}$$

in which R_0 is the Förster distance at which the probability of energy transfer is 1/2. R_0 is a property of the two freely rotating dyes and their surrounding conditions. A measurement of E_{FRET} can be used to estimate or even extract the distance between the dyes [40]. In some cases where the dye orientations are fixed with respect to the molecule, E_{FRET} may also report on changes in orientation of the dyes. Moreover, when a conformational change alters the relative distance between the fluorophores, the FRET value will change accordingly. R_0 has been measured to be in a range between 5.6 nm and 6.5 nm for the Cy5–Cy3 pair [41–44]. With $R_0 = 6.0$ nm, FRET changes from 0.15 to 0.92 when the distance between the dyes decreases from 8 nm to 4 nm. Because this spatial range is comparable to sizes of many macromolecules, the brilliant dye pair has been widely used to study folding and interaction of many molecular systems, particularly in single-molecule FRET (smFRET) experiments.

FRET can be measured by observing changes in the fluorescence intensity or donor lifetime [40]. The former is more commonly used in single-molecule fluorescence experiments, especially when the intent is to measure the dynamic "trajectory" of the molecule. However, combining both measurements provides a robust method for distinguishing between populations of molecules that are otherwise similar [45]. For this reason it is becoming increasingly common that both intensity and lifetime measurements are made simultaneously.

10.2.1 *Folding of a Tetraloop Receptor*

Tertiary interactions between distal domains allow RNAs to fold into compact, three-dimensional structures required for their biological functions [46, 47]. Folding of most RNAs follows a hierarchical order that tertiary interactions form after folding of secondary structures. This feature has been exploited to examine RNA tertiary interaction between folded secondary structures, such as tetraloop receptors, one of the basic tertiary folding motifs [48]. In an smFRET experiment [49], a GAAA tetraloop and a tetraloop receptor linked by a short strand of polyA were used as a simple model system (Fig. 10.1). A pair of Cy5 and Cy3 dyes was labeled at the tetraloop and receptor stems, respectively. When the tetraloop interaction formed, the two dyes were relatively close in the docked conformation and the FRET was ~0.8. In the undocked state, the two stems were farther apart and the FRET was measured to be ~0.3. Hence, smFRET can be used to monitor the tertiary folding of the tetraloop receptor.

Tertiary folding of this RNA critically depends on ionic conditions, particularly divalent metal ions [49]. smFRET trajectories of the tetraloop receptor clearly showed such effects. In the absence of Mg^{2+}, the RNA stayed mostly undocked. In >5 mM Mg^{2+}, the RNA was mostly docked. At intermediate Mg^{2+} concentrations, the RNA transited between the two conformations, as indicated by changes in E_{FRET}. Dwell times at high and low FRET states were pooled to generate a probability density function of lifetimes, from which first-order rate constants for docking (k_{dock}) and undocking (k_{undock}) can be computed. Interestingly, k_{dock} increased by twelvefold as the concentration of Mg^{2+} was raised from 0 mM to 10 mM, whereas k_{undock} decreased by only threefold over the same range of Mg^{2+}

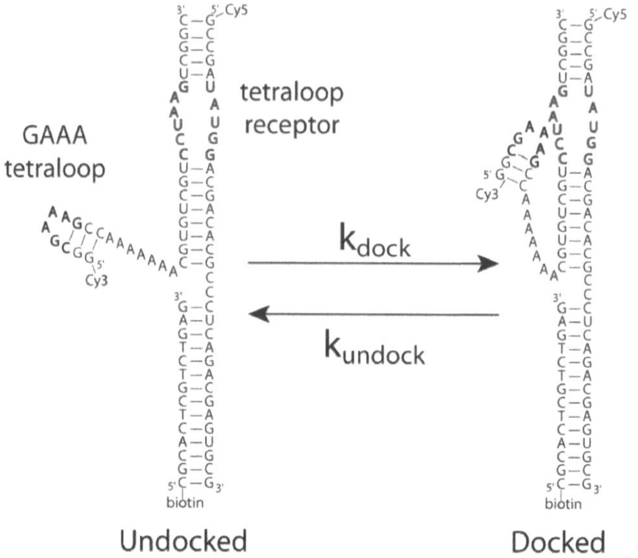

Figure 10.1 smFRET measurement of a tetraloop receptor tertiary interaction. A GAAA tetraloop hairpin and a helix containing a tetraloop receptor were connected by a flexible A_7 linker. Cy3 and Cy5 were placed at the ends of the two helices. The undocked conformation (left) has an average FRET value of 0.22, whereas that of the docked (right) was 0.68. (Adapted from Ref. 49.)

concentration. The equilibrium constant can be computed either by the ratio of the two rate constants or by the ratio of the total dwell times at each state. In subsequent works [50, 51], effects of various mono-, di-, and trivalent metal ions on the tetraloop receptor have been investigated. Monovalent metal ions, such as K^+ and Na^+, also promoted formation of the tetraloop receptor. However, it usually took a few hundred mM of monovalent cations to have similar effect by only a few mM of Mg^{2+}. In addition to metal ions, the length and composition of the linker between the two interacting domains also affect docking kinetics. Particularly, formation of the tetraloop receptor requires a flexible linker between the two stems [50, 52]. When the linker was a rigid duplex, the tertiary folding was greatly reduced.

Most single-molecule experiments were performed at room temperature, which makes it impossible to calculate enthalpy and entropy changes of the reaction. Nesbitt et al. incorporated a temperature-controlled flow cell into smFRET experiments, achieving precisely controlled temperatures up to 45°C [53]. This development has enabled a series of interesting studies on the tetraloop receptor [53–55]. Formation of the tertiary interaction was found to be exothermic and entropically unfavorable. Such measurements can reveal rich thermodynamic information of RNA folding beyond ΔG and are more comparable to bulk thermodynamic characterization.

10.2.2 *Hairpin Ribozymes*

The hairpin ribozyme is a relatively small catalytic RNA that cleaves and ligates RNA substrates reversibly [56]. The RNA can adopt two conformations, docked and undocked, similar to the ones observed in the tetraloop receptor. In the active form, the ribozyme stays at the docked state such that two hairpin loops of the ribozyme interact, whereas in the inactive form, the undocked hairpins costack on each other. The catalytic activity of the RNA depends on the equilibrium between the two conformations, which in turn depends on ionic and temperature conditions. The hairpin ribozyme imposed a challenge different from that of the tetraloop receptor: can catalytic activity of an RNA be studied by smFRET, in addition to conformational changes?

In one elegant study, the hairpin ribozyme was redesigned such that the cleaved strand could not dissociate from the ribozyme, allowing religation to occur quickly (Fig. 10.2a,b) [57]. This scheme ensured that before fluorophores were photobleached, multiple rounds of cleavage and ligation by a single ribozyme molecule can be observed. When the RNA was cleaved, the docked state had similar FRET as the ligated state. However, the cleaved RNA transited frequently between docked and undocked states, as indicated by rapid fluctuation of FRET efficiency (Fig. 10.2b). In contrast, the ligated RNA had more extended lifetimes than the docked state. Therefore, the two conformational states can be distinguished in time domains despite similar FRET values. This distinction

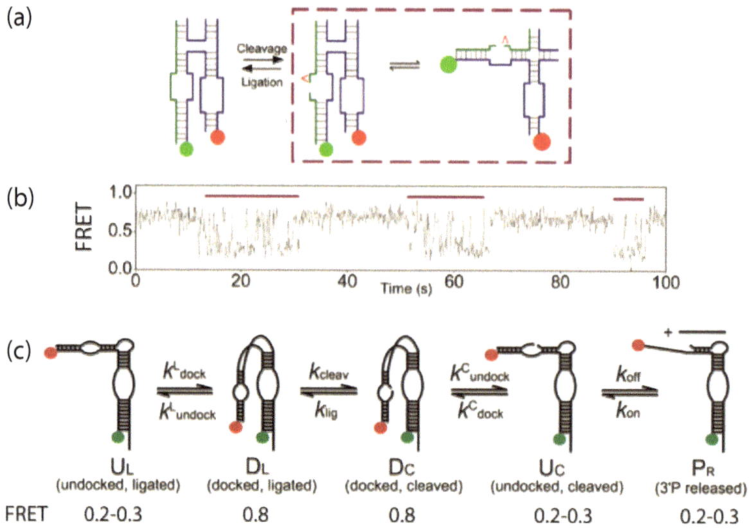

Figure 10.2 Structural dynamics and catalysis of a hairpin ribozyme. (a) Design of a hairpin ribozyme that can undergo multiround catalysis. The cleavage site was marked with. The two colored dots represent fluorophores. (b) FRET trace of a single ribozyme from (a). The purple bar indicates regions that undergo rapid docking/undocking transitions. (Adapted from Ref. 57.) (c) Sequential reaction of a hairpin ribozyme (adapted from Ref. 58).

allows rate constants of individual folding steps to be determined. Furthermore, docking/undocking and cleavage/ligation kinetics were modulated by pH, and the measured folding rates were compared to catalytic rates of the enzyme.

In another study, folding and catalytic steps of a hairpin ribozyme were sequentially analyzed by changing Mg^{2+} and product concentrations [58]. A series of pulse-chase experiments were performed on a single RNA molecule to measure kinetics between different states. Although not all conformations could be distinguished by FRET values (Fig. 10.2c), Mg^{2+}-dependent mean lifetimes were different for intermediates, providing kinetic fingerprints of each folding state.

The two studies both focused on the hairpin ribozymes at the single-molecule level. Multiround catalysis was monitored in one study. In the other one, a single reaction was dissected into substeps. These approaches are reminiscent of classic studies of enzymatic kinetics. However, structural intermediates, particularly the transient ones, are difficult to detect in bulk studies. In both studies, structural intermediates were distinguished and characterized by both FRET values and kinetic fingerprints, making it possible to establish a correlation between structure and function of the hairpin ribozyme.

10.2.3 *Large Ribozymes*

Unlike the hairpin ribozyme with stable folding optimized for catalysis, large ribozymes, such as group I and II introns, have rugged folding energy landscapes with multiple stable on- and off-pathway intermediates [59, 60]. Structural polymorphism and manifold folding pathways of these ribozymes have become ideal a test ground for single-molecule techniques.

10.2.3.1 Group I intron ribozymes

To fluorescently label the large *Tetrahymena thermophila* ribozyme, the RNA was annealed to short oligonucleotides modified by Cy5 and Cy3 [33]. In this scheme, mutations of the ribozyme can be easily incorporated into the system without changing the labeling chemistry. The FRET of the dye pair was used as an indicator to monitor RNA folding in real time and to identify different intermediate states. The overall folding rate was comparable to values obtained from bulk studies. Furthermore, by pinpointing folding intermediates, different folding pathways were identified and their kinetics was measured. For instance, once misfolded, the RNA was significantly slow to reach the native fold. By observing folding trajectories of many single RNA molecules, it was possible to quantify occurrence of different folding pathways.

In a subsequent study, the *Tetrahymena thermophila* ribozyme was shown to display a wide range of structural dynamics and docking/undocking rates [9]. However, when the guanine substrate was added, 94% of all molecules showed cleavage activity despite different conformations. Even more surprisingly, the cleavage rate constants of conformers were similar despite the >800-fold range docking equilibria. These observations strongly support a hypothesis that instead of a single native fold, the ribozyme has multiple active conformations, in addition to many folding intermediates and misfolded states.

The discovery of the rugged energy landscape in the *Tetrahymena thermophila* ribozyme also led to efforts to re-engineer the RNA folding. The P4–P6 domain of the ribozyme exhibited strong heterogeneity in FRET experiments, suggesting long-lived misfolded states. Using a modified purification protocol, Herschlag et al. purified a subpopulation of the ribozyme that was structurally near homogeneous [61]. In smFRET experiments, the newly purified sample showed much reduced heterogeneity as compared to samples prepared using traditional methods. This work provided strong evidence to support a long-standing suspicion that a portion of the RNA sample was chemically damaged during preparation.

10.2.3.2 Group II intron ribozymes

Similar to the *Tetrahymena thermophila* ribozyme, the group II intron ribozymes also have a rugged folding energy landscape [62]. Furthermore, these RNAs often require high concentrations of Mg^{2+} and other molecules to reach the proper folds for catalysis and activity [60]. Because of the stringent reaction conditions, folding of this type of ribozyme was often described using simple models involving a single obligatory intermediate [63]. Using the D135 ribozyme derived from yeast as a model system, single-molecule analysis has revealed that the ribozyme displays three distinctive FRET values, corresponding to extended intermediate (I), fold intermediate (F), and native fold (N) (Fig. 10.3) [64]. A single ribozyme molecule was shown to transit among the three states, depending on Mg^{2+} concentration. The I state dominated at low concentrations of Mg^{2+}, the F state emerged at >10 mM Mg^{2+}, and the N state appeared only above 20 mM Mg^{2+}, which coincided with the Mg^{2+} requirement of the catalysis. Interestingly, even at very high concentrations of Mg^{2+}, the N state existed transiently. This observation led to a suggestion that the F$\rightarrow$N transition was the rate-limiting step of the catalysis. Based on this approach, Rueda

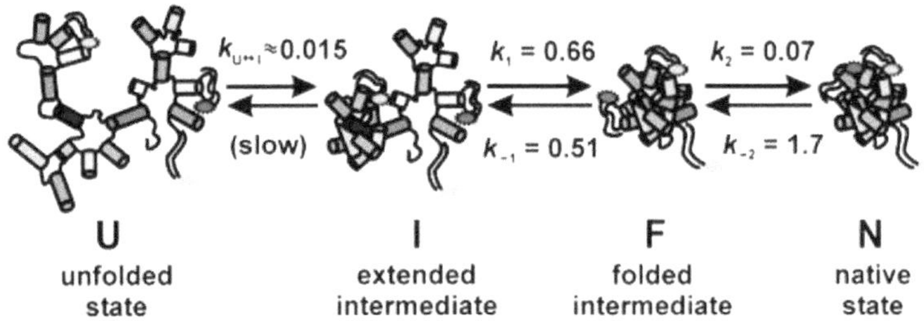

Figure 10.3 Folding pathway of a group II intron ribozyme (adapted from Ref. 64).

et al. further demonstrated that a DEAD-box helicase, Mss166, enhanced group II intron folding by stimulating dynamic sampling between folding intermediates [65].

10.2.4 *Riboswitches*

A riboswitch is a noncoding RNA that regulate biosynthesis of a metabolite in response to its cellular concentration [66, 67]. Consisting of an aptamer domain and an expression platform, the RNA can adopt two mutually exclusive folds, depending on availability of the metabolite. The alternative folding of the riboswitch determines whether its downstream genes will be expressed. Naturally, the ligand-binding and ligand-refolding mechanism of riboswitches becomes a focus of a series of smFRET studies.

In an elegant work, the relatively simple pbuE adenine riboswitch was studied using smFRET [68]. The aptamer domain of the RNA folded in the absence of the ligand, but the folded structure was stabilized by ligand binding. The presence of adenine increased the RNA-folding rate and decreased the unfolding rate. However, in the full riboswitch, folding of the aptamer domain competed poorly with formation of the downstream transcription terminator. These findings suggested an interesting possibility that some riboswtiches function kinetically, that is, folding of a nascent riboswitch has to compete with ligand binding and transcription progression. Similar conclusions were reached previously using bulk kinetics [69, 70].

A pair of fluorophores reflects the relative distance between two positions on a molecule. Such a spatial constraint may not reflect the overall structural dynamics. In addition, local conformational changes can be important for functions of RNA but may not all be revealed by the labeled dyes. To explore these possibilities, Silverman et al. studied the aptamer domain of a guanine riboswitch using bulk and single-molecule FRET [71]. The guanine aptamer has three helical domains. In each experiment, two helices were labeled with a donor and an acceptor, respectively. Repeating the FRET measurement on different pairs of helices yielded three distance constraints for the riboswitch. This comprehensive structural survey revealed complex dynamics of the RNA at various Mg^{2+} and guanine concentrations. More importantly, the powerful multilabeling approach should find more applications in other systems.

10.3 RNA as Part of a Molecular Machinery

A number of molecular machineries, such as spliceosomes and ribosomes, contain RNAs as essential parts of their structures and functions. In these ribonucleoprotein (RNP) complexes, RNAs interact with proteins and other RNAs such that their structures and dynamics have to be understood in the context of the entire assembly. An obvious question is how these RNPs are assembled? Besides smFRET, other single-molecule fluorescent techniques have been employed to study these systems. For instance, a straight-forward application of single-molecule fluorescence microscopy is to count the number of fluorophores in a single-reaction system. By continuously exposing dyes to light, fluorophores can be sequentially photobleached. By counting the number of photobleaching steps/magnitude, it is possible to determine the number of fluorophores associated with a single molecule.

10.3.1 *Counting RNA Molecules in a Bacteriophage-Packaging Motor*

Bacteriophage phi29 encapsidates viral DNA using a ring-shaped packing motor geared by packing RNAs (pRNAs) [72]. Despite several structural studies [73–75], it was still of much debate whether the pRNA ring contains five or six pRNA molecules. In an elegant study, an active viral-packaging motor was reconstructed using pRNA labeled with Cy3 [76]. Fluorescence of individual viral procapsids showed six distinctive photobleaching steps, supporting the model of a six-pRNA ring. The RNA ring was also reconstituted using pRNA dimers, each of which was labeled with a pair of Cy3 and Cy5. In this case, three steps of photobleaching occurred in both Cy3 and Cy5 channels. Significantly, it was demonstrated that under single-molecule experimental conditions, the labeled procapsids were still actively packaging DNA. The hexameric pRNA ring is connected to the procapsid with fivefold symmetry. The confirmation of the hexameric pRNA ring strongly supported the hypothesis that relative motion of the mismatched ring and procapsid generates a force to drive a rotation motor to packaging DNA [77]. Noticeably, the single-molecule technique was further used to monitor assembly and disassembly of the pRNA ring and nanoparticles [76, 78].

10.3.2 *Assembly and Activity of Telomerase*

Eukaryotic telomeres are synthesized by an RNP known as telomerase [79]. Telomerase is responsible for maintaining chromosome stability by adding telomeric DNA repeats at the termini of chromosomes. Abnormal telomerase activity is a hallmark of most cancers [80]. The multicomponent telomerase RNP from *Tetrahymena thermophila* requires at least three components: telomerase RNA, telomerase reverse transcriptase (TERT), and the holoenzyme protein p65 for telomerase activity [81]. Stone et al. investigated the assembly process of this ternary complex using conformation of telomerase RNA as an indicator [82]. Just like in the RNA-folding study, individual telomerase RNAs were labeled with a pair of Cy5 and Cy3 dyes. The free RNA showed a FRET distribution centered at 0.29. Binding of p65 to the RNA shifts the mean FRET value to 0.46, whereas the ternary complex raised the mean FRET to 0.65. Changes in FRET upon protein binding were attributed to conformational changes in the RNA. The distinct FRET values of the three structures were further exploited to monitor assembly of a single ternary complex in real time. An intermediate state between the free RNA

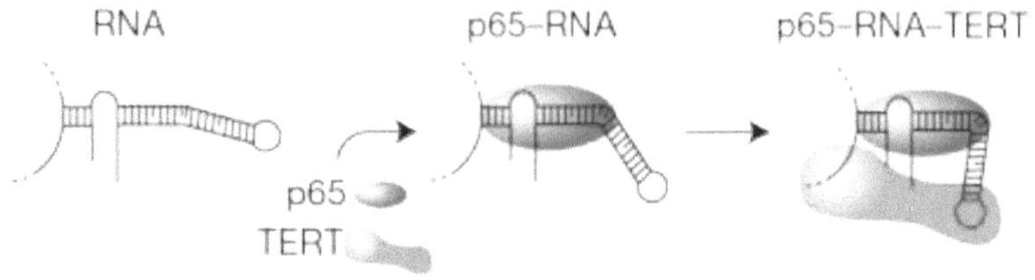

Figure 10.4 A model of sequential binding of p65 and TERT proteins to telomerase RNA. The RNA adopts a different conformation upon binding of p65. (Adapted from Ref. 82.)

and the ternary complex corresponding to the RNA–p65 complex was observed in three-fourths of the trajectories. This observation suggested that in the main hierarchical assembly pathway, p65 induces a conformational change in RNA, allowing further binding of TERT (Fig. 10.4). In addition, a misfolded RNA–TERT complex can be rescued by adding p65.

TERT and telomerase RNA cooperate in reverse transcribing of telomeric DNA. However, the mechanism by which telomerase adds multiple copies of DNA repeats using a single RNA template remains unclear. An "accordion" model was proposed that during reverse transcription, single-stranded RNAs on both sides of the template undergo reciprocal compression and expansion, which leads to reposition the template relative to the transcribed DNA at the end of a catalytic cycle [83]. To test this hypothesis, motion of the telomerase RNA during reverse transcription was monitored using smFRET. The DNA primer was labeled with Cy3, and Cy5 was added to either the 5′ or the 3′ end of the RNA. Binding of the DNA primer increased FRET when Cy5 was labeled at the 3′ end of the RNA, and decreased when the dye was at the 5′ end. The smFRET measurements strongly suggested reciprocal motions of RNAs at the ends of the telomerase RNA. These observations are consistent with the accordion model [83].

10.3.3 *Monitor Splicing in Real Time*

Splicing is indispensable to eukaryotes as it plays an essential role in processing pre-messenger RNA (mRNA) into mRNA [84]. The multi-MegaDalton spliceosome consists of five small nuclear RNAs (snRNAs) (U_1, U_2, U_4, U_5, and U_6) and hundreds of proteins [85]. The complex machinery performs a multistep reaction in which introns are excised and exons are ligated. Details of the assembly of spliceosomes and the splicing reaction remain largely unclear. Single-molecule techniques offer a new perspective into the complex systems.

The U_2/U_6 complex lies in the center of the catalytic core of a spliceosome and plays a critical role in defining substrate specificity of splicing [86, 87]. A smFRET study of the system revealed that the U_2/U_6 complex displayed multiple conformational states indicated by distinctive FRET distributions [88]. More importantly, the structural dynamics of the complex correlate with splicing activation. The mutated complex can be trapped in intermediate conformations, preventing the spliceosome from completing splicing. Further, compensatory mutations can be introduced to restore splicing activity by rescuing these trapped intermediates.

Two single-molecule assays have been established to visualize splicing in vitro. In one experimental setup, single pre-mRNA molecules labeled by a FRET pair, Cy3 on an exon and Alexa647 on an intron, were anchored on the surface [89]. Splicing was induced by adding yeast extracts, Mg^{2+}, and adenosine triphosphate (ATP) to the system (Fig. 10.5a). The cleaved lariat intron RNA diffused away from the surface with Alexa647. By monitoring the number of dyes on each pre-mRNA molecule over time, the kinetics of splicing was measured. It is noticeable that the same molecular design was used in bulk assays, in which the pre-mRNA, intron, and ligated product, each with a different color, were separated by gel electrophoresis. Another study also employed smFRET to monitor splicing on the surface [90]. However, the pre-mRNA was labeled with Cy5 and Cy3 dyes at the two exons (Fig. 10.5b). To determine whether splicing occurred, a DNA oligonucleotide complementary to the intron and RNase H were added. Unreacted pre-mRNAs were cleaved leaving only one exon with Cy5 on the surface, whereas spliced RNA retained both dyes.

Both studies, though still preliminary, have yielded rich information on structural dynamics, catalysis kinetics, and heterogeneous behaviors of spliceosomes. Spliceosomes are large and complex, yet their cellular abundance is low. It is difficult to purify or

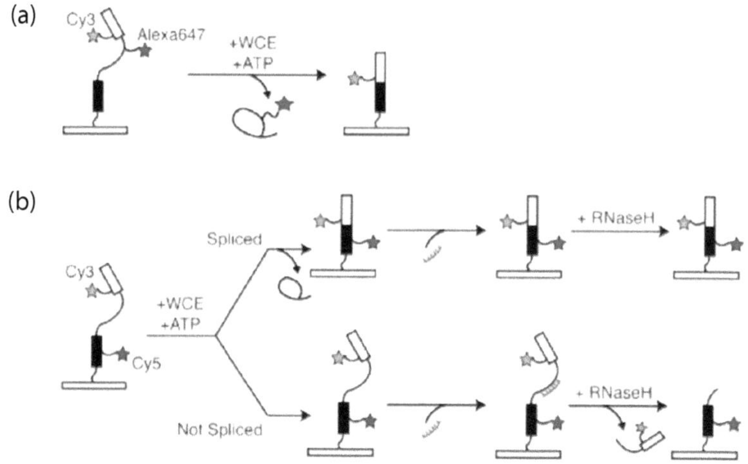

Figure 10.5 Two different experiment setups for single-molecule splicing. (a) A pre-mRNA is tethered to the surface, with a Cy3 on an exon and an Alexa647 on an intron. Adding a WCE to the system induces splicing, leaving only Cy3 on the surface. (b) A pre-mRNA is labeled with Cy5 and Cy3 on two exons. Unspliced introns can be annealed to an olignonucleotide and subsequently be cut by RNase H, resulting in loss of the Cy3. Spliced exons retain both dyes. *Abbreviations*: WCE, whole-cell extract; RNase H, ribonuclease H. (Adapted from Ref. 84.)

reconstitute spliceosomes in vitro. Because whole cell extracts were used in both single-molecule splicing assays, powerful genetic tools, such as mutagenesis of splicing proteins and RNAs, can be used to perturb the reaction. The combination of genetics and single-molecule techniques opens new ways to study RNA splicing.

10.3.4 *Translation by Ribosomes*

Protein synthesis is largely an RNA business. mRNA serves as a template, and tRNAs deliver amino acids to the elongating peptide chain. Most importantly, a ribosome is a ribozyme that catalyzes the translation [91]. A translation cycle involves multiple steps and many factors. Translation is also heavily regulated, and many antibiotics have been developed to target translation [92–94]. Despite the fundamental and medical importance, many details in translation remain unclear. These unresolved questions have stimulated a number of single-molecule studies on protein synthesis.

A major question in translation is how tRNAs move through a ribosome. During translation, an aminoacyl-tRNA first binds to the A site of the ribosome; after passing the amino acid to the nascent peptide chain, it is translocated to the P site; and finally, it exits the ribosome from the E site. To monitor this process, tRNAs labeled with Cy5 or Cy3 were used in a surface-based translation assay, in which co-occupation of tRNAs at the A and P sites was reflected by FRET between the fluorescent tRNAs [95]. Further, binding of aminoacyl-tRNA to the A site was shown to involve at least three conformational states of the tRNA: initial codon recognized, guanosine triphosphatase (GTPase) activated, and fully accommodated [96]. These observations confirmed that correct pairing of codon–anticodon sequences induces a major conformational change that positions the aminoacyl-tRNA toward peptidyl-tRNA at the P site. Further studies revealed how ribosomes select correct aminoacyl-tRNA for next round of incorporation [97, 98]. A tRNA at the A site fluctuates between the codon recognition and GTPase-activated states. A cognate tRNA is more likely to proceed to guanosine triphosphate (GTP) hydrolysis than near-cognate tRNAs, suggesting that substrate selection is achieved by repetitive aligning the bound tRNA to the ribosome before the irreversible GTP hydrolysis occurs. These observations were further confirmed by using various antibiotics that arrest translation in specific stages [99]. Using nanofabrication, a single ribosome can be trapped in a nanoscopic hole, allowing continuous observation of translocation of an tRNA through a ribosome during translation [100].

Translation is a multistep process involving relative motions among ribosomes, mRNA, tRNA, and many translation factors. Single-molecule fluorescence techniques have also been employed to study translation initiation [101], ribosome motions during elongation [102–104], and termination [105]. Most work has so far been focused on prokaryotic translation and relatively simple systems. It can be expected that the single-molecule approach will be extended to study long, perplexing problems such as programmed frame-shifting.

10.4 Conclusion

A few selected examples discussed in this chapter clearly demonstrated the power of single-molecule fluorescence techniques in studying RNA structure and function. Using various clever experimental designs, fluorescence signals are successfully used to reflect conformational dynamics and reactivity of stand-alone RNAs as well as RNAs in large, complex RNPs. These studies have not only confirmed results derived previously from bulk experiments but also revealed novel observations of intermediate states and dynamics hidden from ensemble measurements. The intrinsic structural polymorphism of RNA and the complexity of large RNPs will need many more applications of single-molecule fluorescence techniques in RNA biology in the near future.

References

1. Park, H.Y., Buxbaum, A.R., and Singer, R.H. (2010). Single mRNA tracking in live cells. *Methods Enzymol.*, **472**, 387–406.
2. Li, G.W., and Xie, X.S. (2011). Central dogma at the single-molecule level in living cells. *Nature*, **475**, 308–315.
3. Macklin, J.J., Trautman, J.K., Harris, T.D., and Brus, L.E. (1996). Imaging and time-resolved spectroscopy of single molecules at an interface. *Science*, **272**, 255–258.
4. Tokunaga, M., et al. (1997). Single molecule imaging of fluorophores and enzymatic reactions achieved by objective-type total internal reflection fluorescence microscopy. *Biochem. Biophys. Res. Commun.*, **235**, 47–53.
5. Zheng, H., Goldner, L.S., and Leuba, S.H. (2007). Homebuilt single-molecule scanning confocal fluorescence microscope studies of single DNA/protein interactions. *Methods*, **41**, 342–352.
6. Walter, N.G. (2010). *Single Molecule Tools Part A: Fluorescence Based Approaches*, Vol. 472, Academic Press, San Diego, CA.
7. Walter, N.G. (2010). *Single Molecule Tools Part B: Super-Resolution Particle Tracking Multiparameter and Force Based Methods*, Vol. 475, Academic Press, San Diego, CA.

8. Yanagida, T., and Ishii, Y. (2009). *Single Molecule Dynamics in Life Sciences*, Wiley-VCH Verlag, Weinheim, Germany.
9. Solomatin, S.V., Greenfeld, M., Chu, S., and Herschlag, D. (2010). Multiple native states reveal persistent ruggedness of an RNA folding landscape. *Nature*, **463**, 681–684.
10. Ditzler, M.A., et al. (2008). A rugged free energy landscape separates multiple functional RNA folds throughout denaturation. *Nucleic Acids Res.*, **36**, 7088–7099.
11. Gollnick, P., and Babitzke, P. (2002). Transcription attenuation. *Biochim. Biophys. Acta.*, **1577**, 240–250.
12. Henkin, T.M., and Grundy, F.J. (2006). Sensing metabolic signals with nascent RNA transcripts: the T box and S box riboswitches as paradigms. *Cold Spring Harb. Symp. Quant. Biol.*, **71**, 231–237.
13. Mandal, M., and Breaker, R.R. (2004). Gene regulation by riboswitches. *Nat. Rev. Mol. Cell. Biol.*, **5**, 451–463.
14. Tinoco, I.J., Chen, G., and Qu, X. (2010). RNA reactions one molecule at a time. *Cold Spring Harb. Perspect. Biol.*, **2**, a003624.
15. Noller, H.F. (2005). RNA structure: reading the ribosome. *Science*, **309**, 1508–1514.
16. Eichhorn, C.D., et al. (2012). Unraveling the structural complexity in a single-stranded RNA tail: implications for efficient ligand binding in the prequeuosine riboswitch. *Nucleic Acids Res.*, **40**, 1345–1355.
17. Laederach, A., et al. (2007). Distinct contribution of electrostatics, initial conformational ensemble, and macromolecular stability in RNA folding. *Proc. Natl. Acad. Sci. U S A*, **104**, 7045–7050.
18. Shendure, J., and Ji, H. (2008). Nature biotechnology. *Nat. Biotechnol.*, **26**, 1135–1145.
19. Torkamani, A., Scott-Van Zeeland, A.A., Topol, E.J., and Schork, N.J. (2011). Annotating individual human genomes, *Genomics*, **98**, 233–241.
20. Yandell, M., and Ence, D. (2012). A beginner's guide to eukaryotic genome annotation. *Nat. Rev. Genet.*, **13**, 329–342.
21. Alexander, R.P., et al. (2010). Annotating non-coding regions of the genome, *Nat. Rev. Genet.*, **11**, 559–571.
22. Barash, D., and Gabdank, I. (2010). Energy minimization methods applied to riboswitches: a perspective and challenges. *RNA Biol.*, **7**, 90–97.

23. Bloomfield, V.A., Crothers, D.M., and Tinoco, I., Jr. (2000). Bases, nucleosides, and nucleotides, in *Nucleic Acids: Structures Properties and Functions*, University Science Books, Sausalito, CA, 13–43.
24. Elenko, M.P., Szostak, J.W., and van Oijen, A.M. (2009). Single-molecule imaging of an in vitro-evolved RNA aptamer reveals homogeneous ligand binding kinetics. *J. Am. Chem. Soc.*, **131**, 9866–9867.
25. Raj, A., and Tyagi, S. (2010). Detection of individual endogenous RNA transcripts in situ using multiple singly labeled probes. *Methods Enzymol.*, **472**, 365–386.
26. Bertrand, E., et al. (1998). Localization of ASH1 mRNA particles in living yeast. *Mol. Cell*, **2**, 437–445.
27. Paige, J.S., Wu, K.Y., and Jaffrey, S.R. (2011). RNA mimics of green fluorescent protein. *Science*, **333**, 642–646.
28. Paige, J.S., Nguyen-Duc, T., Song, W., and Jaffrey, S.R. (2012). Fluorescence imaging of cellular metabolites with RNA. *Science*, **335**, 1194.
29. Biteen, J.S., and Moerner, W.E. (2010). Single-molecule and superresolution imaging in live bacteria cells. *Cold Spring Harb. Perspect. Biol.*, **2**, a000448.
30. Patterson, G., Davidson, M., Manley, S., and Lippincott-Schwartz, J. (2010). Superresolution imaging using single-molecule localization. *Annu. Rev. Phys. Chem.*, **61**, 345–367.
31. Lee, N.K., et al. (2007). Three-color alternating-laser excitation of single molecules: monitoring multiple interactions and distances. *Biophys. J.*, **92**, 303–312.
32. Jung, J., et al. (2012). Effect of single-base mutation on activity and folding of 10-23 deoxyribozyme studied by three-color single-molecule ALEX FRET. *J. Phys. Chem. B*, **116**, 3007–3012.
33. Zhuang, X., et al. (2000). A single-molecule study of RNA catalysis and folding. *Science*, **288**, 2048–2051.
34. Greenfeld, M., Pavlichin, D.S., Mabuchi, H., and Herschlag, D. (2012). Single molecule analysis research tool (SMART): an integrated approach for analyzing single molecule data. *PloS ONE*, **7**, e30024.
35. Blanco, M., and Walter, N.G. (2010). Analysis of complex single-molecule FRET time trajectories. *Methods Enzymol.*, **472**, 153–178.
36. Iqbal, A., et al. (2008). Orientation dependence in fluorescent energy transfer between Cy3 and Cy5 terminally attached to double-stranded nucleic acids. *Proc. Natl. Acad. Sci. U S A*, **105**, 11176–11181.

37. Iqbal, A., et al. (2008). The structure of cyanine 5 terminally attached to double-stranded DNA: implications for FRET studies. *Biochemistry*, **47**, 7857–7862.

38. Urnavicius, L., McPhee, S.A., Lilley, D.M., and Norman, D.G. (2012). The structure of sulfoindocarbocyanine 3 terminally attached to dsDNA via a long, flexible tether. *Biophys. J.*, **102**, 561–568.

39. Roy, R., Hohng, S., and Ha, T. (2008). A practical guide to single-molecule FRET. *Nat. Methods*, **5**, 507–516.

40. Clegg, R.M. (1992). Fluorescence resonance energy transfer and nucleic acids. *Methods Enzymol.*, **211**, 353–388.

41. Norman, D.G., Grainger, R.J., Uhrín, D., and Lilley, D.M. (2000). Location of cyanine-3 on double-stranded DNA: Importance for fluorescence resonance energy transfer studies. *Biochemistry*, **39**, 6317–6324.

42. Yasuda, R., et al. (2003). The ATP-waiting conformation of rotating F1-ATPase revealed by single-pair fluorescence resonance energy transfer. *Proc. Natl. Acad. Sci. U S A*, **100**, 9314–9318.

43. Murphy, M.C., et al. (2004). Probing single-stranded DNA conformational flexibility using fluorescence spectroscopy. *Biophys. J.*, **86**, 2530–2537.

44. Rasnik, I., et al. (2004). DNA-binding orientation and domain conformation of the E. coli rep helicase monomer bound to a partial duplex junction: single-molecule studies of fluorescently labeled enzymes. *J. Mol. Biol.*, **336**, 395–408.

45. Kuhnemuth, R., and Seidel, C.A.M. (2001). Principles of single molecule multiparameter fluorescence spectroscopy. *Single Mol.*, **2**, 251–254.

46. Li, P.T.X., Vieregg, J., and Tinoco, I., Jr. (2008). How RNA unfolds and refolds. *Annu. Rev. Biochem.*, **77**(27), 21–24.

47. Tinoco, I.J., and Bustamante, C. (1999). How RNA folds. *J. Mol. Biol.*, **293**, 271–281.

48. Cate, J.H., et al. (1996). Crystal structure of a group I ribozyme domain: principles of RNA packing. *Science*, **273**, 1678–1685.

49. Hodak, J.H., et al. (2005). Docking kinetics and equilibrium of a GAAA tetraloop-receptor motif probed by single-molecule FRET. *Proc. Natl. Acad. Sci. U S A*, **102**, 10505–10510.

50. Downey, C.D., et al. (2006). Metal ion dependence, thermodynamics, and kinetics for intramolecular docking of a GAAA tetraloop and receptor connected by a flexible linker. *Biochemistry*, **45**, 3664–3673.

51. Fiore, J.L., et al. (2012). The role of counterion valence and size in GAAA tetraloop-receptor docking/undocking kinetics. *J. Mol. Biol.*, **423**(2), 198–216.

52. Fiore, J.L., et al. (2008). Monovalent and divalent promoted GAAA tetraloop-receptor tertiary interactions from freely diffusing single-molecule studies. *Biophys. J.*, **95**, 3892–3905.

53. Fiore, J.L., et al. (2009). Enthalpy-driven RNA folding: single-molecule thermodynamics of tetraloop-receptor tertiary interaction. *Biochemistry*, **48**, 2550–2558.

54. Holmstrom, E.D., Fiore, J.L., and Nesbitt, D.J. (2012). Thermodynamic origins of monovalent facilitated RNA folding. *Biochemistry*, **51**, 3732–3743.

55. Fiore, J.L., Holmstrom, E.D., and Nesbitt, D.J. (2012). Entropic origin of Mg2+-facilitated RNA folding. *Proc. Natl. Acad. Sci. U S A*, **109**, 2902–2907.

56. Fedor, M.J. (2000). Structure and function of the hairpin ribozyme. *J. Mol. Biol.*, **297**, 269–291.

57. Nahas, M.K., et al. (2004). Observation of internal cleavage and ligation reactions of a ribozym. *Nat. Struct. Mol. Biol.*, **11**, 1107–1113.

58. Liu, S., Bokinsky, G., Walter, N.G., and Zhuang, X. (2007). Dissecting the multistep reaction pathway of an RNA enzyme by single-molecule kinetic "fingerprinting". *Proc. Natl. Acad. Sci. U S A*, **104**, 12634–12639.

59. Woodson, S.A. (2010). Compact intermediates in RNA folding. *Annu. Rev. Biophys.*, **39**, 61–77.

60. Michel, F., Costa, M., and Westhof, E. (2009). The ribozyme core of group II introns: a structure in want of partners. *Trends. Biochem. Sci.*, **34**, 189–199.

61. Greenfeld, M., and Solomatin, S.V.H., D. (2011). Removal of covalent heterogeneity reveals simple folding behavior for P4-P6 RNA. *J. Biol. Chem.*, **286**, 19872–19879.

62. Pyle, A.M. (2010). The tertiary structure of group II introns: implications for biological function and evolution. *Crit. Rev. Biochem. Mol. Biol.*, **45**, 215–232.

63. Pyle, A.M., Fedorova, O., and Waldsich, C. (2007). Folding of group II introns: a model system for large, multidomain RNAs? *Trends. Biochem. Sci.*, **32**, 138–145.

64. Steiner, M., Karunatilaka, K.S., Sigel, R.K., and Rueda, D. (2008). Single-molecule studies of group II intron ribozymes. *Proc. Natl. Acad. Sci. U S A*, **105**, 13853–13858.

65. Karunatilaka, K.S., Solem, A., Pyle, A.M., and Rueda, D. (2010). Single-molecule analysis of Mss116-mediated group II intron folding. *Nature*, **467**, 935–939.

66. Breaker, R.R. (2011). Prospects for riboswitch discovery and analysis. *Mol. Cell*, **43**, 867–879.

67. Smith, A.M., Fuchs, R.T., Grundy, F.J., and Henkin, T.M. (2010). Riboswitch RNAs: regulation of gene expression by direct monitoring of a physiological signal. *RNA Biol.*, **7**, 104–110.

68. Lemay, J.F., Penedo, J.C., Tremblay, R., Lilley, D.M., and Lafontaine, D.A. (2006). Folding of the adenine riboswitch. *Chem. Biol.*, **13**, 857–868.

69. Wickiser, J.K., Cheah, M.T., Breaker, R.R., and Crothers, D.M. (2005). The kinetics of ligand binding by an adenine-sensing riboswitch. *Biochemistry*, **44**, 13404–13414.

70. Wickiser, J.K., Winkler, W.C., Breaker, R.R., and Crothers, D.M. (2005). The speed of RNA transcription and metabolite binding kinetics operate an FMN riboswitch. *Mol. Cell*, **18**, 49–60.

71. Brenner, M.D., Scanlan, M.S., Nahas, M.K., T. H., and Silverman, S.K. (2010). Multivector fluorescence analysis of the xpt guanine riboswitch aptamer domain and the conformational role of guanine. *Biochemistry*, **49**, 1596–1605.

72. Guo, P., Grimes, S., and Anderson, D. (1986). A defined system for in vitro packaging of DNA-gp3 of the Bacillus subtilis bacteriophage phi29. *Proc. Natl. Acad. Sci. U S A*, **83**, 3505–3509.

73. Ibarra, B., et al. (2000). Topology of the components of the DNA packaging machinery in the phage phi29 prohead. *J. Mol. Biol.*, **298**, 807–815.

74. Simpson, A.A., et al. (2000). Structure of the bacteriophage phi29 DNA packaging motor. *Nature*, **408**, 745–750.

75. Morais, M.C., et al. (2001). Cryoelectron-microscopy image reconstruction of symmetry mismatches in bacteriophage phi29. *J. Struct. Biol.*, **135**, 38–46.

76. Shu, D., Zhang, H., Jin, J., and Guo, P. (2007). Counting of six pRNAs of phi29 DNA-packaging motor with customized single-molecule dual-view system. *EMBO J.*, **26**, 527–537.

77. Bazinet, C., and King, J. (1985). The DNA translocating vertex of dsDNA bacteriophage. *Annu. Rev. Microbiol.*, **39**, 109–129.

78. Shu, D., et al. (2010). Dual-channel single-molecule fluorescence resonance energy transfer to establish distance parameters for RNA nanoparticles. *ACS Nano*, **4**, 6843–6853.

79. Blackburn, E.H., and Collins, K. (2011). Telomerase: an RNP enzyme synthesizes DNA. *Cold Spring Harb. Perspect. Biol.*, **3**, a003558.
80. Blackburn, E.H. (2011). Walking the walk from genes through telomere maintenance to cancer risk. *Cancer Prev. Res.*, **4**, 473–475.
81. Cunningham, D.D., and Collins, K. (2005). Biological and biochemical functions of RNA in the tetrahymena telomerase holoenzyme. *Mol. Cell Biol.*, **25**, 4442–4454.
82. Stone, M.D., et al. (2007). Stepwise protein-mediated RNA folding directs assembly of telomerase ribonucleoprotein. *Nature*, **446**, 458–461.
83. Berman, A.J., Akiyama, B.M., Stone, M.D., and Cech, T.R. (2012). The RNA accordion model for template positioning by telomerase RNA during telomeric DNA synthesis. *Nat. Struct. Mol. Biol.*, **18**, 1371–1375.
84. Hoskins, A.A., Gelles, J., and Moore, M.J. (2011). New insights into the spliceosome by single molecule fluorescence microscopy. *Curr. Opin. Chem. Biol.*, **15**, 864–870.
85. Wahl, M.C., Will, C.L., and Lührmann, R. (2009). The spliceosome: design principles of a dynamic RNP machine. *Cell*, **136**, 701–718.
86. Parker, R., Siliciano, P.G., and Guthrie, C. (1987). Recognition of the TACTAAC box during mRNA splicing in yeast involves base pairing to the U2-like snRNA. *Cell*, **49**, 229–239.
87. Lesser, C.F., and Guthrie, C. (1993). Mutations in U6 snRNA that alter splice site specificity: implications for the active site. *Science*, **262**, 1982–1988.
88. Guo, Z., Karunatilaka, K.S., and Rueda, D. (2009). Single-molecule analysis of protein-free U2-U6 snRNAs. *Nat. Struct. Mol. Biol.*, **16**, 1154–1159.
89. Crawford, D.J., et al. (2008). Visualizing the splicing of single pre-mRNA molecules in whole cell extract. *RNA*, **14**, 170–179.
90. Abelson, J., et al. (2010). Conformational dynamics of single pre-mRNA molecules during in vitro splicing. *Nat. Struct. Mol. Biol.*, **17**, 504–512.
91. Rodnina, M.V., Beringer, M., and Wintermeyer, W. (2007). How ribosomes make peptide bonds. *Trends Biochem. Sci.*, **32**, 20–26.
92. Maguire, B.A. (2009). Inhibition of bacterial ribosome assembly: a suitable drug target? *Microbiol. Mol. Biol. Rev.*, **73**, 22–35.
93. Korostelev, A., and Noller, H.F. (2007). The ribosome in focus: new structures bring new insights. *Trends Biochem. Sci.*, **32**, 434–441.

94. Frank, J., and Gonzalez, R.L.J. (2010). Structure and dynamics of a processive Brownian motor: the translating ribosome. *Annu. Rev. Biochem.*, **79**, 381–412.
95. Blanchard, S.C., et al. (2004). tRNA dynamics on the ribosome during translation. *Proc. Natl. Acad. Sci. U S A*, **101**, 12893–12898.
96. Blanchard, S.C., et al. (2004). tRNA selection and kinetic proofreading in translation. *Nat. Struct. Mol. Biol.*, **11**, 1008–1014.
97. Lee, T.H., et al. (2007). The role of fluctuations in tRNA selection by the ribosome. *Proc. Natl. Acad. Sci. U S A*, **104**, 13661–13665.
98. Geggier, P., et al. (2010). Conformational sampling of aminoacyl-tRNA during selection on the bacterial ribosome. *J. Mol. Biol.*, **399**, 576–595.
99. Gonzalez, R.L.J., Chu, S., and Puglisi, J.D. (2007). Thiostrepton inhibition of tRNA delivery to the ribosome. *RNA*, **13**, 2091–2097.
100. Uemura, S., et al. (2010). Real-time tRNA transit on single translating ribosomes at codon resolution. *Nature*, **464**, 1012–1017.
101. Tsai, A., et al. (2012). Heterogeneous pathways and timing of factor departure during translation initiation. *Nature*, **487**, 390–393.
102. Fei, J., Richard, A.C., Bronson, J.E., and Gonzalez, R.L.J. (2011). Transfer RNA-mediated regulation of ribosome dynamics during protein synthesis. *Nat. Struct. Mol. Biol.*, **18**, 1043–1051.
103. Fei, J., et al. (2009). Allosteric collaboration between elongation factor G and the ribosomal L1 stalk directs tRNA movements during translation. *Proc. Natl. Acad. Sci. U S A*, **106**, 15702–15707.
104. Fei, J., Kosuri, P., MacDougall, D.D., and Gonzalez, R.L.J. (2008). Coupling of ribosomal L1 stalk and tRNA dynamics during translation elongation. *Mol. Cell*, **30**, 348–359.
105. Sternberg, S.H., et al. (2009). Translation factors direct intrinsic ribosome dynamics during translation termination and ribosome recycling. *Nat. Struct. Mol. Biol.*, **16**, 861–868.

Chapter 11

Rational Design of RNA Nanoparticles and Nanoarrays

Arkadiusz Chworos
Centre of Molecular and Macromolecular Studies, Polish Academy of Sciences in Lodz, Sienkiewicza 112, 90363 Lodz, Poland
achworos@cbmm.lodz.pl

11.1 Introduction: Development of DNA Nanotechnologies

In modern days we tend to change the structure of science, introducing a variety of new disciplines, and classical early divisions, such as mathematics, physics, chemistry, and biology, undergo modern branching into biochemistry, biophysics, and bioinformatics, which evolve into new technologies, such as biotechnology, structural biology, or nanobiotechnology. But they all share a common purpose—to comprehend the surrounding world and living organisms and to applying existing knowledge for new developments and useful technologies.

A similar motivation drove the crystallographer Nadrian Seeman to the idea of building three-dimensional (3D) scaffolds made of

RNA Nanotechnology
Edited by Bin Wang

ISBN 978-981-4411-64-6 (Hardcover), 978-981-4411-65-3 (eBook)
www.panstanford.com

deoxyribonucleic acid (DNA) molecules. At that time (mid-1980s) we had already extensive knowledge about the function and role of DNA in living organisms. Moreover the structure of nucleic acids was known for more than 30 years due to the meticulous work and ingenious observation of Rosalind Franklin passed to Maurice Wilkins and the finalizing work done by Watson and Crick, published in 1953 [1, 2].

Seeman's idea was revolutionary, and despite all the knowledge of the atom's structural coordinates in DNA and prediction of its folding, it took many years to prove its validity [3–5]. This was partially caused by the lack of sufficient methodology that allows analysis of such fine objects. Here one has to realize that emerging of DNA nanotechnology, and consequently ribonucleic acid (RNA) bionanotechnology, would not be possible without advancements in two fundamental steps in these materials' development—synthesis of DNA/RNA molecules and nanoparticle analysis. Fabrication of artificial DNA objects started to be feasible due to automated solid-support DNA synthesis. This was, by many, believed to be the greatest achievement of modern structural genetics. Initiated in the middle of the 1960s by Letsinger and perfected in the early 1980s by the work of Marvin Caruthers, who introduced highly reactive nucleoside phosphoramidities as building blocks undergoing nucleophilic substitution by the 5'-hydroxyl group of the nucleotides, which are immobilized on the solid support of micron-size glass or polymer bids [6, 7], this methodology opened new possibilities for the chemical synthesis of DNA fragments of any sequence of the length of 120–150 nt. Short DNA strands can be further ligated into even longer constructs. It is worth noting that the power of this technique is enhanced by the ability to introduce modified nucleotides at any preselected position of a growing chain, which has been used in later stages of DNA nanoparticle development.

In terms of DNA bioscaffold analysis two other techniques were crucial for its success, namely, scanning probe microscopy (SPM) and low-energy electron micrograph of a frozen sample. Scanning force microscopy, otherwise called atomic force microscopy (AFM), is the implementation of tunneling probe microscopy, invented by two Nobel Prize laureates, Gerd Binnig and Heinrich Rohrer, in the early 1980s, distinguished for studies on near-surface tunneling effects on conductive materials. This powerful technology allows the visualization of flat objects with remarkable resolution below 1 nm in air and under biologically native buffer conditions. These days this is a primary technique used to study self-assembly processes for DNA and RNA scaffolds as well as surface protein–nucleic acid interactions [8, 9]. These technical and methodological advances enabled scientists to use DNA as a medium to create marvelous nanostructures and form a new field of DNA nanobiotechnology [10–13].

11.2 RNA Modular Functions

RNA plays an important role in all living organisms. Because of its functions RNA can be divided into two groups, coding and noncoding RNAs. Coding RNAs include one class of molecules—messenger RNA (mRNA). The second group, noncoding RNAs, also called functional RNAs, does not undergo translation processes but serves as a regulatory molecule. This group includes RNA molecules of different sizes:

- Ribosomal RNA (rRNA) constitutes about 80% of RNA in the cell.
- Transfer RNA (tRNA), 74–95-nucleotide-long molecules, is responsible for the transport of amino acids from the cytoplasm to the ribosome (approximately 15% of total RNA in a cell).
- Small (noncoding) RNA (sRNAs).

The sRNAs are responsible for a variety of biological functions in cells: small nuclear RNA (snRNA) supports the folding of pre-mRNA during maturation, and small nucleolar RNAs (snoRNA) is involved in chemical modification of rRNA. Small regulatory RNA (srRNA), such as microRNA (miRNA) and small interfering RNA (siRNA), has relatively short RNA fragments involved in the regulation of gene expression [14].

However, most important for current advancements in the modular functions of RNA is the link between comparative sequence analysis, enzymatic function, and RNA structure. It was known that RNA is not only an information transfer medium between DNA and biosynthesis of proteins; also crucial was the discovery of the catalytic activity and structural function of RNA, by Sidney Altman and Thomas R. Czech in the 1980s. Later, when self-splicing mechanisms were discovered and structural constrains started to be reported for folded RNA molecules, RNA bioengineering became possible. These fundamental aspects of RNA functions and their role in cellular biology are discussed by Christian W. Zwieb in chapter 2 of this book.

11.3 Vocabulary of RNA Motifs

Concurrent to the biochemical analysis of RNA, two great scientists in another part of the world, Leontis and Westhof, took upon a methodical approach to the problem of understanding the RNA language and dissecting the long-range noncanonical RNA–RNA interactions into a set of vocabulary rules [15].

When we consider possible interactions between two nucleotides in the structure of RNA, in addition to Watson–Crick binding, one can find other noncanonical interactions, which can be present in the same helix or between distant motifs [16]. An example of such interactions is the hydrogen bonds involving guanine and uracil, the so-called G:U wobble. The G:U base pair may occur in the helix without causing a major disruption in its secondary structure, but it makes the helix a little broader. Other noncanonical binding plays an important role in accommodating the function of RNA. The presence of such interaction can disrupt the regularity of the helix structure, exposing the functional group, which creates site recognition for proteins, other RNAs, ligands, or binding sites of ions. An example of this is four purine–purine bonds within the regular A type of helix within the 5S rRNA E-loop, which lead to an increase in the width of the major groove and the formation of large, unique hydrogen bonds in a shallow groove. These can cause the formation of a recognition site for other RNAs or, in some instances, for ribosomal proteins. Among the noncanonical bonds the homopurine pairs are very important, since they can form a strong 2–3 hydrogen bond interaction between RNA

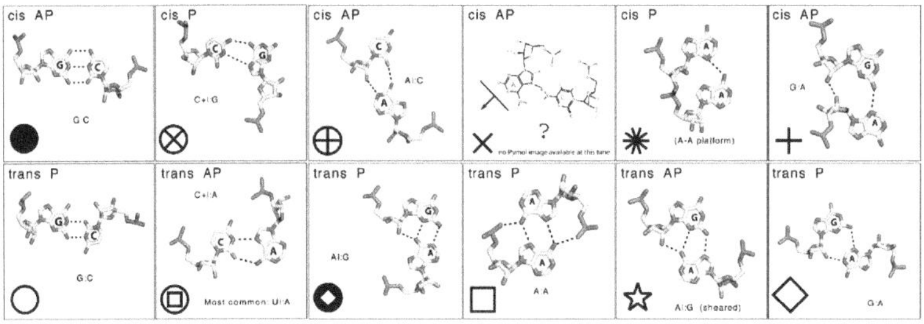

Figure 11.1 Noncanonical base pair interactions [15].

nucleotides. Another conservative elements present in the tertiary structures of many RNAs are tetraloops, such as the GNRA loop (G—guanine; N—any nucleotide; R—purine, A—adenine), typical in structures of ribozymes and fragments of rRNA [17]. Metal ions and water molecules (as a hydrogen bridge) often participate in the stabilization of such interactions.

To categorize noncanonical interactions in the RNA secondary structure Leontis and Westhof consider three edges for each nucleotide: Watson and Crick (WC), Hoogsten (H), and Sugar (SC). In addition, interactions between bases can be either in *cis* or *trans* orientation. Such an assumption identifies 12 major groups of interactions between the nucleotides formed by at least two hydrogen bonds (Fig. 11.1) [15, 18, 19].

11.4 Rational Design of the First RNA Self-Assembly Units

Understanding of structural constraints and folding of RNA led to the idea of creating de novo artificial RNA modules built on the basis of RNA motifs. Initially interhelical interactions present in the natural P4-P6 domain of the group I intron (11 nt motif receptor for the GAAA tetraloop) has been selected to mediate specific, high-affinity intermolecular RNA interactions [20]. At first a hairpin-loop molecule has been designed in a way that it can self-dimerize in the presence of magnesium ions, which stabilize specific loop-receptor interactions (Fig. 11.2a,b). Replacement of motifs and chemical probing (lead cleavage) were used to prove that dimerization was indeed mediated by specific tertiary interactions and not by complementary Watson–Crick base pairing between two monomers. This was one of the first examples for directed RNA one-dimensional self-assembly. Soon after it was determined that RNA motifs are modular and can be exchanged and reorganized. On the basis of the same principles of modularity a set of similar dimerizing tecto-RNA units has been constructed, including heterodimers with or without flexible helix hinges [21]. Using similar principles Geary and Jaeger designed an RNA switch molecule, which can either form a rod structure that self-assembles into a long RNA wire or forms a flexible self-dimerizing unit [22].

A few years later, the structure of the self-assembling dimer was investigated using nuclear magnetic resonance (NMR) spectroscopy techniques and led to the refinement of the computer model (Fig. 11.2c) showing positioning of magnesium ions in the tetraloop receptor pocket (PDB id:2ADT) [23].

11.5 The Concept of the RNA "Lego" Playground

Introduced by Seeman's idea of creating geometrical shapes with DNA on the one hand and multidimensionality and modularity of RNA motifs on the other hand, a new approach has been initiated. It was commonly described as a reverse design of tecto-RNA objects, since indeed the procedure is opposite to the common structural prediction. In a typical biocomputational approach the primary sequence is folded into a secondary structure, whose formation is primarily driven by Watson–Crick hydrogen bonds. The secondary structure for relatively short sequences can be predicted with high accuracy. This calculation is possible on the basis of a semiempirical model of the nearest neighbor, first published in 1986 [24] and later refined and used as a based algorithm for secondary structure prediction available on many servers [25, 26]. With the secondary structure in hand different approaches can be taken to predict folding of the tertiary structure, which in RNA, similarly to protein, is based on (1) initial metal ion condensation, (2) helix assembly collapse, and, finally, (3) conformation search and final structure formation [27]. These tertiary structure predictions are based

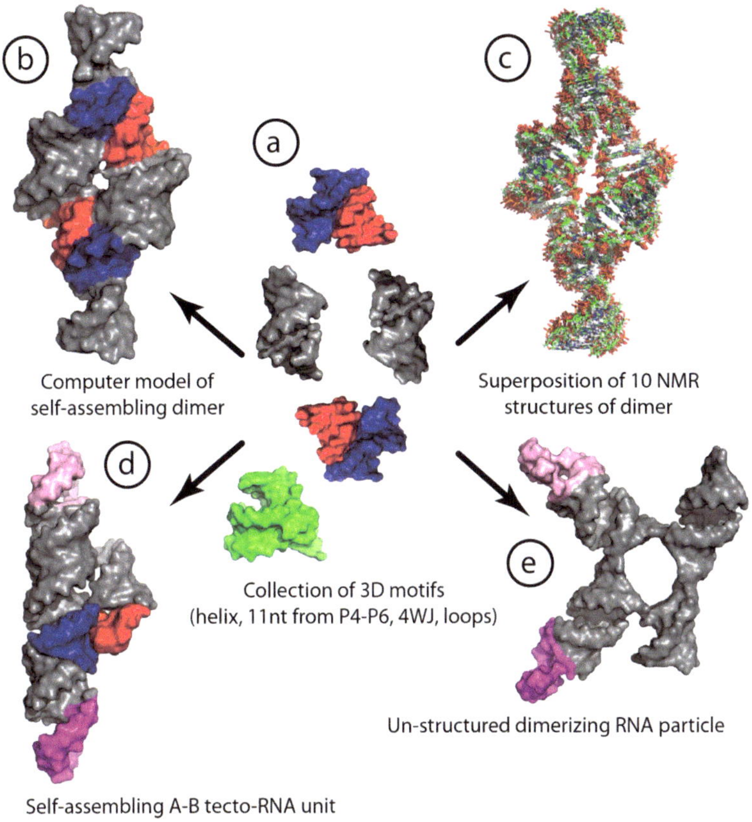

Figure 11.2 Modular self-assembly of tecto-RNA molecules. (a) An 11-nulceotide loop receptor motif extracted from the X-ray crystal structure of the P4-P6 domain of the group I intron (GAAA tetraloop in red, 11 nt receptor in blue), (b) a computer model of a self-dimerizing tecto-RNA unit [20], (c) superposition of 10 NMR structures corresponding to dimer formation in solution [23], and (d, e) computer models of novel self-assembly units with rigid or flexible junctions [22].

either on a database search for a sequence consensus of larger (10–20 nt) RNAs with known secondary structures (FRABASE) or on the division of larger structural motifs into submotifs and searching structural databases for specific sequence signatures [28]. Yet another approach can be undertaken where dynamic modeling simulation is employed to search for the most stable (most probable) structure. Considering a constant increase of accessible computer power new discoveries are being made every day. Many of these are comprehensively covered in other chapters of this book. Overall the whole process of RNA structure prediction follows the general trend from primary sequence to secondary helix to tertiary structure analysis [29, 30].

In a methodology called rational design of tecto-RNA or RNA architectonics, the reverse approach was undertaken [31]. Here the tertiary structure of the final, desirable unit is the platform for computer modeling. Modeling, or rather more appropriately called "molding" of tecto-RNAs, is the method where small structural RNA motifs are virtually extracted from an already solved crystal structure, rearranged according to the initial model, and connected with other motifs according to rules of directionality, distance spacing, and the angle between the connecting base pairs (Fig. 11.3). Simple motifs are deposited in RNA motif databases; others can be extracted from the crystal structure of ribozymes, ribosomes (resolved with remarkably high resolution [32–34]), or other smaller RNA structures available in the protein data bank (PDB) database. Occasionally deletion or rather substitution of a nucleotide or base pair is necessary to accommodate the proper spacing. The sequence of a helix-motif-helix consensus can be read from the model creating a secondary structure blueprint. At this point base pair substitution can be made at the level of the helix. However, to facilitate proper folding of the molecules, noncanonical interactions present in the structure should to be conserved, although canonical base pairs present in the helixes can be replaced. Such substitution can facilitate higher energy of helix formation (A-T into G-C) or higher sequence specificity. In some instances base pairs adjacent to the motif also have to be maintained. For instance a G-U wobble is commonly present at the second base pair position in the loop–loop interaction or in the tetraloop (GNRA). In the next step of tecto-RNA engineering the secondary structure is converted into a primary sequence (Fig. 11.3). If the position of the loop–loop interaction or a tail placement in the molecule has to be changed, circular permutation can be employed. After computer design and secondary structure stability analysis, molecules can be manufactures, folded, and assembled according to the hierarchical

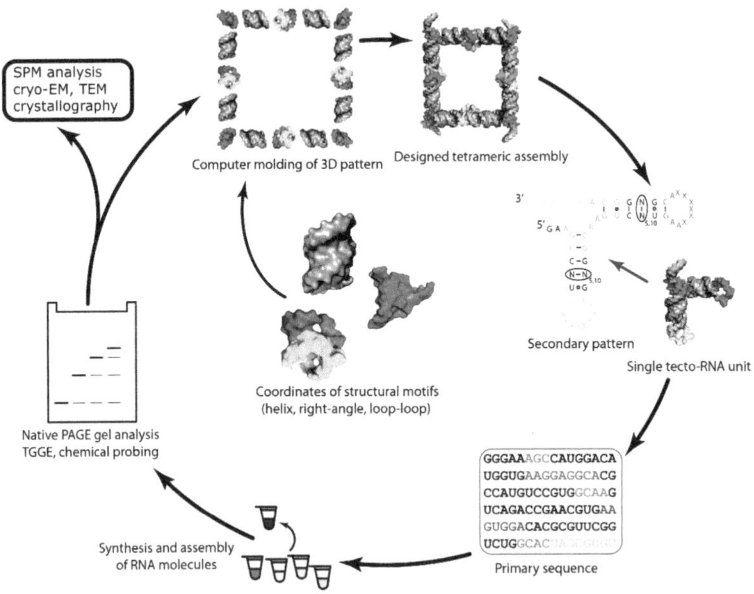

Figure 11.3 Rational design scheme for a tecto-RNA tetramer. Starting from a set of structural motifs (center) and used for relative positioning and building of a computer model for a tecto-RNA tetramer (top). The model is used for deriving the secondary structure and primary sequence blueprint for a single tecto unit (right). RNA molecule synthesis, assembly, and gel analysis (left). Gel results are used for model (structure) improvement or further analysis.

design of the final object or pattern. This methodology appears to be versatile, and it has been shown that multiple designs of RNA structures can be made using this method. Since RNA motifs can be connected at the helix point with any other RNA motif, these motifs resemble and were referred to as "Lego RNA pieces," and as such modeling makes it a "Lego" RNA playground.

11.6 Creating tecto-RNA Squares Using RNA Junction Motifs

The initial idea was to create the simplest flat, geometrical shape that would be enough arbitrary, but it cannot be formed accidently.

Therefore a square-shaped structure has been chosen as a model. For this purpose four tecto-RNA molecules have been designed in a way that they are able to form a self-assembled tetramer (Figs. 11.3 and 11.4). a loop–loop interaction, structurally identical to the dimerization initiation side of the human immunodeficiency

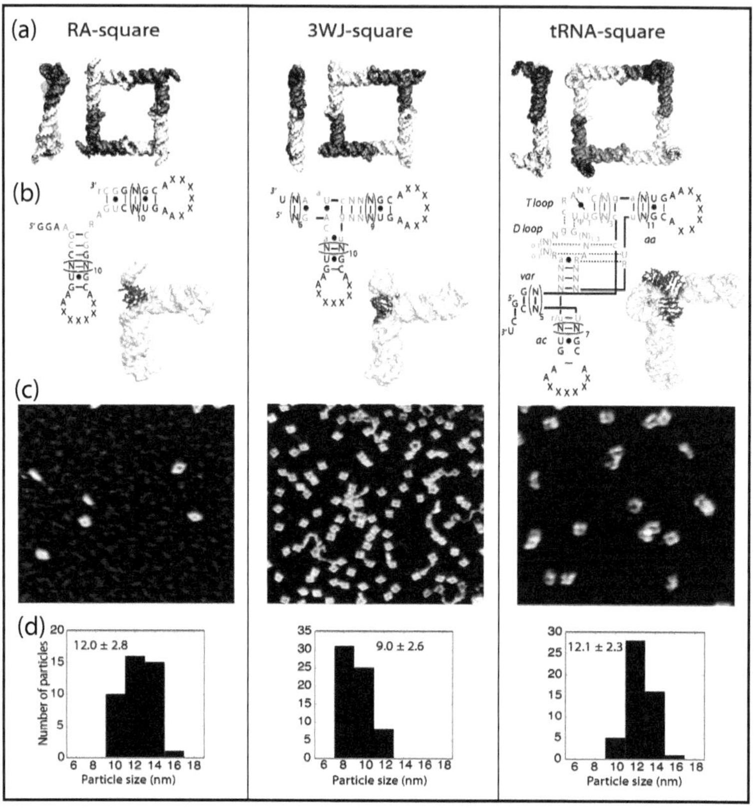

Figure 11.4 Three different multiway junction motifs have been used to design tecto-RNA units, which are able to form square-shaped tetramers. (a) Computer models of tecto-RNA squares, respectively, for right-angle, 3WJ, and tRNA motif, show how these objects are not always planar and how this is dictated by the sequence constraint at the level of the motif. (b, c) The squares' relative planarity is expressed in the amount of the material deposited on the surface analyzed by surface scanning techniques, (d) and the size of RNA particles has been confirmed, as theoretically predicted (for details read Ref. 36).

virus (HIV), has been selected as an interaction side between RNA monomers [35]. To ensure rigidity of the structure as well as to accommodate a sharp turn between helixes in the corner junction, the right-angle motif has been selected. This naturally occurring RNA motif accommodates a cabin-log type of arrangement with adjacent helixes stabilized by six noncanonical base pair interactions. Small and large types of tecto-RNA squares with respective sizes of 124 Å and 87 Å have been assembled. AFM has been used to provide evidence that indeed a tetramer is a square-shaped object [9].

Later two other types of RNA tecto-squares have been constructed [36]. The first was designed on the basis of the three-way junction (3WJ) motif that belongs to the family of so called UA_handle motifs, which resembles an almost perfect T-shaped multi-way junction fold with two helixes stack in-line on the top of each other, and the third protruding at close to a 90° angle [37]. Another tecto-RNA square was designed using an energetically stable tRNA motif. This four- or five-helix junction motif is stabilized by very conservative internal loop–loop interactions (T-D). Thermal stability of these three types of assemblies, studied using temperature-gradient gel electrophoresis, showed a large contribution of motifs to the overall stability of squares and shows that the tRNA square is the most stable among all three squares.

Assuming a relatively small size of those objects, and the fact that RNA assemblies are made of bioorganic material, the only suitable method for its characterization is SPM, a method commonly used for biomaterial characterization. It has been demonstrated, using high-resolution SPM, that, indeed, tetramers formed in solution using four tecto-RNA units resemble a square-shaped object when deposited on a flat mica surface [9, 36]. Remarkable is the fact that RNA structures deposited on a mica surface can sustain physical forces during multiple scanning based only entirely on electrostatic interaction. In this particular example of three squares, the 3D structure of a single unit mediates the strength of the RNA–mica interaction and, consecutively, transposes into the amount of the material deposited on the surface. Respectively, the relative flat arrangement of three helixes in the 3WJ motif made the RNA tetramer highly interactive with flat surfaces such as the crystal surface of mica, and this translates to the high amount of material deposited on the surface. The next most abandoned RNA square on the surface was a tRNA one (flat only on one side) followed by the RA-square, which structure resembles cabin-log type of arrangement, where one helix is on the top of the other (Fig. 11.4).

More recently crystallography techniques have been used successfully to resolve the structure of another square-shaped tetramer [38]. In these studies the structure of the IIa bulge in the internal ribosome entry site (IRES) of the hepatitis C virus RNA has been incorporated in every corner of the square to accommodate a right-angle bend and close the whole structure. The IIa domain, which is a potential target of antiviral drugs that block hepatitis C virus (HCV) protein synthesis, adopts a sharp bend that is necessary to accommodate spatial positioning of the HCV IRES domain for binding with host cell ribosomes. Structural analysis of the IIa domain shows that this motif constitutes one of the most compact L-shaped arrangement [39]. Molecular modeling based on the IRES IIa domain and self-complementary tail–tail interaction assumes fourfold symmetry of the tetramer; however, crystal structure analysis shows that the RNA square is not symmetrical. This is probably caused by internal tension in the structure (Fig. 11.5).

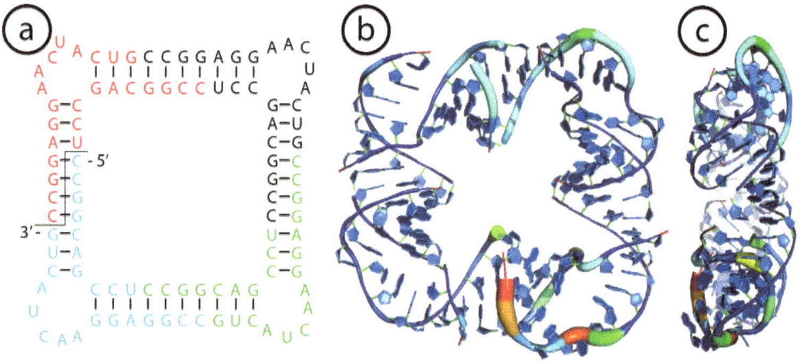

Figure 11.5 Crystal structure determination of the smallest square-shaped nano-object made with dsRNA [38]. (a) Secondary structure pattern for a self-assembly tetramer, one molecule with different colors for clarity; (b, c) top and side views of a crystal structure model showing the asymmetry of the tetramer and flexibility at the level of the tail (red, yellow). *Abbreviation*: dsRNA, double-stranded RNA.

11.7 From Model Tetramers to Multidimensional Assemblies

Formation of the square-shaped nanoparticle made entirely with artificial RNA molecules was important for developing the field of RNA nanotechnology, but this type of assembly alone has limited applications. The obvious next step was the formation of multidimensional assemblies based on tecto-RNA squares. The tetramer was designed in such a way that the conjunction of both termini of RNA molecules forms a stem, which can be used for functionalization. With strong loop–loop interaction inside the single unit (Figs. 11.5 and 11.6) implementation of single-stranded RNA (ssRNA) tail–tail interaction seems to be most appropriate. Since a structurally stabilized kissing complex is stabler (in the presence of magnesium [40]) than tail–tail interaction of the same length, this energy difference enables stepwise assembly [31]. Following this method, in the first step, tecto-RNA units are being

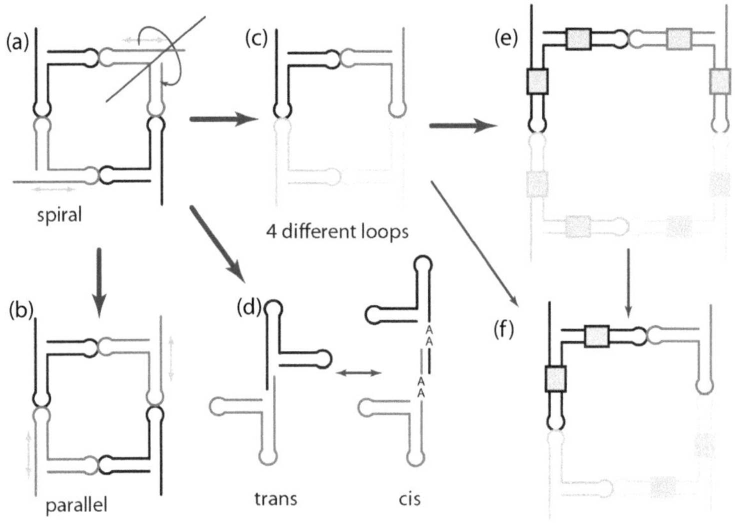

Figure 11.6 Relative orientation of tecto-RNA units. (a, b) Relative arrangement corners in the tetramer causes spiral vs. parallel orientation and (c) self-assembly of the tetramer using four different RNA molecules with four specific loop–loop interactions.

assembled in the presence moderate amount of divalent ions into tetramer. Once tetramers are made, they can be associated together through weaker tail–tail interactions into larger 2D structures.

The orientation of the protruding stem can be either vertical or horizontal with respect to the square, and therefore we can generate different types of tetramers with respect to the relative orientation of tails. This is controlled by relative programmation of loop sequences (Fig. 11.6). Let us take, for example, two same-size squares and orient them with respect to each other. They can be arranged in a *trans* type of conformation, with one connector between each of two units, or in *cis* type, where two tetramer units are connected through two contiguous helixes with two complementary sequences and the length of the stem between two squares controls the arrangement; half of the helical turn rotates the object by 180°. Additionally we can substitute a single square unit with a larger unit, and this adds another variable to the library of possible two-dimensional (2D) array patterns. In summary, with 4 kissing complexes, 2 stem orientations (W-E or S-N), and 2 lengths of the stems and assuming 16 specific tail–tail interactions and 2 sizes of squares (small, large), we potentially can make more than 500 different molecules, which can be assembled to over 2,000 fully programmable one-dimensional (1D) units, and each can be connected specifically with the other, making a large variety of 2D arrays. To prove this concept an ensemble of different patterns was created and analyzed using AFM (Fig. 11.7). With a high number of specific tail–tail interactions, also a fully addressable pattern of 4×4 squares has been made.

All scaffolds described so far are relatively flat because motifs used for these assemblies are flat (loop loop, RA motif, 3WJ motif, etc.). A different approach has to be undertaken to make an RNA object that is three-dimensional. In other words, to make a rectangular type of structure the tRNA motif had been selected. Assuming proper folding of the tRNA motif, stabilized with a D-T kissing complex, two stems (imaginary anticodon and amino acid stem) are oriented in such a way that they form an L-shaped structure with a variable stem protruding from the basic plane [41]. This potentially enables rational bioengineering of the 3D objects made entirely with RNA. Initially the cuboid structure was designed,

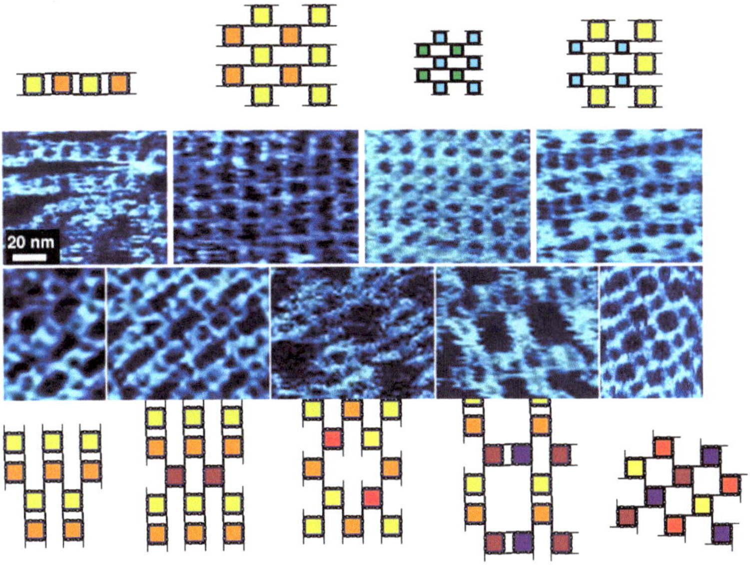

Figure 11.7 Collection of tecto-RNA arrays made by assembly of different types of small (blue, green) and large (warm colors) of tetramers. (Inside) AFM visualization of RNA arrays deposited on mica surface (adapted from Ref. 9).

with all angles at the corner to be at around 90°; however, it turned out that the variable stem accommodates a tilted conformation and the entire structure forms an antiprism type of object (Fig. 11.8). Since this is not a flat object and the SPM technique is not sufficient to unambiguously prove its structure, a different method has to be employed. Collected AFM data strongly suggests that the tRNA octamer favors the formation of a closed, compact molecular complex. However, the most convincing structural evidence for the antiprism formation came from the cryo-EM experiment with data analyzed by single-particle reconstruction [42].

11.8 Using RNA Nanoparticles toward Gene Regulation

The idea of using RNA nanoparticles for gene regulation was first published in 2005 [43]. The same principles of rational

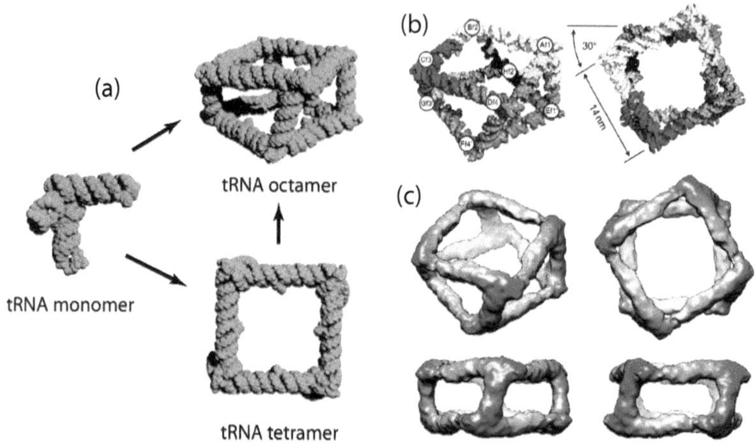

Figure 11.8 Assembly design and visualization of an RNA polyhedral with a permutated tRNA motif [42]. (a) Formation of an RNA tetramer (four specific kissing complexes) and an octamer (four specific tail–tail interactions) using tecto-RNA units based on a four-way junction tRNA motif, (b) 3D computer model of a tecto-RNA antiprism, and (c) single-particle image reconstruction from cryo-EM analysis of the antiprism. *Abbreviation*: cryo-EM, cryo–electron microscopy.

design were used to create an RNA trimer consisting of a CD4 recognition aptamer sequence, an siRNA fragment, and a fluorescein isothiocyanate (FITC)-labeled probe strand. This three-domain construct allows specific binding and fluorescent localization of the whole construct in cells overexpressing the CD4 receptor [43]. In the following paper a similar design of the packaging RNA (pRNA) motif and kissing complex was used to construct bipartite pRNA/siRNA particles [44]. These sequence-directed particles were successfully used for silencing the endogenous enhanced green fluorescent protein (eGFP), coded by a plasmid that was cotransfected with an RNA complex [45].

A different approach for gene regulation can also be undertaken, where an "inverse kissing complex" is used to design a hexamer particle with specific properties [46]. Let us consider a loop–loop interaction with the bend between helixes to be 120°. In such a configuration we can design six specific interactions to form a stable hexameric nanoring. If we want to embedded short interfering

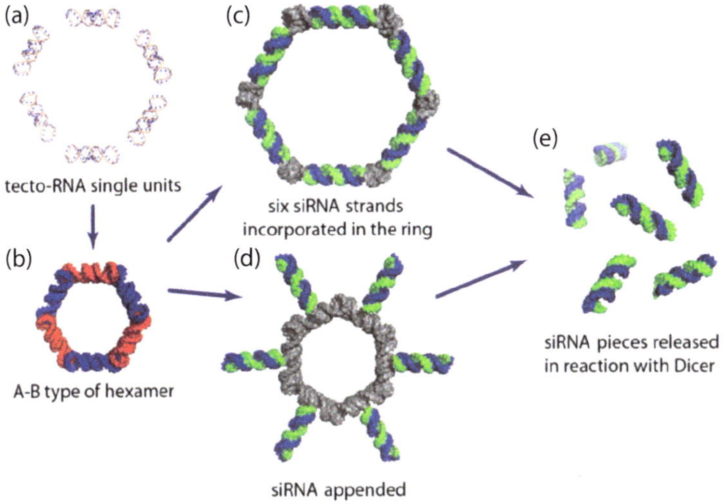

Figure 11.9 Design of a symmetric RNA hexamer using inverted ColE1 loop–loop interaction that creates a 120° bend. (a, b) Formation of the hexamer using alternate A-B molecules (blue, red), (c, d) incorporation of short dsRNA sequences (green, blue) into the structure of the RNA hexamer; dsRNA is either within the hexamer or appended by a the 3WJ to the sides, and (e) six fragments of 22 nt long siRNAs can be released by the enzymatic reaction with Dicer [46].

dsRNAs, two orthogonal strategies can be used to incorporate siRNAs into nanorings, and apparently both can facilitate the release of short dsRNAs upon digestion with human Dicer [46]. Each strategy allows up to six different siRNA duplexes to be incorporated into the hexamer. One strategy involves encoding siRNA sequences within the helical stems of each nanoring strand (Fig. 11.9). In the second strategy, the siRNA stem can be connected to the nanoring in the 3WJ fashion, appended at each side of the hexamer, also allowing up to six different silencing fragments to be released upon delivery. It has been proven that this type of close arrangement is less susceptible to enzymatic degradation as compared to unstructured RNA. However, despite nuclease resistivity both types of hexamers are substrates for Dicer, a small endoribonuclease class III, which is able to process long ds or pre-microRNA into short fragments of siRNA or miRNA.

11.9 Future Directions and Applicability of RNA Technologies: The Promise of siRNA

Considered by many scientists as a molecule of life, RNA is a very versatile medium. In nature it plays an essential role in the information transfer from a DNA sequence into protein biosynthesis. But it also has multiple enzymatic functions, such as self-processing, gene regulation, and protein synthesis, in the RNA-based factory called ribosomes. All this knowledge can be applied to future applications, which are part of the newly discovered field of RNA bionanotechnology. The principles we acquired during the journey of rational design of RNA nanoparticles can be used in different scientific endeavors. Essentially looking at the naturally occurring processes involving RNA we can learn about not only molecular folding, interactions, and enzymatic activity but also selection pressure and evolution. We can extract this virtual knowledge, redesign the structure of RNA according to our needs, and apply to novel, targeted objects (Fig. 11.10). Objects like smart therapeutics will perform specific tasks, such as search for unwanted cells and destroy them or bind with specific tissue, virtually like minimized robots in Isaac Asimov's novel *Fantastic Voyage* (1966).

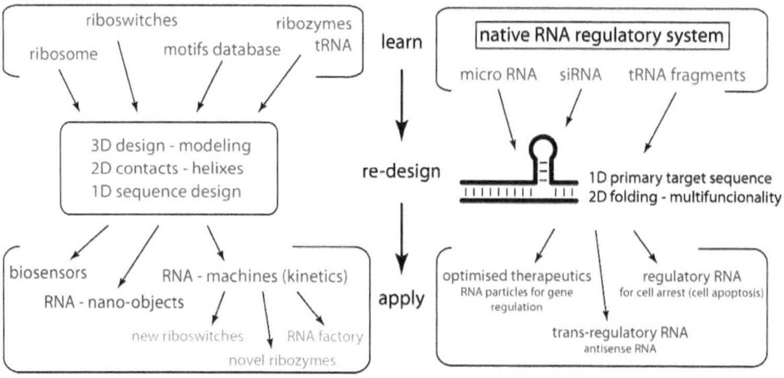

Figure 11.10 Comprehensive approach scientists can take toward design and application of novel RNA nano-objects, new machines, and therapeutics. Inverse to the conventional 1D-2D-3D computational approach one can model the target 3D structure or multicomponent object in the first step and the helix and 1D sequence in the next step.

This is a multidisciplinary approach bridging different fields of biosciences: crystallographers try to solve structures of large RNAs with superhigh resolution, computer scientists try to predict the proper folding and structural determinates of existing or new structures, biochemists provide evidence for integrity and stability of molecules, and bioengineers apply all this knowledge into a practical package that can be successfully delivered into specific and preselected places in the human body. One can imagine that in the near future we will be able to build an entire system using RNA molecules, and even if it needs to be packaged in a cage, such as liposomes, polymeric vesicles, protein capsids, or organic cages (dextran), the entire engine of therapeutics can be made of RNA. However, to be able to use this strategy of "learn—redesign—apply" we need a deeper understanding of processes of molecular uptake, recognition, and enzymatic degradation in a cellular environment.

Acknowledgments

This contribution was supported by NCN grant N302 643740 and is dedicated to two women who never stop believing in me, Walentyna (Dmitruk) Chworos and Gabriela (Czulba) Chworos. I also thank Wojciech J. Stec for critical reading of the manuscript.

References

1. Franklin, R.E., and Gosling, R.G. (1953). Evidence for 2-chain helix in crystalline structure of sodium deoxyribonucleate. *Nature*, **172**, 156–157.
2. Watson, J.D., and Crick, F.H. (1953). Molecular structure of nucleic acids; a structure for deoxyribose nucleic acid. *Nature*, **171**, 737–738.
3. Seeman, N.C. (1982). Nucleic acid junctions and lattices. *J. Theory Biol.*, **99**, 237–247.
4. Seeman, N.C. (1996). The design and engineering of nucleic acid nanoscale assemblies. *Curr. Opin. Struct. Biol.*, **6**, 519–526.
5. Yang, X., et al. (1998). Torsional control of double-stranded DNA branch migration. *Biopolymers*, **45**, 69–83.

6. Caruthers, M.H. (1985). Gene synthesis machines: DNA chemistry and its uses. *Science*, **230**, 281–285.
7. Matteucci, M.D., and Caruthers, M.H. (1992). Synthesis of deoxyoligonucleotides on a polymer support. 1981. *Biotechnology*, **24**, 92–98.
8. Rothemund, P.W. (2006). Folding DNA to create nanoscale shapes and patterns. *Nature*, **440**, 297–302.
9. Chworos, A., et al. (2004). Building programmable jigsaw puzzles with RNA. *Science*, **306**, 2068–2072.
10. Niemeyer, C.M. (2000). Self-assembled nanostructures based on DNA: towards the development of nanobiotechnology. *Curr. Opin. Chem. Biol.*, **4**, 609–618.
11. Feldkamp, U., and Niemeyer, C.M. (2006). Rational design of DNA nanoarchitectures. *Angew. Chem., Int. Ed. Engl.*, **45**, 1856–1876.
12. Seeman, N.C. (2003). At the crossroads of chemistry, biology, and materials: structural DNA nanotechnology. *Chem. Biol.*, **10**, 1151–1159.
13. Chen, X., and Ellington, A.D. (2010). Shaping up nucleic acid computation. *Curr. Opin. Biotechnol.*, **21**, 392–400.
14. Chapman, E.J., and Carrington, J.C. (2007). Specialization and evolution of endogenous small RNA pathways. *Nat. Rev. Genet.*, **8**, 884–896.
15. Leontis, N.B., and Westhof, E. (2001). Geometric nomenclature and classification of RNA base pairs. *RNA*, **7**, 499–512.
16. Leontis, N.B., and Westhof, E. (2003). Analysis of RNA motifs. *Curr. Opin. Struct. Biol.*, **13**, 300–308.
17. Hermann, T., and Patel, D.J. (1999). Stitching together RNA tertiary architectures. *J. Mol. Biol.*, **294**, 829–849.
18. Leontis, N.B., Stombaugh, J., and Westhof, E. (2002). The non-Watson-Crick base pairs and their associated isostericity matrices. *Nucleic Acids Res.*, **30**, 3497–3531.
19. Leontis, N.B., and Westhof, E. (2002). The annotation of RNA motifs. *Comp. Funct. Genomics*, **3**, 518–524.
20. Jaeger, L., and Leontis, N.B. (2000). Tecto-RNA: one-dimensional self-assembly through tertiary interactions. *Angew. Chem., Int. Ed. Engl.*, **39**, 2521–2524.
21. Jaeger, L., Westhof, E., and Leontis, N.B. (2001). TectoRNA: modular assembly units for the construction of RNA nano-objects. *Nucleic Acids Res.*, **29**, 455–463.
22. Geary, C., Chworos, A., and Jaeger, L. (2011). Promoting RNA helical stacking via A-minor junctions. *Nucleic Acids Res.*, **39**, 1066–1080.

23. Davis, J.H., Foster, T.R., Tonelli, M., and Butcher, S.E. (2007). Role of metal ions in the tetraloop-receptor complex as analyzed by NMR. *RNA*, **13**, 76–86.
24. Freier, S.M., et al. (1986). Improved free-energy parameters for predictions of RNA duplex stability. *Proc. Natl. Acad. Sci. U S A*, **83**, 9373–9377.
25. Sugimoto, N., et al. (1995). Thermodynamic parameters to predict stability of RNA/DNA hybrid duplexes. *Biochemistry*, **34**, 11211–11216.
26. Zuker, M. (2003). Mfold web server for nucleic acid folding and hybridization prediction. *Nucleic Acids Res.*, **31**, 3406–3415.
27. Woodson, S.A. (2005). Metal ions and RNA folding: a highly charged topic with a dynamic future. *Curr. Opin. Chem. Biol.*, **9**, 104–109.
28. Popenda, M., et al. (2010). RNA FRABASE 2.0: an advanced web-accessible database with the capacity to search the three-dimensional fragments within RNA structures. *BMC Bioinf.*, **11**, 231.
29. Das, R., and Baker, D. (2007). Automated de novo prediction of native-like RNA tertiary structures. *Proc. Natl. Acad. Sci. U S A*, **104**, 14664–14669.
30. Cruz, J.A., et al. (2012). RNA-puzzles: a CASP-like evaluation of RNA three-dimensional structure prediction. *RNA*, **18**, 610–625.
31. Jaeger, L., and Chworos, A. (2006). The architectonics of programmable RNA and DNA nanostructures. *Curr. Opin. Struct. Biol.*, **16**, 531–543.
32. Ban, N., et al. (2000). The complete atomic structure of the large ribosomal subunit at 2.4 A resolution. *Science*, **289**, 905–920.
33. Ben-Shem, A., Jenner, L., Yusupova, G., and Yusupov, M. (2010). Crystal structure of the eukaryotic ribosome. *Science*, **330**, 1203–1209.
34. Klinge, S., et al. (2011). Crystal structure of the eukaryotic 60S ribosomal subunit in complex with initiation factor 6. *Science*, **334**, 941–948.
35. Lodmell, J.S., Ehresmann, C., Ehresmann, B., and Marquet, R. (2000). Convergence of natural and artificial evolution on an RNA loop-loop interaction: the HIV-1 dimerization initiation site. *RNA*, **6**, 1267–1276.
36. Severcan, I., et al. (2009). Square-shaped RNA particles from different RNA folds. *Nano Lett.*, **9**, 1270–1277.
37. Jaeger, L., Verzemnieks, E.J., and Geary, C. (2009). The UA_handle: a versatile submotif in stable RNA architectures. *Nucleic Acids Res.*, **37**, 215–230.
38. Dibrov, S.M., McLean, J., Parsons, J., and Hermann, T. (2011). Self-assembling RNA square. *Proc. Natl. Acad. Sci. U S A*, **108**, 6405–6408.

39. Dibrov, S.M., Johnston-Cox, H., Weng, Y.H., and Hermann, T. (2007). Functional architecture of HCV IRES domain II stabilized by divalent metal ions in the crystal and in solution. *Angew. Chem., Int. Ed. Engl.*, **46**, 226–229.

40. Jossinet, F., et al. (2001). Identification of the in vitro HIV-2/SIV RNA dimerization site reveals striking differences with HIV-1. *J. Biol. Chem.*, **276**, 5598–5604.

41. Biou, V., Yaremchuk, A., Tukalo, M., and Cusack, S. (1994). The 2.9 A crystal structure of T. thermophilus seryl-tRNA synthetase complexed with tRNA(Ser). *Science*, **263**, 1404–1410.

42. Severcan, I., et al. (2010). A polyhedron made of tRNAs. *Nat. Chem.*, **2**, 772–779.

43. Khaled, A., Guo, S., Li, F., and Guo, P. (2005). Controllable self-assembly of nanoparticles for specific delivery of multiple therapeutic molecules to cancer cells using RNA nanotechnology. *Nano Lett.*, **5**, 1797–1808.

44. Shu, D., et al. (2011). Thermodynamically stable RNA three-way junction for constructing multifunctional nanoparticles for delivery of therapeutics. *Nat. Nanotechnol.*, **6**, 658–667.

45. Shu, Y., et al. (2011). Assembly of therapeutic pRNA-siRNA nanoparticles using bipartite approach. *Mol. Ther.*, **19**, 1304–1311.

46. Grabow, W.W., et al. (2011). Self-assembling RNA nanorings based on RNAI/II inverse kissing complexes. *Nano Lett.*, **11**, 878–887.

Chapter 12

Tailor-Made RNAs: From Self-Folding RNAs to Ribonucleoproteins

Kazuo Harada

Department of Life Sciences, Tokyo Gakugei University, Koganei, Tokyo 184-8501, Japan

harada@u-gakugei.ac.jp

12.1 Introduction

Transfer ribonucleic acid (tRNA) has been described as "nature's attempt to make RNA do the job of a protein" [1]. Indeed, the structures of both RNA and proteins consist of secondary structural elements that are brought together by tertiary interactions to form complex functional structures. However, in the case of proteins, the energies that stabilize secondary and tertiary structures are similar, so the existence of a secondary structure is highly context dependent. On the other hand, the RNA structure is hierarchical in that the secondary structure is much more stable than the tertiary structure, so the secondary and tertiary structures can be distinguished, for example, by the presence or absence of Mg^{2+} [2]. As a consequence, functional RNA structures can be divided into individual secondary structural units, which can then be reconstituted into the functional

RNA Nanotechnology
Edited by Bin Wang

ISBN 978-981-4411-64-6 (Hardcover), 978-981-4411-65-3 (eBook)
www.panstanford.com

structure through tertiary interactions, thereby providing unique opportunities for the construction of RNA-based nanostructures [3].

In the biological context, RNA is usually associated with proteins in the form of ribonucleoprotein (RNP) complexes. RNPs such as the ribosome and the spliceosome are thought to be remnants of the RNA world where the RNA components could not be replaced by proteins during the transition to the contemporary genetic system, suggesting that there are functional roles that may only be achieved by RNPs Therefore, an increased understanding of the structure and function of RNPs should be useful in expanding the repertoire of functions capable by RNA-based nanostructures.

12.2 Assembly of Self-Folding RNA Structures and RNA Assemblies Using Tertiary Interactions

Over the last decade, RNA tertiary interactions have been shown to greatly expand the repertoire of structures and functions that may be created for nanobiotechnology [4, 5]. In this section, a few representative examples will be described.

12.2.1 *RNA Tertiary Interactions*

Structural studies of RNA have revealed the presence of a number of unique secondary structural motifs, which in turn are involved in tertiary interactions [4, 5]. Common RNA secondary structural motifs are shown in Fig. 12.1. Other unique secondary structural motifs include the U-turn motif found in tRNAs, GNRA tetraloops, and the adenosine platform. RNA tertiary interactions are stabilized by canonical and noncanonical base pairing, stacking interactions, and interactions involving the ribose and phosphate moieties and can be classified into three types: (1) interactions between helical motifs, (2) interactions between helical and nonhelical motifs, and (3) interactions between nonhelical motifs [6, 7]. Representative tertiary interactions are listed in Table 12.1, and tertiary interactions observed in the group I intron ribozyme are shown in Fig. 12.2.

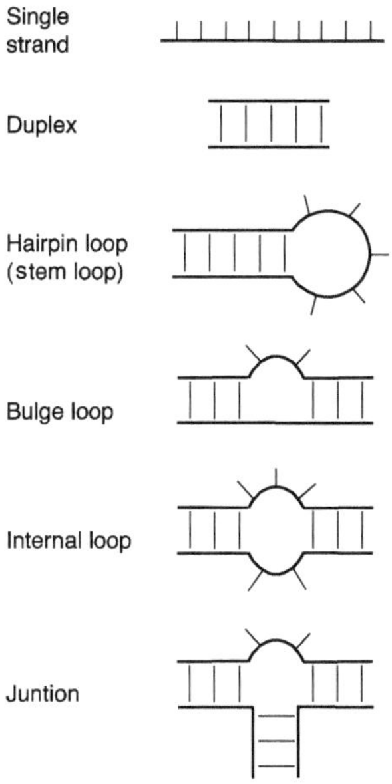

Figure 12.1 Common RNA secondary structural elements.

Table 12.1 Representative RNA tertiary interactions

Interactions between helical motifs
Coaxial stacking, Adenosine platform (③)
Interactions between helical and unpaired motifs
Base triples and triplexes (④), Tetraloop-tetraloops receptor motif (①) ribose zipper. A minor motif
Interactions between unpaired motifs
Loop-loop (kissing) interactions (②), Pseudoknot motif

12.2.2 *RNA Assembly Using the Tetraloop–Tetraloop Receptor Interaction*

The first attempts to construct specific RNA structures and assemblies utilized the tetraloop–tetraloop receptor interaction. In this interaction, the tetraloop receptor forms two G-C base pairs, a reverse Hoogsteen A-U base pair, an adenosine platform, and a G-U base pair. The three adenines in the 5′-GAAA-3′ tetraloop stack with

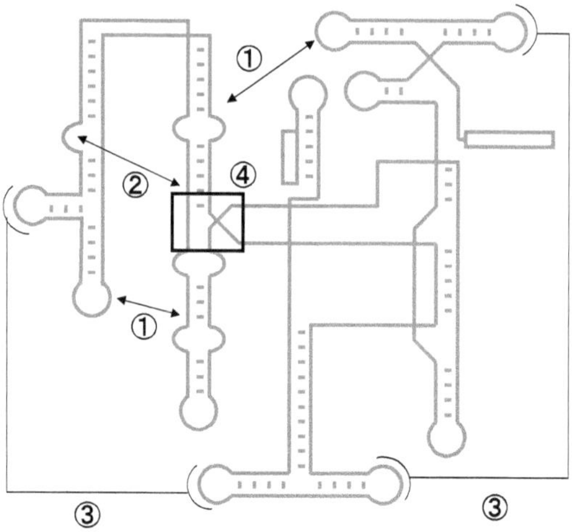

Figure 12.2 The secondary structure of the *Tetrahymena* group I intron and representative secondary structures (numbers shown with arrows correspond to numbers in Table 12.1).

the adenosine platform while forming hydrogen bonds to the minor groove of the tetraloop receptor.

Inoue et al. showed how RNA tertiary interactions could be used to design functional self-folding RNA structures in a rational manner (Fig. 12.3). First, a novel self-folding RNA was constructed

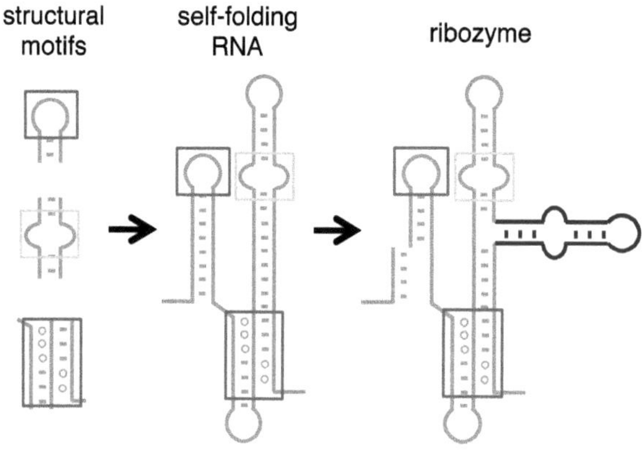

Figure 12.3 Stepwise construction of a novel functional self-folding RNA.

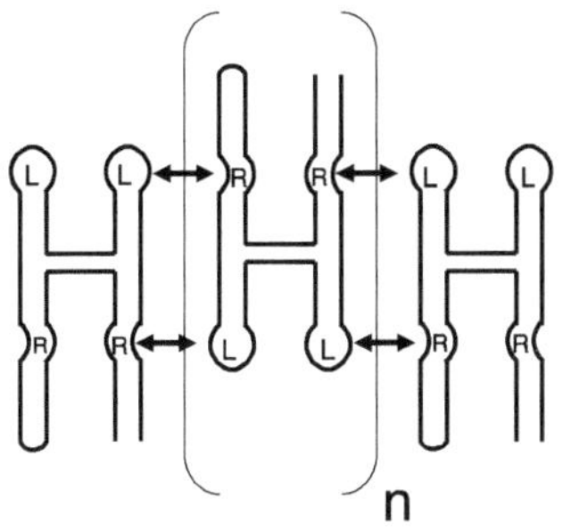

Figure 12.4 Construction of a one-dimensional RNA assembly using the tetraloop (L)–tetraloop receptor (R) interaction.

by modeling a tetraloop, a tetraloop receptor, and a triple-helical scaffold in a rational manner [8]. Formation of a stable RNA fold was found to be magnesium dependent by polyacrylamide gel electrophoresis (PAGE). This artificial RNA structure was then used as a scaffold to select for an RNA structural unit, by in vitro selection, that catalyzed a selfligation reaction [9].

Jaeger et al. showed that RNA substrates containing both a tetraloop and a tetraloop receptor can self-assemble to form one-dimensional assemblies [10]. This was accomplished by utilizing the four-way junction from the hairpin ribozyme to construct a building block containing two tetraloops (L) and two receptor structures (R), resulting in two coaxially stacked stems with a 60° angle (Fig. 12.4). Assembly of the "Tecto RNA" building blocks was shown to be magnesium dependent, with polymeric assemblies being observed at 9 mM magnesium by PAGE.

12.2.3 *RNA Assembly Using the Loop–Loop (Kissing) Interaction*

Loop–loop interactions are observed in many prokaryotic antisense control systems [11], as well as in RNA folding [12] and in the dimerization of retroviral genomic RNAs [13]. The dimerization initiation site (DIS) of HIV has a nine-nucleotide loop containing a self-complementary hexanucleotide 5′-GCGCGC-3′ flanked by purine residues that are thought to be important in stabilizing the kissing complex through stacking interactions [14]. As in many prokaryotic

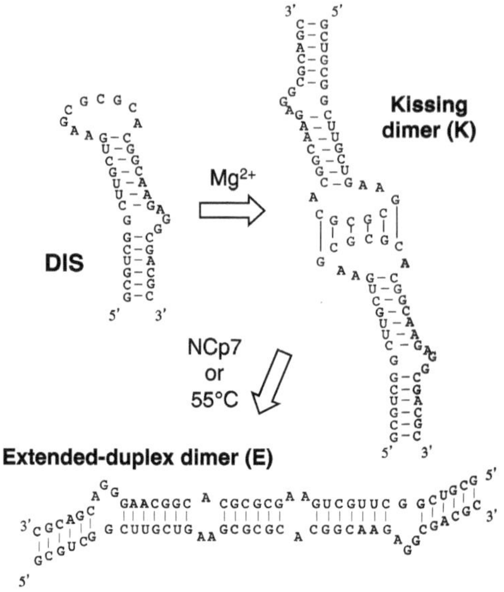

Figure 12.5 Mg^{2+}-dependent loop–loop interaction of the HIV DIS and isomerization to an extended duplex structure.

antisense control systems, the loop–loop complex isomerizes to a stable duplex (Fig. 12.5).

The DIS of HIV genomic RNA has been shown to be a particularly useful starting point for the design of building blocks because the strength of the interaction can be adjusted in a predictable manner [15]. To construct specific linear and circular arrangements of RNA building blocks through loop–loop interactions, several pairs of non-self-complementary loop sequences were designed, and their stabilities were found to correspond well with the thermodynamic stability calculated using the expanded nearest-neighbor parameters for the formation of RNA duplexes [16].

RNA building blocks containing two hairpin loops, based on the DIS of HIV RNA, connected by a short linker were used to construct large RNA assemblies through hairpin loop–loop ("kissing") interactions [15]. Using the non-self-complementary loop sequences, we showed that specific linear and circular assemblies can be constructed in a magnesium-dependent manner, as analyzed by PAGE (Fig. 12.6). Because the assembly and disassembly of

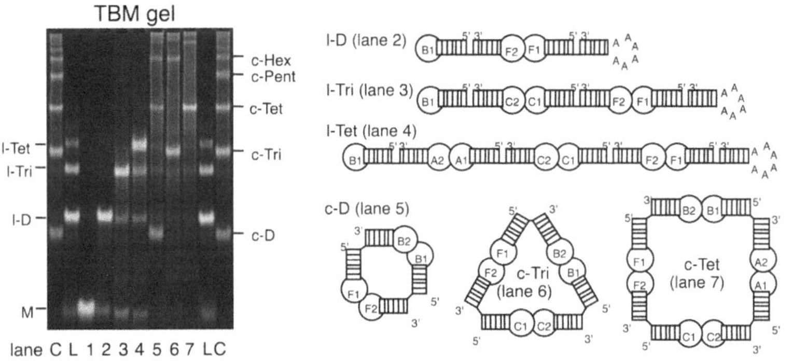

Figure 12.6 Formation of specific linear and circular RNA assemblies using RNA building blocks with two DIS-like stem loops. (Left) PAGE analysis on a Tris–boric acid gel containing Mg^{2+}. Lanes C and L are markers of circular and linear assemblies, respectively. (Right) Schematic representation of the linear and circular assemblies with the loop sequences A1-A2 to F1-F2 indicated in the loops.

RNA building blocks could be controlled in a Mg^{2+}-dependent manner, the process was referred to as "RNA LEGO." Furthermore, dendrimeric structures could be constructed using RNA building blocks with three DIS-based stem loops [17].

Jaeger et al. used RNA building blocks containing two hairpin loops at a fixed 90° angle to construct square-shaped RNA "jigsaw puzzle pieces," which could be further assembled into two-dimensional sheet structures [18]. Guo et al. have used the loop–loop interaction from the bacteriophage phi29 to construct nanostructures with multiple functional domains (small interfering RNA [siRNA], aptamers, etc.), which may be useful as therapeutics [19].

12.3 Modular Assembly of Ribonucleoprotein Complexes

It has been speculated that the first step in the transition from the RNA world to the contemporary deoxyribonucleic acid (DNA)/RNA/protein world may have occurred by primitive translation of short, positively charged peptides [20]. These positively charged peptides, in the form of ribonucleopeptide complexes, would have assisted in RNA folding and helped to expand the structural and functional repertoire of RNA, eventually leading to the evolution of sophisticated RNPs such as the ribosome and spliceosome. An understanding of this process should provide us with additional tools for constructing functional RNA-based nanostructures using RNA as a scaffold. A number of studies that may be considered to mimic this process will be introduced in this section.

12.3.1 *Conversion of a Ribozyme into a Ribonucleoprotein Enzyme*

In an attempt to convert the group I intron ribozyme into an RNP enzyme, an RNA–polypeptide interaction was used to replace an RNA tetraloop–tetraloop receptor interaction (Fig. 12.7) [21]. Modification of the group I intron so that the tetraloop in the P5b element and the tetraloop receptor in the P6ab element were replaced by the bacteriophage λ boxB RNA and the HIV RRE, respectively, thereby disrupting an important tertiary interaction,

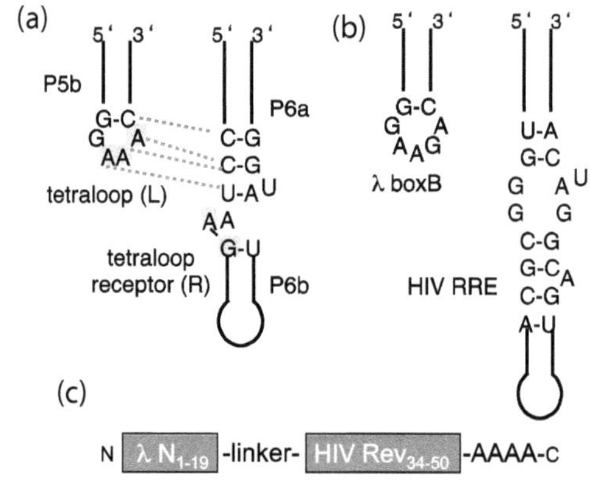

Figure 12.7 Evolution of an RNA enzyme into a RNP enzyme. (A) The interaction between the P5b and P6b domains of the group I intron. (B) Replacement of the two domains by the λ boxB and HIV RRE RNAs leads to a loss of the tertiary interaction. (C) The structure of the λ N and Rev peptide fusion designed to bridge boxB and RRE. *Abbreviation*: RRE, Rev-response element.

resulted in a decrease in ribozyme activity, as would be expected However, when a polypeptide containing the binding partners of boxB and RRE, the λ N and Rev peptides that was modeled so as to bridge the two RNAs, was added, a recovery in activity was observed This study shows how readily functional RNA can be converted to functional RNP, possibly reflecting the adaptability of RNA as a scaffold for constructing nanostructures.

12.3.2 *Construction of Ribonucleoprotein Complexes with Novel Functions*

The assembly and function of RNP complexes such as the ribosome and the spliceosome has been shown to be accompanied by large RNA conformational changes [22, 23]. These conformational changes can be classified into two types (1) rearrangement of the spatial orientation of structural elements and (2) isomerization to alternative secondary structures. As an attempt to demonstrate how the formation of an RNP complex may be regulated by an external signal, a thermal RNA switch utilizing the isomerization of the loop–loop complex to a stable complex was constructed [24]. As shown in Fig. 12.5, the HIV DIS loop–loop complex isomerizes to a stable conformation upon treatment with the HIV nucleocapsid protein NCp7 or by heating to 55°C. In a manner similar to the RNA LEGO assemblies described above, RNA building blocks containing two DIS-like hairpin loops connected by a short linker were used to construct a dimeric assembly through loop–loop interactions so that thermal isomerization leads to the efficient unmasking of the peptide-binding domain, HIV RRE (Fig. 12.8). The relative ease

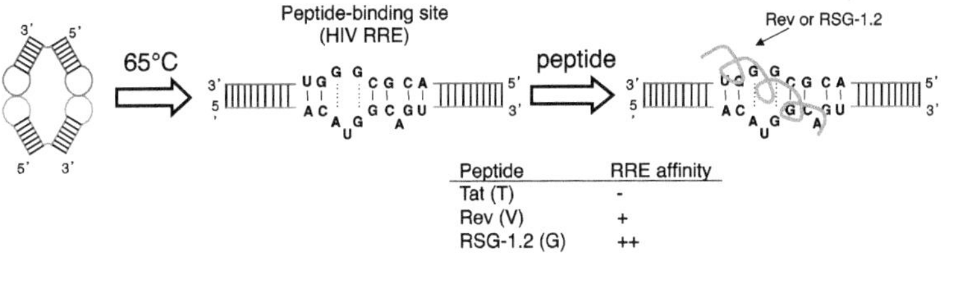

Figure 12.8 A thermal peptide-binding RNA switch.

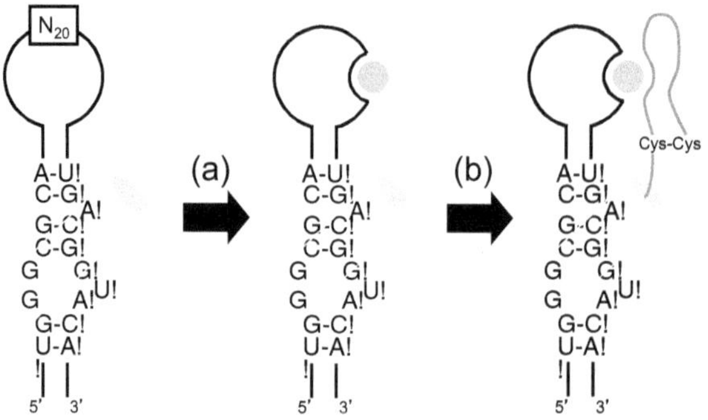

Figure 12.9 Stepwise selection of a ribonucleopeptide receptor for ATP.

with which the RRE could be engineered into the RNA substrates suggested that a variety of functional RNA structures may be introduced for different applications.

In an attempt to demonstrate how RNA-binding peptides may help to improve or expand the function of RNA, in vitro selection of an adenosine triphosphate (ATP)-binding ribonucleopeptide complex has been carried out (Fig. 12.9) [25]. A random 25-mer was introduced adjacent to the HIV RRE, and selection of sequences that bind to ATP was carried out in the presence of the HIV Rev peptide. The selected RNA-peptide receptor molecule bound specifically to the adenine moiety of ATP with a Kd of 33 μM. In the absence of the Rev peptide, ATP binding was completely lost. In the next step, a random 7-mer peptide was introduced to the N-terminal of the Rev peptide. Further selection resulted in the isolation of ribonucleopeptides that bound specifically to the ribonucleotide.

12.4 Conclusions

As described above, RNA is an attractive biopolymer for nanostructural design, in that it possesses programmability through tertiary interactions, in addition to that by Watson–Crick base pairing. The functional repertoire of RNA-based nanostructures should be greatly expanded by the incorporation of proteins to form functional RNPs. The relative ease with which novel

RNA structures and assemblies can be constructed using tertiary interactions presumably reflects the versatility of RNA as a material for biotechnology.

References

1. Crick, F.H.C. (1996). The genetic code—yesterday, today, and tomorrow *Cold Spring Harb. Symp. Quant. Biol.*, **31**, 1–9.
2. Tinoco, Jr., I., and Bustamante, C. (1999). How RNA folds. *J. Mol. Biol.*, **293**, 271–281.
3. Doudna, J.A., and Cech, T.R. (1995). Self-assembly of a group I intron active site from its component tertiary structural domains. *RNA*, **1**, 36–45.
4. Jaeger, L., and Chworos, A. (2006). The architectonics of programmable RNA and DNA nanostructures. *Curr. Opin. Struct. Biol.*, **16**, 531–543.
5. Guo, P. (2010). The emerging field of RNA nanotechnology. *Nat. Nanotechnol.*, **5**, 833–842.
6. Batey, R.T., Rambo, R.P., and Doudna, J.A. (1999). Tertiary motifs in RNA structure and folding. *Angew. Chem. Int. Ed.*, **38**, 2326–2343.
7. Butcher, S.E., and Pyle, A.M. (2011). The molecular interactions that stabilize RNA tertiary structure: RNA motifs, patterns, and networks. *Acc. Chem. Res.*, **44**, 1302–1311.
8. Ikawa, Y., et al. (2002). Design, construction, and analysis of a novel class of self-folding RNA. *Structure*, **10**, 527–534.
9. Ikawa, Y., Tsuda, K., Matsumura, S., and Inoue, T. (2004). De novo synthesis and development of an RNA enzyme. *Proc. Natl. Acad. Sci. U S A*, **101**, 13750–13755.
10. Jaeger, L., and Leontis, N.B. (1999). TectoRNA: modular assembly units for the construction of RNA nano-objects. *Angew. Chem., Int. Ed.*, **39**, 2521–2524.
11. Zeiler, B.N., and Simons, R.W. (1998). *RNA Structure and Function*, Simons, R.W., and Grunberg-Manago, M. (eds.), Cold Spring Harbor Laboratory Press, 437–464.
12. Lehnert, V., Jaeger, L., Michel, F., and Westhof, E. (1996). New loop-loop tertiary interactions in self-splicing introns of subgroup IC and ID: a complete 3D model of the Tetrahymena thermophila ribozyme. *Chem. Biol.*, **3**, 993–1009.

13. Paillart, J.C., et al. (1996). A loop-loop "kissing" complex is the essential part of the dimer linkage of genomic HIV-1 RNA. *Proc. Natl. Acad. Sci. U S A*, **93**, 5572–5577.
14. Jossinet, F., et al. (1999). Dimerization of HIV-1 genomic RNA of subtypes A and B: RNA loop structure and magnesium binding. *RNA*, **5**, 1222–1234.
15. Horiya, S., et al. (2003). RNA LEGO: magnesium-dependent formation of specific RNA assemblies through kissing interactions. *Chem. Biol.*, **10**, 645–654.
16. Xia, T., et al. (1998). Thermodynamic parameters for an expanded nearest-neighbor model for formation of RNA duplexes with Watson-Crick base pairs. *Biochemistry*, **37**, 14719–14735.
17. Fujiya, I., and Harada, K. (2004). Formation of dendrimeric RNA assemblies though RNA loop-loop interactions. *Nucleic Acids Symp. Ser.*, **48**, 97–98.
18. Chworos, A., et al. (2004). Building programmable jigsaw puzzles with RNA. *Science*, **306**, 2068–2072.
19. Khaled, A., Guo, S., Li, F., and Guo, P. (2005). Controllable self-assembly of nanoparticles for specific delivery of multiple therapeutic molecules to cancer cells using RNA nanotechnology. *Nano Lett.*, **5**, 1797–1808.
20. Noller, H.F. (2005). Evolution of ribosomes and translation from an RNA world, in *The RNA World*, 3rd ed., Cold Spring Harbor Laboratory Press, 287–307.
21. Atsumi, S., Ikawa, Y., Shiraishi, H., and Inoue, T. (2001). Design and development of a catalytic ribonucleoprotein. *EMBO J.*, **20**, 5453–5460.
22. Shajani, Z., Sykes, M.T., and Williamson, J.R. (2011). Assembly of bacterial ribosomes. *Annu. Rev. Biochem.*, **80**, 501–526.
23. Will, C.L., and Luhrmann, R. (2011). Spliceosome structure and function. *Cold Spring Harb. Perspect. Biol.*, **3**, a003707.
24. Li, X., Horiya, S., and Harada, K. (2006). An efficient thermally induced RNA conformational switch as a framework for the functionalization of RNA nanostructures. *J. Am. Chem. Soc.*, **128**, 4035–4040.
25. Sato, S., et al. (2005). Stepwise molding of a highly selective ribonucleopeptide receptor. *J. Am. Chem. Soc.*, **127**, 30–31.

Chapter 13

Small RNA–Mediated Transcriptional Modulation

Thomas C. Roberts and Matthew J. A. Wood
Department of Physiology, Anatomy and Genetics, University of Oxford, South Parks Road, Oxford, OX1 3QX, UK
thomas.roberts@dpag.ox.ac.uk

The utility of small RNAs to modulate the transcription of specific target genes in mammalian cells is an exciting recent development with implications for our understanding of endogenous transcriptional control, the development of novel RNA-based therapeutics, and the study of gene function. Both transcriptional gene silencing (TGS) and transcriptional gene activation (TGA) have been reported. In the case of TGS, small RNAs targeting promoter regions either recruit epigenetic modifying complexes to a promoter-associated RNA in order to induce silent-state chromatin modifications and DNA methylation or sterically inhibit RNA polymerase procession via direct interaction with chromosomal DNA. Conversely, in the case of TGA, targeting antisense RNA transcripts relieves endogenous epigenetic silencing, leading to an increase in target gene expression. Noncoding RNA transcripts appear to be involved in both transcriptional silencing and activation processes. Here we

RNA Nanotechnology
Edited by Bin Wang

ISBN 978-981-4411-64-6 (Hardcover), 978-981-4411-65-3 (eBook)
www.panstanford.com

describe the evidence that these processes occur in mammalian cells, and discuss their utility in the treatment of human disease.

13.1 Introduction

Ribonucleic acid (RNA) interference (RNAi) by post-transcriptional gene silencing (PTGS) utilizes small RNA effector molecules that direct the RNA-induced silencing complex (RISC) to homologous messenger RNAs (mRNAs) [1]. The target transcript is then cleaved by the "slicer" activity of the RISC component Argonaute2 (AGO2) [2]. This endogenous gene regulation pathway has been used extensively in the study of gene function and holds much promise as a therapy for a wide range of diseases. However, the effect of silencing by PTGS is short lived, typically declining over a period of $\sim$7 days, with peak silencing observed between 24 and 72 hours [3, 4]. Silencing is also dependent on the presence of effector RNA molecules, the concentration of which is reduced by the activity of cellular ribonucleases (RNases) and diluted by successive rounds of cell division.

Multiple studies now suggest that an alternative regulatory pathway, transcriptional gene silencing (TGS), is operative in mammalian cells. TGS is a potent and specific RNA-mediated, homology-dependent gene-silencing pathway in which small RNA molecules homologous to gene promoters either induce epigenetic changes at the targeted locus or sterically inhibit procession of RNA polymerase, thus achieving silencing at the transcriptional level. Conversely, small RNAs targeting promoters can also induce transcriptional gene activation (TGA). Although the mechanistic details of this process are less well understood, several reports suggest that TGA is operative through targeting of promoter-overlapping antisense RNA (asRNA) transcripts, which leads to the reversal of endogenous epigenetic silencing.

The potential to control transcription through the use of small RNA molecules provides new opportunities for both the development of novel therapeutics and the study of gene function. In particular, the epigenetic control of gene expression by small and long noncoding RNAs (lncRNAs) presents the opportunity for both silencing and activation of specific target genes that are stable and heritable and so, in theory, permanent. Here we discuss recent developments in the field of RNA-targeted transcriptional control and prospects for therapeutic applications.

13.2 Mechanisms of Transcriptional Gene Silencing

Although TGS is relatively well understood in *Schizosaccharomyces pombe* (*S. pombe*), drosophila, and Arabidopsis (reviewed in [5–8]), until recently it was not known to occur in mammals. Multiple studies now show that small RNAs complementary to target promoters silence gene expression at the level of transcription. Two competing models have emerged to explain these observations. In the RNA:RNA model, the small RNA effector molecule targets a sense promoter-associated RNA (pRNA) and recruits an epigenetic remodeling complex to the target promoter in order to induce heterochromatin formation and promoter deoxyribonucleic acid (DNA) methylation. Conversely, the RNA:DNA model posits that the small RNA effector molecule interacts directly with chromosomal DNA. The binding of small RNAs to DNA occurs either by DNA:DNA:RNA triplex formation or by binding to single-stranded sequences that are susceptible to transient melting, most notably at the site of transcription initiation where RNA polymerase II (RNAPII) mediates local DNA unwinding. Although these two models were initially in competition, it now seems that both models may be true under different specific circumstances. Here we refer to silencing by the RNA:RNA model as epigenetic-TGS and silencing by the RNA:DNA model as antigene-TGS.

13.2.1 *Epigenetic-TGS*

Epigenetics is the study of mitotically and meiotically heritable changes in gene expression that are not coded in the DNA itself [9]. Here we are primarily concerned with two kinds of epigenetic modification, direct chemical modification of DNA nucleotides (specifically, methylation of cytosine at CpG dinucleotides, which is associated with gene silencing in mammals) and post-translational modifications of accessory proteins that regulate the accessibility of DNA to the transcriptional machinery. Eukaryotic genomic DNA is packaged with histone proteins to form the dynamic polymer chromatin. Genomic DNA wraps around a histone protein octamer in 1.67 superhelical turns to form a structure called a nucleosome. The N-terminal tails of histones H3 and H4 protrude from the nucleosome core structure and undergo extensive post-translational modification. The pattern of histone tail modifications regulates the transition between transcriptionally active (euchromatin) and transcriptionally silent (heterochromatin) conformations and is known as the histone code. For example, the methylation of lysine 9 (H3K9me2) and lysine 27 (H3K27me3) on the N-terminal tail of histone H3 results in chromatin compaction, which reduces the accessibility of DNA to transcription factors and RNA polymerase [10]. Consequently, the histone code greatly expands the information content of the genome [11]. The dual mechanisms of DNA methylation and silent-state chromatin formation are functionally linked such that methylation of histone H3K9 directs DNA methylation (as in the case of pericentromeric chromatin) [12], and conversely, the polycomb protein enhancer of zeste homolog 2 (EZH2) (which trimethylates H3K27) has been shown to directly influence DNA methylation [13]. The patterns of CpG methylation and histone tail modifications are inherited as somatic cells divide and, in some cases, in the germ line [14].

Epigenetic-TGS (RNA:RNA model) is a homology-dependent gene-silencing pathway in which small RNA effector molecules targeting low-copy-number promoter transcripts recruit RITS (RNA-Induced Transcriptional Silencing complex) to the target promoter. RITS acts to induce silent-state chromatin formation and, in some cases, de novo promoter DNA methylation. These epigenetic modifications at the target promoter potentially result in long-term, stable gene silencing (Fig. 13.1). The first study to report small RNA–mediated TGS by Morris et al. showed that targeting the eukaryotic elongation factor 1α (*EEF1A1*) with a small interfering RNA (siRNA) resulted in silencing of both an integrated GFP transgene driven by the *EEF1A1* promoter and endogenous *EEF1A1* expression. Silencing was sensitive to the histone deacetylase (HDAC) inhibitor trichostatin A (TSA) and the DNA methyltransferase inhibitor

5′-azacytidine (5-azaC), which implied that epigenetic remodeling at the *EEF1A1* promoter was responsible for the observed silencing effect. Methylation of the *EEF1A1* promoter was also detected by methyl-specific digestion [15]. Subsequent studies showed that treatment with promoter-targeting siRNAs results in enrichment of the silent-state chromatin modifications H3K9me2 and H3K27me3 as measured by chromatin immunoprecipitation (ChIP) both for *EEF1A1* [16] and for other promoters (Table 13.1).

The induction of targeted epigenetic changes enables long-term therapeutic gene silencing. To this end, Hawkins et al. showed that targeting the ubiquitin C (*UbC*) promoter for two to three days with a tetracycline-induced small hairpin RNA (shRNA) was sufficient to induce long-term gene silencing (~30 days) [17]. Similarly, Suzuki et al. showed that transient transfection of an siRNA targeting the human immunodeficiency virus type 1 (HIV-1) 5′–long terminal repeat (LTR) was able to suppress HIV replication for 31 days in the HeLa-derived MAGIC-5 cells [18], and building on this work, Yamagishi et al. were able to silence HIV mRNA transcription and viral replication for up to 1 year using a retrovirus-expressed shRNA in a T-cell line (Molt-4) [19]. Collectively, these studies suggest that long-term, targeted gene suppression by TGS may be possible in a therapeutic context.

Cytosine methylation has been reported in the case of a number of mammalian promoters targeted for epigenetic-TGS [15, 17, 20–22]. Conversely, other studies have shown that TGS can occur in the absence of DNA methylation [23]. A study by Kim et al. demonstrated TGS of transforming growth factor β receptor II (*Tgfbr2*) in rat hepatic stellate cells (SBC10) following lentiviral transduction. shRNAs targeting the *Tgfbr2* promoter induced dense methylation of CpGs and non-CpG cytosines. DNA methylation and silencing were alleviated in the presence of 5-azaC, and the strength of silencing was found to correlate with the degree of promoter methylation. *Tgfbr2* inhibition was only observed seven days after transduction, suggesting that promoter DNA methylation is a prerequisite for silencing [20]. This relatively late appearance of DNA methylation may explain some of the discrepancies between studies. Alternatively, persistent targeting may be required to modulate DNA methylation as an shRNA, but not an equivalent

Table 13.1 Studies reporting epigenetic-TGS. Studies in which small RNA–based effectors silence gene expression with epigenetic changes reported. "Enrichment" and "loss of" indicate that the described proteins were enriched or reduced, respectively, at the effector target site as measured by ChIP. "Dependent" indicates that the reported effect was abolished or reduced when the described genes were down-regulated by siRNAs or antisense oligonucleotides. "Sensitive" indicates that the reported effect was abolished or reduced when cultures were treated with the chemical inhibitors described (i.e., TSA and sodium butyrate inhibit HDACs, 5-azaC and 5-aza-dC inhibit DNA methyltransferases, BIX inhibits HMTs specifically targeting H3K9, and 3′hExo digests siRNAs)

Promoter	Effector	Cell line	Evidence	Ref.
EEF1A1	siRNA	HEK293FT	TSA/5-azaC sensitive	[15, 16, 46]
			Promoter methylation	
			H3K9me2 enrichment	
			pRNA dependence	
RASSF1A	shRNA (stable transfection)	HeLa	AGO1, EZH2, and TARBP2 enrichment	[27, 36]
			Promoter methylation	
CCR5	siRNA	HEK293T	H3K9me2 and AGO1 enrichment	[27]
			Promoter methylation	
CDH1	siRNA	HCT116, MCF-7	H3K9me2 enrichment	[23]
			Silencing in methylation deficient cell line	
			No promoter methylation	
UBC	siRNA	HEK293GT	H3K9me2, H3K7me3, and AGO1 enrichment	[17]
	shRNA (stable transfection)		G9a, DNMT3A, HDAC1, and AGO1 dependence	
			pRNA dependence	
			Promoter methylation	

Tgfbr2	shRNA (lentiviral transduction)	SBC10	5-aza-dC sensitivity Promoter methylation	[20]
HIV1 5′-LTR	siRNA asRNA shRNA (retroviral transduction)	MAGIC-5, IG5, TZMbl	H3K9me2, H3K27me3, HDAC1 enrichment Reduced NF-κB and Sp1 binding TSA and α-amanitin sensitive Promoter methylation AGO1, DNMT3a, and HDAC1 dependent	[16, 30, 18, 22, 19]
	siRNA miRNA	HEK293, TZMbl	TSA sensitive HDAC1 enrichment	[50]
SIV 5′-LTR	siRNA	MAGIC-5, CEMx174	H3K9me2 and H3K27me3 enrichment 5-azaC and TSA sensitive	[51]
PLAU	siRNA	PC3, DU145	Promoter methylation TSA/5-azaC sensitive	[21]
VEGF-A	shRNA	C166, Intramuscular injection	H3K9me2, H3K27me3, and nucleosome enrichment	[52]
Txnip	siRNA (combination of two)	retinal EC, Intravitreal injection	Reduced association of p300, suggesting closed chromatin formation	[53]
HPV16 E6/E7	siRNA	SiH3, CaSki Intratumoral injection	H3k9me3 enrichment No promoter methylation detected	[54–56]
IL2	shRNA	jurkat	Promoter methylation 5-azaC sensitivity Loss of H3Ac	[57]

(*Continued*)

Table 13.1 *(Continued)*

Promoter	Effector	Cell line	Evidence	Ref.
eNOS	siRNA (19 and 27nt)	HAEC	cDNA microarray confirming specificity of silencing	[58]
β–actin-Ig-μ minigene	proposed endogenous siRNA	HeLa	H3Kme enrichment Sodium butyrate and TSA sensitive Sensitive to 3′hExo	[59]
RD-INK4/ARF	siRNA	HEK293T	H3K9me3, AGO1/AGO2 enrichment pRNA dependent	[60, 61]
MTAP	shRNA	MEF		[60]
CDKN2A (p16)	siRNA	HEK293T	H3K9me3 enrichment	[62]
RUNX3	shRNA (stable transfection)	SGC7901	Not 5-azaC sensitive	[63]
MYC	siRNA shRNA (stable transfection)	MCF-7, SiHa, Caco-2, U87MG	Promoter methylation 5-azaC sensitive	[37]
CDKN1A (p21)	tiRNA mimic	MCF-7, THP-1	Loss of CTCF binding	[64]
PGR	siRNA	T47D, MCF-7	H3K27me3 enrichment Loss of RNAPII	[65]
FN1	siRNA	Hep3B, HeLa	H3K9me2 and H3K27me3, enrichment TSA, 5-aza-dC, and BIX sensitive AGO1, AGO2, and HP1α dependent	[66]
POLR3D	miR-320	HEK293	H3K27me3, AGO1, and EZH2 enrichment	[67]

HOXD4	siRNA miR-10a	MCF-7, MDA-MB-231, MCF10A, HepG2, HeLa, A549	H3K27me3 enrichment. Dicer, AGO1, and AGO3 dependent Promoter methylation pRNA involvement 5azaC sensitive	[68]
TBCEL	miR-17-5p miR-20a	HCT116	AGO1/AGO2 enrichment	[62]
RASA2	miR-17-5p miR-20a	HCT116	AGO1/AGO2 enrichment	[62]
RHPN2	miR-17-5p miR-20a	HCT116	AGO1/AGO2 enrichment	[62]
WHSC1	miR-17-5p miR-20a	HCT116	AGO1 enrichment	[62]
CDCA8	miR-let-7f	WI38	H3K27me3 and AGO2 enrichment	[69]
CDC2	miR-let-7f	WI38	H3K27me3 and AGO2 enrichment	[69]

Abbreviations: PGR, progesterone receptor; SIV, simian immunodeficiency virus; CCR5, chemokine receptor 5; VEGF-A, vascular endothelial growth factor A; MTAP, methylthioadenosine phosphorylase; PLAU, plasminogen activator urokinase; HPV, human papillomavirus.

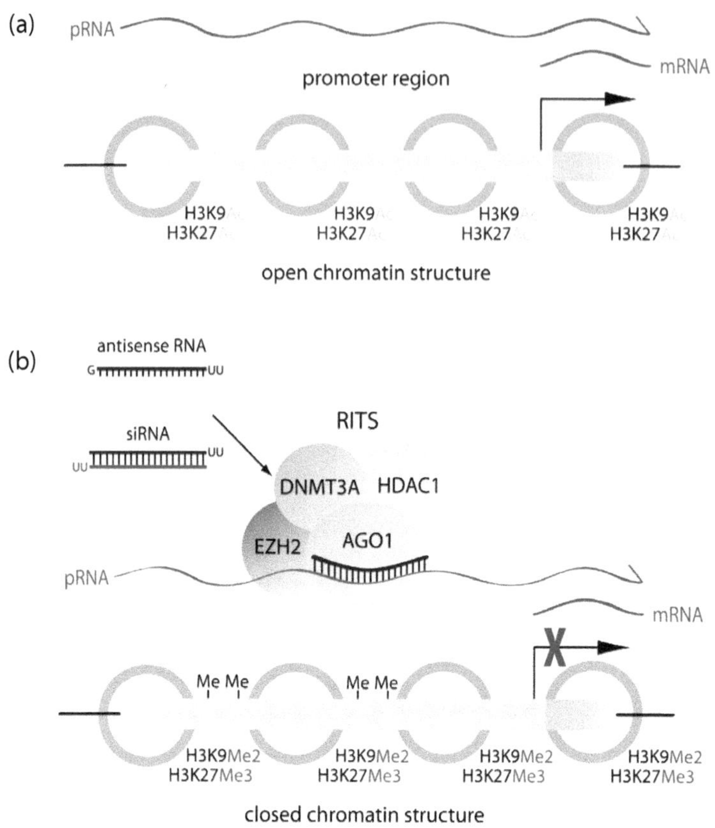

Figure 13.1 A model of transcriptional gene silencing in mammalian cells. (a) pRNA is transcribed from the promoter region of a hypothetical gene locus. (b) A promoter-targeting siRNA or asRNA recruits the RITS complex, which includes HDAC1, an HMT (EZH2), and a de novo DNA methyltransferase (DNMT3A) to the targeted promoter. Replacement of H3K9 and H3K27 acetyl groups with di- and trimethyl modifications (and in some cases methylation of promoter DNA) results in transcriptional silencing of the target locus. *Abbreviation*: HMT, histone methyltransferase.

siRNA, induced cytosine methylation at the *Tgfbr2* promoter [20]. Conversely, Ting et al. showed that siRNAs targeting the *CDH1* promoter (E-cadherin) induced TGS in the absence of promoter methylation, as demonstrated by bisulphite sequencing. In addition, the same effect could be reproduced in cells deficient in the de novo DNA methylation machinery, suggesting that DNA methylation is not required for TGS [23]. Taken together, these studies suggest that there is redundancy between the epigenetic mechanisms of TGS (i.e., heterochromatin formation and promoter DNA methylation). Consequently, the ability to induce targeted promoter methylation may be cell-type dependent or gene specific.

13.2.1.1 RITS components

The RITS complex that facilitates epigenetic-TGS is analogous to the RISC that is involved in post-transcriptional gene silencing and contains the RNA-binding protein AGO2 [2, 24]. In Arabidopsis and fission yeast, AGO proteins are involved in TGS and constitute part of the RITS complex. Notably, in *S. pombe* the RITS complex contains AGO1 [25, 26]. These observations led researchers to hypothesize a role for AGO1 in mammalian TGS. Subsequently, Kim et al. showed that the presence of AGO1 is required for TGS in HEK293T cells [27]. AGO1, a close homolog of AGO2, lacks the catalytic amino acid residues (the disulfide-directed hairpin [DDH] motif [28]) required for "slicer" functionality and thus does not cleave target RNA [29]. However, several studies have detected enrichment of AGO2 at the promoters of genes targeted for TGS by ChIP, and the RISC component TARBP2 was enriched at the *RASSF1A* promoter following treatment with a promoter-targeting siRNA, suggesting the involvement of other components of the RNAi machinery in mammalian TGS [27]. Initiation of TGS requires the deacetylation of histone N-terminal tails. The involvement of HDACs has been demonstrated indirectly through sensitivity of silencing to treatment with TSA and directly through ChIP and RNAi depletion experiments [17, 30]. Similarly, the HMTs, EHMT2 (formerly G9a) and EZH2, which methylate H3K9 and H3K27, respectively, have been similarly implicated in mammalian TGS [17, 27]. The de novo DNA methyltransferase DNMT3A coimmunoprecipitates with H3K27me3 [16], HDAC1 [31, 32], the HMT SUV39H [33], and EZH2 [34] and binds siRNAs in vitro [35]. Furthermore, inhibition of DNA methyltransferases by treatment with 5-azaC or an siRNA against DNMT3A abrogates TGS [17, 21, 23, 36, 37]. Expression of DNMT1 was required for maintenance of long-term silencing [17].

13.2.1.2 Promoter-associated RNA

Next-generation sequencing and RNA-tiling array technologies have revealed that transcription from mammalian genomes is much more prevalent than once thought [38, 39]. Transcription is observed in both sense and antisense orientations, and divergent transcription is frequently observed at promoters [40–43]. Noncoding RNA (ncRNA) transcripts that overlap the 5′ and 3′ termini of genes have been described previously as pRNAs and termini-associated RNAs, respectively [44]. Furthermore, depletion of the exonucleolytic RNA exosome reveals ubiquitous unstable promoter-associated transcripts [45]. Han et al. showed the existence of a pRNA at the *EEF1A1* promoter, which was spliced and polyadenylated and could be considered an mRNA transcript with an extended 5′-untranslated region (5′-UTR). The *EEF1A1* pRNA could be recovered after streptavidin pull-down when HEK293T cultures were treated with a biotinylated siRNA, thus demonstrating direct physical association between the TGS effector siRNA and the pRNA target transcript [46]. Epigenetic-TGS has been shown to be dependent on the presence of pRNA for a number of different promoters as determined by antisense oligonucleotide and RNAi depletion experiments (Table 13.1). There are several examples of pRNAs acting as *cis*-regulatory elements that influence the transcription of their respective downstream promoters [47–49]. As such, pRNAs may directly associate with the chromatin from which they are transcribed. Alternatively, pRNAs may be indirectly tethered to their respective promoter chromatin by association with RNAPII rather than by direct binding to chromatin. In either case the pRNA is retained in the proximity of the promoter for the downstream gene and acts as an "RNA signature" for the associated locus.

13.2.1.3 Non-promoter-targeting TGS

Several studies have demonstrated TGS or TGS-like effects through targeting nonpromoter regions and these are noteworthy "exceptions that prove the rule." The observation that siRNAs can direct promoter chromatin remodeling raises the question as to whether similar effects can be seen in nonpromoter regions. Several studies have now shown that this is, in fact, the case. A study by Bühler et al. showed that a small RNA species produced from intron 4 of an immunoglobulin (Ig) minigene driven by the β-actin (*ACTB*) promoter induced transcriptional silencing in HeLa cells [59]. Enrichment of methylated H3K9 was observed at the minigene promoter, and the silencing was sensitive to treatment with TSA and sodium butyrate, suggesting the involvement of HDACs. Interestingly, a neomycin resistance gene found on the same plasmid as the Ig minigene was also silenced. This suggested that the induced heterochromatic region had spread upstream from the small RNA target site in intron 4 to silence the minigene promoter and the upstream neomycin cassette.

In some cases nonpromoter transcripts may be present in close proximity to promoters as a result of the formation of higher-order chromatin structures [70]. A recent study by Yue et al. showed that the human progesterone receptor (*PGR*) could be silenced by siRNAs that target downstream of the 3′ terminus of the gene and are therefore not complementary to either the mRNA or the promoter. Chromosome conformation capture (3C) analysis showed that the locus formed a gene loop structure whereby the promoter region and the 3′ terminal region were found to be in close proximity in three-dimensional space. Additionally, ncRNA species were identified at both termini and were found to associate with AGO2, as determined by RNA immunoprecipitation (RIP) and ChIP. Consequently, the siRNA was indirectly targeting the *PGR* promoter through interaction with a promoter-proximal ncRNA [71].

Epigenetic control of splicing has also been shown in a study by Alló et al. In this case, an intron-targeting siRNA induced local heterochromatin formation within the coding region of the fibronectin gene (*FN1*) and consequently influenced alternative splicing [66]. The formation of a local closed chromatin region resulted in slowed RNAPII procession and preferential inclusion of an alternatively spliced exon proximal to the siRNA target site.

13.2.2 *Antigene-TGS*

Antigene-TGS (RNA:DNA model) is the use of oligonucleotides to inhibit transcription through direct interaction of the effector molecule with chromosomal DNA by either DNA:DNA:oligonucleotide triplex formation or by the oligonucleotide binding to exposed single-stranded DNA (e.g., at the transcription start site (TSS)) [72, 73]. The effector oligonucleotide acts to sterically hinder RNA polymerase procession and consequently induce transcriptional gene silencing [74]. Janowski et al. showed that siRNAs (which they term "antigene RNAs" [agRNAs] as they are designed to target DNA) that target the TSS of the *PGR* B-promoter induce transcriptional inhibition. Functional agRNAs were found to overlap the −9 to +2 region that was predicted to form an open complex during transcription initiation. This region constitutes an oligonucleotide target site where the two strands of DNA are transiently separated in order to permit transcription initiation. Inhibition by agRNAs was highly potent with an IC50 of 2.5 nM, and similar results were seen in the case of the major vault protein (*MVP*), the androgen receptor (*AR*), and cyclo-oxygenase-2 (*COX2*) genes, indicating the effect is observable for TATA-containing and TATA-less promoters. No promoter DNA methylation was observed (up to five days post-transfection) for any of these genes studied [75]. Similarly, for an agRNA targeting just upstream of *PGR* TSS (−26 to −7) no TSA [76] or 5-azaC [75] sensitivity, DNA methylation, or substantial changes in histone H3K4, H3K9, or H3K27 methylation were observed [77]. Taken together these results suggest a nonepigenetic mechanism of silencing. Peptide nucleic acid (PNA) [78], locked nucleic acid (LNA) [79, 80], and other chemically modified mixmer oligonucleotides [81] were also shown to silence *PGR* in a similar manner.

Similar results were reported in a study by Napoli et al. who targeted the TSS of the *MYC* promoter with an siRNA [82] (Table 13.2). Interestingly, this study also showed that the presence of a promoter transcript was required for silencing. The role of this transcript in the mechanism of silencing is unclear. Despite early reports of antigene-TGS failing to show any evidence of epigenetic involvement in silencing, a recent study targeting the *PGR* promoter did show enrichment of silent-state chromatin modifications [71]. Additionally, the expression of AGO2 was shown to be required for PGR silencing, which points toward an RNA:RNA interaction [83]. Consequently, the degree to which agRNAs target chromosomal DNA or RNA remains an open question.

Table 13.2 Studies reporting antigene-TGS. Studies in which small RNA–based effectors targeted to TSSs silence gene expression in the absence of epigenetic changes. "Enrichment" and "loss of" indicate that the described proteins were enriched or reduced, respectively, at the effector target site, as measured by ChIP. "Dependent" indicates that the reported effect was abolished or reduced when the described genes were down-regulated by siRNAs or antisense oligonucleotides. "Sensitive" indicates that the reported effect was abolished or reduced when cultures were treated with the chemical inhibitors described (i.e., TSA inhibits HDACs)

Promoter	Effector	Cell line	Evidence	Ref.
PGR	agRNA agPNA agLNA agENA	T47D	Not TSA sensitive; no histone changes; no promoter methylation	[75, 84, 76, 85, 81, 79, 80]
			AGO1 and AGO2 dependent	[71]
			AGO1 and AGO2 enrichment AGO1 and AGO2 associating with a PGR antisense transcript, as measured by RIP	
			Conflicting evidence of epigenetic changes	
AR	agRNA agLNA agENA	T47D	AGO1 and AGO2 dependent No promoter methylation	[75, 76, 80]
MVP	agRNA	T47D		[75]
COX2	agRNA	T47D		[75]
HTT	agRNA	T47D	AGO1 and AGO2 dependent	[76]
MYC	siRNA	PC3, DU145 and LNCaP	No histone changes	[82]
			pRNA dependent	
Hpa	siRNA shRNA	PC3 EJ SGC-7901	Loss of RNAPII and TFIIB	[86]
			No histone changes or DNA methylation detected	
			AGO1 and AGO2 dependent	

Abbreviations: agENA, antigene ethylene-bridged nucleic acid; HTT, human serotonin transporter; Hpa, heparanase.

13.2.3 *Transcriptional Gene Activation*

Small RNAs have also been shown to activate gene expression (Table 13.3). Although some groups have termed this phenomenon "RNA activation" (RNAa), here we use the term "transcriptional gene activation," as recent reports suggest that the effect is closely related to epigenetic-TGS. The first evidence that small RNAs could activate gene transcription came from the Functional Annotation of the Mammalian genome (FANTOM) consortium transcriptome project [42]. In this study, pairs of overlapping sense and antisense transcripts were targeted with siRNAs and both concordant (i.e., both transcripts were silenced) and discordant (i.e., the target transcript was silenced, but the expression of the overlapping transcript increased) modes of regulation were observed. Separately, Li et al. showed that siRNAs targeting the *CDH1* (E-cadherin) promoter induced sequence-specific, long-lasting (∼13 days) gene activation in human cells. Two- to tenfold induction was observed, and 1 nM siRNA was sufficient to induce silencing. RNAi experiments demonstrated that AGO2 was indispensable for gene activation and silencing was associated with a loss of H3K9 methylation. Similarly, *CDKN1A* (p21) and *VEGF-A* could also be activated following transient transfection with promoter-targeting RNA duplexes [87]. Morris et al. investigated these results further and demonstrated that siRNAs targeting the p21 promoter induced PTGS of an antisense transcript, leading to gene activation through the reversal of an endogenous TGS mechanism. In this case the antisense transcripts themselves were shown to be directing epigenetic silencing [88] (Fig. 13.2). Similarly, Modarresi et al. have shown that targeting an asRNA overlapping the brain-derived neurotrophic factor (*BDNF*) gene using antisense oligonucleotide gapmers (which degrade the target RNA) results in abrogation of epigenetic silencing with consequent gene activation [89].

Building on previous work targeting the *PGR* promoter, the Corey lab demonstrated that when the sequence of a silencing siRNA was shifted by a single nucleotide, it induced potent and specific activation of PGR mRNA and protein expression in T47D and MCF7 breast cancer cells. Importantly, PGR is differentially expressed in these two cell lines. In T47D cells, which express high levels of PGR,

Table 13.3 Studies reporting TGA. Studies in which small RNA–based effectors induce transcriptional gene activation. "Enrichment" and "loss of" indicate that the described proteins were enriched or reduced, respectively, at the effector target site, as measured by ChIP. "Dependent" indicates that the reported effect was abolished or reduced when the described genes were down-regulated by RNA interference. "Sensitive" indicates that the reported effect was abolished or reduced when cultures were treated with the chemical inhibitors described (i.e., TSA inhibits HDACs). "antagoNAT" refers to antinatural antisense transcript oligonucleotides (i.e., LNA gapmers and LNA mixmers)

Promoter	Effector	Cell line	Comments	Ref.
CD97	siRNA	Hepa1-6		[42]
CDH1	siRNA	PC-3, DU-145, 5637, COS1, WES, MCF-7, MDA-MB-453	AGO2 dependent	[87, 90–92]
CDKN1A	siRNA	PC-3, Hela, COS1, WES, MCF-7	Bidirectional transcription regulating silencing and activation	[87, 88, 91]
PR	siRNA	TF7D, MCF-7	AGO2 enrichment AGO2 dependent	[85, 93]
MVP	siRNA	MCF-7		[93]
LDLR	siRNA	HepG2		[94]
VEGF-A	siRNA shRNA (lentiviral tranduction)	Hela, COS1, WES, CCSMC, C166, IM injection	In vivo transcriptional up-regulation	[52, 87, 91, 95]
IL10	siRNA	THP1	Enhancer targeting	[96]
IL24	miR-205	PC3		[97]
IL32	miR-205	PC3		[97]
CDH1	miR-373	PC3		[98]

(Continued)

Table 13.3 (*Continued*)

Promoter	Effector	Cell line	Comments	Ref.
CSDC2	siRNA	PC3, HCT116		[98]
OCT4	siRNA	MCF-7		[99]
BDNF	siRNA antagoNAT	N2a	Loss of H3K27me3,	[89]
		HEK293T	EZH2	
		ICV injection	In vivo transcriptional up-regulation	
GDNF	antagoNAT	HEK293T		[89]
EPHB2	antagoNAT	HEK293T		[89]
p53	siRNA	COS1, WES		[91]
PAR4	siRNA	WES		[91]
NKX3-1	siRNA	COS1, WES		[91]
Ccnb1	siRNA miR-744 miR-1186 miR-366d-3p	NIH/3T3, TRAMP C1	H3Kme3, RNAPII, and AGO1 enrichment	[91, 100]
CXCR4	siRNA	Primary rat stem cells		[91]
KLF4	siRNA	DuPro, PC3, DU145		[101]
NANOG	siRNA	NCCIT		[102]
WT1	siRNA	HepG2		[103]

Abbreviations: GDNF, glial cell-derived neurotrophic factor; CSDC2, cold-shock domain-containing protein C2.

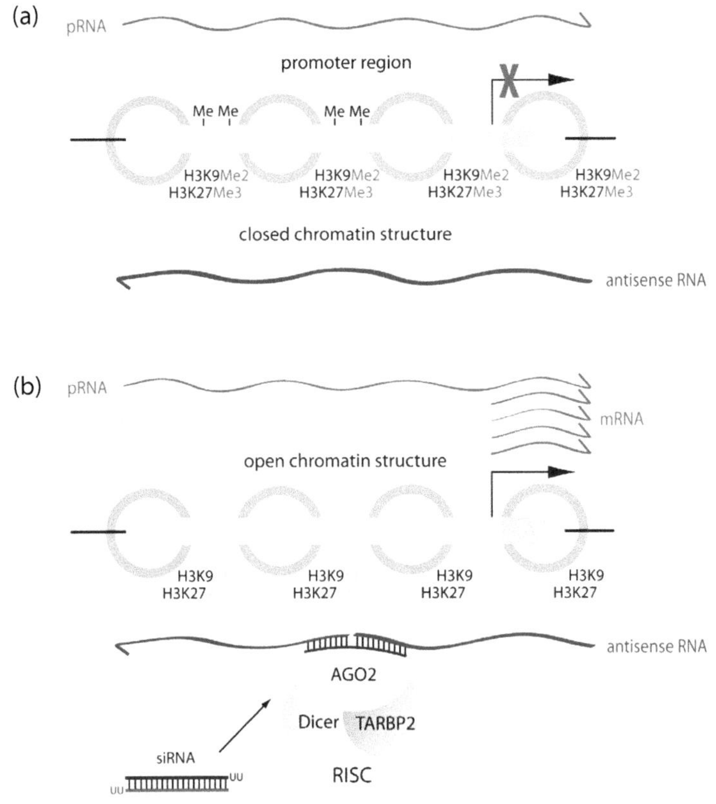

Figure 13.2 A model of transcriptional gene activation in mammalian cells. (a) An antisense transcript directs epigenetic silencing at a hypothetical promoter. (b) Targeted degradation of the antisense transcript (e.g., by siRNA-directed RISC cleavage) alleviates endogenous epigenetic repression and results in transcriptional activation.

gene activation was relatively modest. However, in MCF7 cells, which express low levels of PGR, the observed activation was as much as 18-fold. These results suggest that the gene activation is dependent on cellular context and that cells that already express the targeted gene at high levels may be less susceptible to TGA. Gene activation was reversed by treatment with TSA and resulted in reduced H3K9 and H3K14 acetylation and increased di- and trimethylation of H3K4 implicating chromatin conformation changes in gene activation.

Similarly, a fourfold induction in *MVP* gene expression was observed following treatment with siRNAs [93]. Further work showed that antisense transcripts overlapping the *PGR* promoter are the targets for activating siRNAs. These antisense transcripts are spliced and polyadenylated, and they span a 70,000 kb region of genomic DNA and are present at levels 10- to 1,000-fold lower than the PGR mRNA. Direct association of the siRNAs was determined by biotin pull-down, and pan-AGO RIP showed recruitment of AGO to the asRNA [83]. A combination of RNAi depletion and RIP for all four human AGO proteins demonstrated that AGO2 is involved in siRNA-mediated *PGR* transcriptional silencing and activation [87].

13.2.4 *Endogenous Epigenetic Modulation*

The observation that exogenously introduced small RNAs can utilize cellular machinery to induce epigenetic changes in gene expression implied that microRNAs (miRNAs) might be endogenous epigenetic triggers of epigenetic silencing. Kim et al. utilized a bioinformatic search for miRNA target sites in human promoters 200 bp upstream of the TSS, and 10 completely complementary miRNA target sites were identified. One candidate, miR-320, was found to target the *POLR3D* promoter and induced transcriptional gene silencing, including promoter enrichment of AGO1, EZH2, and H3K27me3. Low levels of sense transcription were also detected at the *POLR3D* promoter [67]. Conversely, Place et al. showed miR-373 induces activation of *CDH1* and cold-shock domain-containing protein C2 (*CSDC2*) in PC3 cells [98]. Gene activation was induced by both mature miR-373 and expressed pre-miR-373 synthetic double-stranded RNA mimics, was Dicer dependent, and resulted in enrichment of RNAPII at the *CDH1* and *CSDC2* promoters. Interestingly, an miRNA with only partial complementarity to its target promoters has also been shown to induce TGA, suggesting there may be numerous endogenous transcriptional modulation events [97]. These studies suggest new functions for miRNAs in the regulation of mammalian transcription, in addition to PTGS by transcript degradation and translational repression.

Endogenous epigenetic regulation is also mediated by long (>200 nucleotides in length) noncoding RNAs (lncRNAs). lncRNAs are implicated in a wide variety of cellular processes, including X chromosome inactivation [104], mono-allelic expression of imprinted loci [105], and determination of cellular differentiation [106, 107]. lncRNAs are thought to operate as molecular guides or scaffolds [47, 49, 108] that mediate associations between epigenetic remodeling complexes and chromatin. (Mechanisms of gene regulation by lncRNAs are reviewed here [109].) As such there are clear parallels with small RNA–mediated transcriptional regulation. For example, Yu et al. identified an antisense transcript at the p15 promoter. Overexpression of portions of the p15 asRNA led to silent-state chromatin formation and long-term gene silencing of the p15 tumor suppressor. Levels of the asRNA were also found to be inversely correlated with p15 expression in a variety of cancers. Silencing was reversed by treatment with TSA and 5-azaC. The same effect was also observed in a Dicer-deficient cell line, suggesting that the asRNA is not processed into smaller siRNA effectors [110]. Additionally, ectopic transcription of an asRNA through the CpG island of the *HBA2* (α-globin) promoter results in transcriptional silencing and α-thalassemia in a patient with a rare chromosomal translocation, suggesting that epigenetic regulation by lncRNAs is relevant to human disease [111].

13.3 The Practical Application of Transcriptional Modulation

Methods for facilitating epigenetic-TGS and TGA are discussed here [112] and the use of agRNAs and agPNAs here [113]. Typically, the target gene promoter is targeted within 200 bp upstream of the TSS. Some studies have specifically targeted CpG dinucleotides [51]. There is, as yet, no consensus on TGS/TGA effector design criteria as some researchers have utilized conventional siRNA design rules [87] to good effect, whereas other have shown highly functional siRNAs that score poorly [18] based on Reynolds criteria [114].

Small RNA–mediated transcriptional modulation requires delivery to the nucleus, which can be achieved in a number of ways. In actively cycling cells it is likely that dissolution of the nuclear membrane during cell division is sufficient to allow small

RNA effector molecules to enter the nucleus. Similarly, lentiviral transduction facilitates nuclear membrane permeabilization [115]. Generally, mammalian TGS studies have used relatively high (>50 nM) siRNA concentrations, presumably to achieve effective nuclear concentrations of siRNA. This suggests that in the absence of a dedicated nuclear-targeting strategy, nuclear delivery may be achieved through passive diffusion from the cytoplasm to the nucleus (thus high oligonucleotide concentrations in the cytoplasm are required). Nuclear targeting has also been achieved through noncovalent complexing of siRNA with nuclear-targeting peptides such as MPG [16, 116].

TGS/TGA can be induced by siRNA molecules or viral/plasmid-expressed shRNA precursors [36]. The antisense strand of an siRNA is alone sufficient to induce TGS [16], and to this end ~21-mer-expressed asRNAs have also been shown to induce TGS [16, 22]. The use of expressed asRNAs may be preferable to shRNAs in two respects. Firstly, there is no corresponding sense strand to mediate off-target effects, and secondly, transcription alone generates the mature antisense species in the nucleus, whereas shRNAs must be transcribed, exported, diced, and imported back into the nucleus for functionality.

Recently, Hwang et al. demonstrated that a hexanucleotide motif is sufficient to direct miR-29b to the nucleus. The same nuclear-targeting effect was observed when the sequence was incorporated into a synthetic siRNA targeting luciferase [117]. The potential for this sequence motif to direct siRNAs to the nucleus may be highly useful in the design of TGS effectors. However, a word of caution is warranted as Liao et al. detected no nuclear enrichment of RNAs carrying the hexanucleotide motif using a deep sequencing approach [118].

Chemical modification of TGS/TGA effector oligonucleotides offers a potential means of enhancing nuclear delivery. For example, phosphorothioate-substituted oligonucleotides are primarily located in the nucleus following transient transfection [119]. Similarly, a recent study has shown that chemical addition of a 2,2,7-trimethylguanosine cap (m_3G-CAP) to antisense oligonucleotides improves their nuclear accumulation [120]. These simple modifications may enhance nuclear delivery, although it remains to be seen as to whether chemical alteration reduces the functionality of the TGS/TGA effector molecule.

13.3.1 *Potential for Off-Target Effects*

The main technologies utilized to induce TGS and TGA (primarily siRNAs and shRNAs) are chemically identical to that used for conventional RNAi and so can potentially mediate similar off-target effects. Indeed, the requirement for high concentrations of siRNA to induce TGS/TGA means that potential off-target effects must be considered carefully [16, 116]. siRNAs with complete or partial complementarity to nontarget mRNAs will induce mRNA slicing or translational repression respectively [121]. Furthermore, siRNAs and shRNAs can, in some circumstances, induce nonspecific gene silencing by activation of the interferon response [122, 123]. The presence of long double-stranded RNA [124], specific immunostimulatory sequence motifs [125, 126], or 5′-triphosphates that result from in vitro transcribed RNAs [127] are markers of viral infection and so have the potential to induce the interferon response. As with conventional RNAi, care must be taken to avoid these immunostimulatory effects when designing oligonucleotides for TGS/TGA.

Weinberg et al. showed that an asRNA targeting the HIV-1 5′-LTR, which activated expression of proviral mRNA transcription, was actually functioning via the off-target down-regulation of a general transcriptional suppressor [128]. This study indicates that care must be taken to validate the specificity of interactions in studies of transcriptional modulation. Similarly, a study by Moses et al. showed that the intended transcriptional gene silencing of *VEGF-A* was, in fact, a sequence-dependent off-target effect [129]. This study also provided evidence that TGS in a previous report [16] may also be due to nonspecific effects. The authors emphasize the need to mutate the small-RNA target sequence within the targeted promoter as a means of ensuring specificity. However, experiments of this type, whilst potentially informative, suffer from two technical difficulties. Firstly, mutations in the promoter sequence may adversely affect target gene transcription. And secondly, cloning the target promoter upstream of reporter genes may result in loss of endogenous ncRNA regulation and thereby not accurately model the transcriptional landscape of the target gene.

Conversely, a thorough study by Suzuki et al. showed highly specific TGS of HIV-1 replication by a promoter-targeting shRNA. Scrambled and mismatched controls did not suppress viral transcription, and other genes involved in viral entry showed no change in expression. Furthermore, the shRNA failed to silence HIV-2 (which does not contain the complementary shRNA target site), and genes with similar regulatory elements in their promoters (i.e., NF-κB-binding sites) were similarly unaffected. Additionally, no interferon induction was observed. Taken together this study shows that sequence-specific on-target TGS is possible in the case of HIV-1, with no observed off-target effects [130].

One strategy to enhance the potency of conventional PTGS is to utilize a pool of siRNAs (typically three or four) targeting different regions of the target mRNA, so-called combinatorial RNAi. This approach has the advantage of exhibiting high levels of knockdown with minimal off-target effects as the siRNAs converge on the same target mRNA but are individually at relatively low concentration. To this effect, Ting et al. showed that combining two promoter-targeting siRNAs induced higher levels of knockdown relative to treatment with each individual siRNA alone [23]. This suggests that combinatorial TGS may be a means of maximizing potency while minimizing off-target effects.

Heterochromatin has been shown to propagate following induction by small-RNA effectors [30]. Spreading of heterochromatin to upstream genes has been observed in at least one report of mammalian TGS [59]. This presents an additional source of potential off-target effects as heterochromatin spreading may result in the silencing of genes adjacent to the target locus. In addition, small RNAs with complementarity to nontarget promoters may mediate off-target TGS and/or TGA. Equally, siRNAs intended to silence genes by RNAi may potentially elicit off-target TGS/TGA effects.

13.3.2 *Therapeutic Applications of Transcriptional Modulation*

13.3.2.1 Human immunodeficiency virus

HIV is a particularly promising target for gene-silencing therapies. PTGS has been utilized to target the HIV RNA genome and transcribed viral mRNA in order to inhibit viral replication. The high rate of viral turnover and the error-prone nature of HIV reverse transcriptase contribute to the rapid accumulation of escape mutants to conventional small-molecule inhibitor drugs and RNA interference effectors [131–133]. von Elje et al. showed that the ability of HIV to accumulate escape mutations was restricted by targeting shRNAs to conserved regions of the viral genome [134]. Consequently, nucleic acid–based therapeutics is a promising anti-HIV approach as modifying effector molecules to account for escape mutants is relatively trivial. A combination therapy whereby multiple gene-silencing approaches target several viral genes may limit the capacity of the virus to mutate, analogous with highly active antiretroviral therapy (HAART) [135]. TGS offers an additional advantage by silencing the integrated proviral DNA, therefore inhibiting the viral life cycle before the reverse transcription stage and thus limiting the possibility of escape mutations occurring [18]. An alternative antiviral strategy is to target host factors that are required for viral replication or entry, as these are not subject to the same evolutionary pressure as viral genomes. For example, Kim et al. showed that the HIV-1 coreceptor *CCR5* promoter is amenable to silencing by TGS [27].

One of the first reports of HIV TGS was by Suzuki et al., who utilized 21 nt duplex siRNAs targeting tandem NF-κB motifs in the U3 region of the HIV-1 5′-LTR. Transfection in MAGIC-5 cells (CCR5-CXCR4-expressing Hela/CD4^{+} cell line infected with HIV) resulted in LTR cytosine methylation and the suppression of viral replication for at least 30 days [18]. Four LTR-targeting duplexes were found to induce differential CpG methylation, with the degree and density of methylation correlating with the strength of viral suppression. Two subsequent studies showed that (1) treatment with an LTR-targeting siRNA was accompanied by sustained recruitment of RITS components AGO1 and HDAC1, enrichment of the silent chromatin modification H3K9me2, and rearrangement of nucleosome positioning around the TSS within the integrated provirus [30] and (2) that viral suppression could be maintained for up to one year [19]. The same group also showed inhibition of SIV (SIVmac251) replication mediated by siRNAs targeting the 5′-LTR upstream of the TSS in two cell lines (MAGIC-5 and CEMx174; human lymphoid cells). The silencing effect induced DNA methylation and enrichment of H3K9me2 and H3K27me3 and was partially reversed by treatment with TSA and 5-azaC [51].

In parallel, Weinberg et al. used asRNAs targeting the U3 region of HIV-1 5′-LTR in IG5 cells (which contain an integrated tat-inducible firefly luciferase transgene driven by the HIV-1 5′-LTR [136]). asRNAs were found to be comparable to equivalent double-stranded siRNAs and directed H3K27 trimethylation in an RNAPII-dependent manner [16]. Building on this work, Turner et al. developed a mobilization-competent lentivector system to express these HIV-targeting asRNAs. Suppression of viral replication was observed for up to one month and TGS verified by nuclear run-on analysis, enrichment of silent-state chromatin modifications, and CpGs methylation at the HIV-1 5′-LTR. In addition silencing was found to be dependent on the presence AGO1, HDAC1, and DNMT3A [22].

13.3.2.2 Cancer

Cancer refers to a broad group of diseases characterized by the formation of masses of cells that have lost cell cycle regulation and thus divide uncontrollably (i.e., tumors). Tumorigenesis is a multistep process in which the various cellular mechanisms that restrain cell proliferation are progressively lost [137]. Activating mutations in oncogenes (i.e., genes that promote proliferation) leading to ectopic expression is one such step. Conversely, loss-of-function mutation or epigenetic silencing of tumor suppressor genes (i.e., genes that act to protect against cancer progression) are common features of many tumors. The silencing of oncogenes and the activation of tumor suppressor genes, by TGS and TGA, respectively, have thus been the focus of numerous transcriptional modulation studies [21, 23, 27, 36, 37, 54, 55, 60–63, 87, 88, 93].

One promising anticancer TGS target is the plasminogen activator urokinase (*PLAU*), a gene involved in tumor invasion and metastasis. DNA demethylation at the *PLAU* promoter leads to increased *PLAU* expression, which is observed in many cancers and correlates with poor survival [138]. In a study by Pulukuri and Rao, PC3 cells were transfected with siRNAs that target the *PLAU* promoter and implanted into immunodeficient mice. (The lung is a common metastatic site of prostate cancers, and implantation of PC3 human prostate cancer cells into mice induces lung metastasis.) Mice implanted with *PLAU* promoter–targeting siRNA-treated cells showed long-term *PLAU* silencing, suppressed primary tumor growth, and reduced lung metastasis relative to those implanted with control siRNA-treated cells [21]. This ex vivo study provided the first evidence that TGS may be a relevant therapeutic strategy in vivo.

13.3.2.3 Transcriptional modulation in vivo

To date, four studies have demonstrated transcriptional modulation in vivo. A landmark study by Turunen et al. demonstrated both TGS and TGA of *VEGF-A* following lentiviral delivery of shRNAs in a mouse hind limb ischemia model [52]. Inhibition of *VEGF-A* has the potential to suppress metastasis by inhibiting tumor vascularization, whereas activation of *VEGF-A* could be used to promote neovascularization as a therapy for myocardial or peripheral ischemias [139]. The silencing shRNA induced reductions in H3K4 methylation and acetylation and increases in methylation of H3K9 and H3K27, consistent with epigenetic silencing. Conversely, treatment with the activating shRNA resulted in an increase in H3K4 trimethylation and reductions in H3K9 and H3K27 methylation, consistent with transcriptional activation [52]. Other in vivo TGS studies have utilized local injections of promoter-targeting siRNAs against the human papillomavirus (HPV) E6/E7 and thioredoxin-interacting protein (*Txnip*) promoters to induce in vivo TGS in mouse xenograft tumors [54] and rat retina [53], respectively, although these reports did not include investigations of epigenetic remodeling. More recently, in vivo TGA of *BDNF* was shown by intracerebroventricular (ICV) injection of siRNAs and antisense oligonucleotides (which they term "antagoNATs") in the mouse brain. The oligonucleotides targeted a natural antisense transcript, which overlaps with the *BDNF* locus and recruits EZH2. Following treatment, a loss of H3K27me3 and EZH2 was observed at the *BDNF* promoter by ChIP [89].

13.4 Conclusions

An appraisal of recent discoveries reveals a new vision of genome biology in which RNA plays a central role. ncRNA transcripts are found throughout the genome and act to regulate the expression of genes at the epigenetic level. Targeting these transcripts with small RNAs presents exciting opportunities for the development of molecular therapies and the study of gene function. In particular, the use of small RNA molecules to activate target genes in a sequence-specific manner is a pharmacological novelty as the mode of action of conventional small-molecule drugs and RNAi approaches is limited to the inhibition of therapeutic targets. Consequently, the development of epigenetic-TGS and TGA approaches has opened up the epigenome to molecular medicine.

The therapeutic application of epigenetic-TGS presents several advantages over existing gene-silencing approaches. Conventional PTGS is dependent on the presence of effector molecules that are subject to degradation by cellular nucleases and are progressively dispersed between daughter cells, following cell division. Consequently, repeat administration of effector molecules is required in order to attain persistent silencing. In contrast, TGS effectors can induce long-term silencing that is inherited following cell division, and maintenance of silencing is not dependent on the presence of TGS effector molecules [17]. A single treatment may be sufficient to elicit stable gene silencing, hence removing the need for repeat administration [17, 18], limiting toxicity and off-target effects, lowering the risk of saturating endogenous RNA-processing pathways [140], and greatly reducing material costs. Targeting a gene promoter effectively down-regulates all splice isoforms of that target gene, whereas a PTGS approach is limited to only those splice variants that contain the homologous target sequence. Similarly,

TGS allows for the targeting of multiple genes derived from the same promoter, as in the case of the HIV-1 5′-LTR and HPV E6/E7 promoter. Silencing at the level of transcription is also preferable in the therapeutic targeting of viral genes prone to escape mutations. For example, in the case of HIV infection, targeting early in the viral life cycle reduces the possibility of escape mutations occurring.

Although tentative models of TGS and TGA have been proposed, many mechanistic details are still poorly understood. Unresolved issues such as the differential involvement of AGO proteins, the role of pRNA, and the extent to which effector oligonucleotides interact directly with chromosomal DNA need to be addressed. It is possible that cell-type or gene-specific differences are responsible for the discrepancies reported in the literature, and as the number of reports of TGS and TGA increases, a clearer picture is likely to emerge. In the case of TGS we have described two distinct modes of action in this review, epigenetic-TGS and antigene-TGS. Given certain similarities between these two mechanisms it is possible that there is an overlap between them. This is exemplified by a recent report of a pRNA from the ribosomal RNA (rRNA) locus that not only acts to recruit an epigenetic modifying factor (DNMT3B) to the rRNA promoter but also interacts directly with the promoter DNA by forming a triplex structure [47].

The successful translation of transcriptional modulation from cell culture studies to the in vivo setting may be dependent on similar expression levels of the ncRNA target transcripts between cultured cells and tissues. The low abundance of these target RNAs may be a limiting factor for treatment efficacy, given that many are highly unstable [45]. Although initial reports of transcriptional modulation in vivo are highly encouraging [52–54, 89], more studies are required to demonstrate their general applicability as therapeutic modalities.

References

1. Filipowicz, W. (2005). RNAi: the nuts and bolts of the RISC machine. *Cell*, **122**, 17–20.

2. Meister, G., et al. (2004). Human Argonaute2 mediates RNA cleavage targeted by miRNAs and siRNAs. *Mol. Cell*, **15**, 185–197.
3. Dykxhoorn, D.M., Novina, C.D., and Sharp, P.A. (2003). Killing the messenger: short RNAs that silence gene expression. *Nat. Rev. Mol. Cell Biol.*, **4**, 457–467.
4. Capodici, J., Karikó, K., and Weissman, D. (2002). Inhibition of HIV-1 infection by small interfering RNA-mediated RNA interference. *J. Immunol.*, **169**, 5196–5201.
5. Creamer, K.M., and Partridge, J.F. (2011). RITS-connecting transcription, RNA interference, and heterochromatin assembly in fission yeast. *Wiley Interdiscip. Rev. RNA*, **2**, 632–646.
6. Verdel, A., Vavasseur, A., Le Gorrec, M., and Touat-Todeschini, L. (2009). Common themes in siRNA-mediated epigenetic silencing pathways. *Int. J. Dev. Biol.*, **53**, 245–257.
7. Grewal, S.I.S., and Elgin, S.C.R. (2007). Transcription and RNA interference in the formation of heterochromatin. *Nature*, **447**, 399–406.
8. Matzke, M.A., Matzke, A.J., Pruss, G.J., and Vance, V.B. (2001). RNA-based silencing strategies in plants. *Curr. Opin. Genet. Dev.*, **11**, 221–227.
9. Egger, G., Liang, G., Aparicio, A., and Jones, P.A. (2004). Epigenetics in human disease and prospects for epigenetic therapy. *Nature*, **429**, 457–463.
10. Strahl, B.D., and Allis, C.D. (2000). The language of covalent histone modifications. *Nature*, **403**, 41–45.
11. Jenuwein, T., and Allis, C.D. (2001). Translating the histone code. *Science*, **293**, 1074–1080.
12. Lehnertz, B., et al. (2003). Suv39h-mediated histone H3 lysine 9 methylation directs DNA methylation to major satellite repeats at pericentric heterochromatin. *Curr. Biol.*, **13**, 1192–1200.
13. Viré, E., et al. (2006). The Polycomb group protein EZH2 directly controls DNA methylation. *Nature*, **439**, 871–874.
14. Paro, R. (1995). Propagating memory of transcriptional states. *Trends Genet.*, **11**, 295–297.
15. Morris, K.V., Chan, S.W.-L., Jacobsen, S.E., and Looney, D.J. (2004). Small interfering RNA-induced transcriptional gene silencing in human cells. *Science*, **305**, 1289–1292.

16. Weinberg, M.S., et al. (2006). The antisense strand of small interfering RNAs directs histone methylation and transcriptional gene silencing in human cells. *RNA*, **12**, 256–262.

17. Hawkins, P.G., et al. (2009). Promoter targeted small RNAs induce long-term transcriptional gene silencing in human cells. *Nucleic Acids Res.*, **37**, 2984–2995.

18. Suzuki, K., et al. (2005). Prolonged transcriptional silencing and CpG methylation induced by siRNAs targeted to the HIV-1 promoter region. *J. RNAi Gene Silencing*, **1**, 66–78.

19. Yamagishi, M., et al. (2009). Retroviral delivery of promoter-targeted shRNA induces long-term silencing of HIV-1 transcription. *Microbes Infect.*, **11**, 500–508.

20. Kim, J.-W., et al. (2007). Short hairpin RNA causes the methylation of transforming growth factor-beta receptor II promoter and silencing of the target gene in rat hepatic stellate cells. *Biochem. Biophys. Res. Commun.*, **359**, 292–297.

21. Pulukuri, S.M.K., and Rao, J.S. (2007). Small interfering RNA directed reversal of urokinase plasminogen activator demethylation inhibits prostate tumor growth and metastasis. *Cancer Res.*, **67**, 6637–6646.

22. Turner, A.-M.W., De La Cruz, J., and Morris, K.V. (2009). Mobilization-competent lentiviral vector-mediated sustained transcriptional modulation of HIV-1 expression. *Mol. Ther.*, **17**, 360–368.

23. Ting, A.H., Schuebel, K.E., Herman, J.G., and Baylin, S.B. (2005). Short double-stranded RNA induces transcriptional gene silencing in human cancer cells in the absence of DNA methylation. *Nat. Genet.*, **37**, 906–910.

24. Liu, J., et al. (2004). Argonaute2 is the catalytic engine of mammalian RNAi. *Science*, **305**, 1437–1441.

25. Lippman, Z., et al. (2003). Distinct mechanisms determine transposon inheritance and methylation via small interfering RNA and histone modification. *PLoS Biol.*, **1**, E67.

26. Verdel, A., et al. (2004). RNAi-mediated targeting of heterochromatin by the RITS complex. *Science*, **303**, 672–676.

27. Kim, D.H., Villeneuve, L.M., Morris, K.V., and Rossi, J.J. (2006). Argonaute-1 directs siRNA-mediated transcriptional gene silencing in human cells. *Nat. Struct. Mol. Biol.*, **13**, 793–797.

28. Rivas, F.V., et al. (2005). Purified Argonaute2 and an siRNA form recombinant human RISC. *Nat. Struct. Mol. Biol.*, **12**, 340–349.

29. Tolia, N.H., and Joshua-Tor, L. (2007). Slicer and the argonautes. *Nat. Chem. Biol.*, **3**, 36–43.
30. Suzuki, K., et al. (2008). Closed chromatin architecture is induced by an RNA duplex targeting the HIV-1 promoter region. *J. Biol. Chem.*, **283**, 23353–23363.
31. Fuks, F., et al. (2001). Dnmt3a binds deacetylases and is recruited by a sequence-specific repressor to silence transcription. *EMBO J.*, **20**, 2536–2544.
32. Datta, J., et al. (2003). Biochemical fractionation reveals association of DNA methyltransferase (Dnmt) 3b with Dnmt1 and that of Dnmt 3a with a histone H3 methyltransferase and Hdac1. *J. Cell. Biochem.*, **88**, 855–864.
33. Fuks, F., Hurd, P.J., Deplus, R., and Kouzarides, T. (2003). The DNA methyltransferases associate with HP1 and the SUV39H1 histone methyltransferase. *Nucleic Acids Res.*, **31**, 2305–2312.
34. Viré, E., et al. (2006). The Polycomb group protein EZH2 directly controls DNA methylation. *Nature*, **439**, 871–874.
35. Jeffery, L., and Nakielny, S. (2004). Components of the DNA methylation system of chromatin control are RNA-binding proteins. *J. Biol. Chem.*, **279**, 49479–49487.
36. Castanotto, D., et al. (2005). Short hairpin RNA-directed cytosine (CpG) methylation of the RASSF1A gene promoter in HeLa cells. *Mol Ther.*, **12**, 179–183.
37. Mehndiratta, M., et al. (2011). CpG Hypermethylation of the C-myc promoter by dsRNA results in growth suppression. *Mol. Pharm.*, **8**, 2302–2309.
38. Clark, M.B., et al. (2011). The reality of pervasive transcription. *PLoS Biol.*, **9**, e1000625; discussion e1001102.
39. Kapranov, P., et al. (2005). Examples of the complex architecture of the human transcriptome revealed by RACE and high-density tiling arrays. *Genome Res.*, **15**, 987–997.
40. Dahary, D., Elroy-Stein, O., and Sorek, R. (2005). Naturally occurring antisense: transcriptional leakage or real overlap? *Genome Res.*, **15**, 364–368.
41. Lapidot, M., and Pilpel, Y. (2006). Genome-wide natural antisense transcription: coupling its regulation to its different regulatory mechanisms. *EMBO Rep.*, **7**, 1216–1222.

42. Katayama, S., et al. (2005). Antisense transcription in the mammalian transcriptome. *Science*, **309**, 1564–1566.
43. Seila, A.C., Core, L.J., Lis, J.T., and Sharp, P.A. (2009). Divergent transcription: a new feature of active promoters. *Cell Cycle*, **8**, 2557–2564.
44. Kapranov, P., et al. (2007). RNA maps reveal new RNA classes and a possible function for pervasive transcription. *Science*, **316**, 1484–1488.
45. Preker, P., et al. (2008). RNA exosome depletion reveals transcription upstream of active human promoters. *Science*, **322**, 1851–1854.
46. Han, J., Kim, D., and Morris, K.V. (2007). Promoter-associated RNA is required for RNA-directed transcriptional gene silencing in human cells. *Proc. Natl. Acad. Sci. U S A*, **104**, 12422–12427.
47. Schmitz, K.-M., Mayer, C., Postepska, A., and Grummt, I. (2010). Interaction of noncoding RNA with the rDNA promoter mediates recruitment of DNMT3b and silencing of rRNA genes. *Genes Dev.*, **24**, 2264–2269.
48. Martianov, I., et al. (2007). Repression of the human dihydrofolate reductase gene by a non-coding interfering transcript. *Nature*, **445**, 666–670.
49. Sun, B.K., Deaton, A.M., and Lee, J.T. (2006). A transient heterochromatic state in Xist preempts X inactivation choice without RNA stabilization. *Mol. Cell*, **21**, 617–628.
50. Klase, Z., et al. (2007). HIV-1 TAR element is processed by Dicer to yield a viral micro-RNA involved in chromatin remodeling of the viral LTR. *BMC Mol. Biol.*, **8**, 63.
51. Lim, H.G.W., Suzuki, K., Cooper, D.A., and Kelleher, A.D. (2008). Promoter-targeted siRNAs induce gene silencing of simian immunodeficiency virus (SIV) infection in vitro. *Mol. Ther.*, **16**, 565–570.
52. Turunen, M.P., et al. (2009). Efficient regulation of VEGF expression by promoter-targeted lentiviral shRNAs based on epigenetic mechanism: a novel example of epigenetherapy. *Circ. Res.*, **105**, 604–609.
53. Perrone, L., et al. (2010). Inhibition of TXNIP expression in vivo blocks early pathologies of diabetic retinopathy. *Cell Death Dis.*, **1**, e65.
54. Zhou, J., et al. (2011). Transcriptional gene silencing of HPV16 E6/E7 induces growth inhibition via apoptosis in vitro and in vivo. *Gynecol. Oncol.*, **124**, 296–302.

55. Hong, D., et al. (2009). Gene silencing of HPV16 E6/E7 induced by promoter-targeting siRNA in SiHa cells. *Br. J. Cancer*, **101**, 1798–1804.
56. Palanichamy, J.K., et al. (2010). Silencing of integrated human papillomavirus-16 oncogenes by small interfering RNA-mediated heterochromatization. *Mol. Cancer Ther.*, **9**, 2114–2122.
57. Murayama, A., et al. (2006). A specific CpG site demethylation in the human interleukin 2 gene promoter is an epigenetic memory. *EMBO J.*, **25**, 1081–1092.
58. Zhang, M.-X., et al. (2005). Regulation of endothelial nitric oxide synthase by small RNA. *Proc. Natl. Acad. Sci. U S A*, **102**, 16967–16972.
59. Bühler, M., Mohn, F., Stalder, L., and Mühlemann, O. (2005). Transcriptional silencing of nonsense codon-containing immunoglobulin minigenes. *Mol. Cell*, **18**, 307–317.
60. Gonzalez, S., Pisano, D.G., and Serrano, M. (2008). Mechanistic principles of chromatin remodeling guided by siRNAs and miRNAs. *Cell Cycle*, **7**, 2601–2608.
61. Gonzalez, S., et al. (2006). Oncogenic activity of Cdc6 through repression of the INK4/ARF locus. *Nature*, **440**, 702–706.
62. Wang, X., et al. (2007). The proximal GC-rich region of p16(INK4a) gene promoter plays a role in its transcriptional regulation. *Mol. Cell. Biochem.*, **301**, 259–266.
63. Feng, X.-Z., et al. (2008). Investigation of transcriptional gene silencing and mechanism induced by shRNAs targeted to RUNX3 in vitro. *World J. Gastroenterol.*, **14**, 3006–3014.
64. Taft, R.J., Hawkins, P.G., Mattick, J.S., and Morris, K.V. (2011). The relationship between transcription initiation RNAs and CCCTC-binding factor (CTCF) localization. *Epigenet. Chromatin*, **4**, 13.
65. Younger, S.T., and Corey, D.R. (2011). Transcriptional regulation by miRNA mimics that target sequences downstream of gene termini. *Mol. Biosyst.*, **7**, 2383–2388.
66. Alló, M., et al. (2009). Control of alternative splicing through siRNA-mediated transcriptional gene silencing. *Nat. Struct. Mol. Biol.*, **16**, 717–724.
67. Kim, D.H., Saetrom, P., Snøve, O., and Rossi, J.J. (2008). MicroRNA-directed transcriptional gene silencing in mammalian cells. *Proc. Natl. Acad. Sci. U S A*, **105**, 16230–16235.
68. Tan, Y., et al. (2009). Transcriptional inhibiton of Hoxd4 expression by miRNA-10a in human breast cancer cells. *BMC Mol. Biol.*, **10**, 12.

69. Benhamed, M., et al. (2012). Senescence is an endogenous trigger for microRNA-directed transcriptional gene silencing in human cells. *Nat. Cell Biol.*, **14**, 266–275.
70. Dean, A. (2011). In the loop: long range chromatin interactions and gene regulation. *Brief Funct. Genomics*, **10**, 3–10.
71. Yue, X., et al. (2010). Transcriptional regulation by small RNAs at sequences downstream from 3′ gene termini. *Nat. Chem. Biol.*, **6**, 621–629.
72. Holstege, F.C., Fiedler, U., and Timmers, H.T. (1997). Three transitions in the RNA polymerase II transcription complex during initiation. *EMBO J.*, **16**, 7468–7480.
73. Milne, L., Xu, Y., Perrin, D.M., and Sigman, D.S. (2000). An approach to gene-specific transcription inhibition using oligonucleotides complementary to the template strand of the open complex. *Proc. Natl. Acad. Sci. U S A*, **97**, 3136–3141.
74. Janowski, B.A., et al. (2005). Inhibiting gene expression at transcription start sites in chromosomal DNA with antigene RNAs. *Nat. Chem. Biol.*, **1**, 216–222.
75. Janowski, B.A., et al. (2005). Inhibiting gene expression at transcription start sites in chromosomal DNA with antigene RNAs. *Nat. Chem. Biol.*, **1**, 216–222.
76. Janowski, B.A., et al. (2006). Involvement of AGO1 and AGO2 in mammalian transcriptional silencing. *Nat. Struct. Mol. Biol.*, **13**, 787–792.
77. Janowski, B.A., et al. (2005). Inhibiting gene expression at transcription start sites in chromosomal DNA with antigene RNAs. *Nat. Chem. Biol.*, **1**, 216–222.
78. Janowski, B.A., et al. (2005). Inhibiting transcription of chromosomal DNA with antigene peptide nucleic acids. *Nat. Chem. Biol.*, **1**, 210–215.
79. Beane, R.L., et al. (2007). Inhibiting gene expression with locked nucleic acids (LNAs) that target chromosomal DNA. *Biochemistry*, **46**, 7572–7580.
80. Beane, R., et al. (2008). Recognition of chromosomal DNA inside cells by locked nucleic acids. *Biochemistry*, **47**, 13147–13149.
81. Watts, J.K., et al. (2010). Effect of chemical modifications on modulation of gene expression by duplex antigene RNAs that are complementary to non-coding transcripts at gene promoters. *Nucleic Acids Res.*, **38**, 5242–5259.

82. Napoli, S., et al. (2009). Promoter-specific transcriptional interference and c-myc gene silencing by siRNAs in human cells. *EMBO J.*, **28**, 1708–1719.

83. Schwartz, J.C., et al. (2008). Antisense transcripts are targets for activating small RNAs. *Nat. Struct. Mol. Biol.*, **15**, 842–848.

84. Janowski, B.A., et al. (2005). Inhibiting transcription of chromosomal DNA with antigene peptide nucleic acids. *Nat. Chem. Biol.*, **1**, 210–215.

85. Chu, Y., et al. (2010). Involvement of argonaute proteins in gene silencing and activation by RNAs complementary to a non-coding transcript at the progesterone receptor promoter. *Nucleic Acids Res.*, **38**, 7736–7748.

86. Jiang, G., et al. (2012). Small RNAs targeting transcription start site induce heparanase silencing through interference with transcription initiation in human cancer cells. *PLoS ONE*, **7**, e31379.

87. Li, L.-C., et al. (2006). Small dsRNAs induce transcriptional activation in human cells. *Proc. Natl. Acad. Sci. U S A*, **103**, 17337–17342.

88. Morris, K.V., et al. (2008). Bidirectional transcription directs both transcriptional gene activation and suppression in human cells. *PLoS Genet.*, **4**, e1000258.

89. Modarresi, F., et al. (2012). Inhibition of natural antisense transcripts in vivo results in gene-specific transcriptional upregulation. *Nat. Biotechnol.*, **30**(5), 453–459.

90. Mao, Q., et al. (2008). Up-regulation of E-cadherin by small activating RNA inhibits cell invasion and migration in 5637 human bladder cancer cells. *Biochem. Biophys. Res. Commun.*, **375**, 566–570.

91. Huang, V., et al. (2010). RNAa is conserved in mammalian cells. *PLoS ONE*, **5**, e8848.

92. Junxia, W., et al. (2010). Double strand RNA-guided endogeneous E-cadherin up-regulation induces the apoptosis and inhibits proliferation of breast carcinoma cells in vitro and in vivo. *Cancer Sci.*, **101**, 1790–1796.

93. Janowski, B.A., et al. (2007). Activating gene expression in mammalian cells with promoter-targeted duplex RNAs. *Nat. Chem. Biol.*, **3**, 166–173.

94. Matsui, M., et al. (2010). Activation of LDL receptor expression by small RNAs complementary to a noncoding transcript that overlaps the LDLR promoter. *Chem. Biol.*, **17**, 1344–1355.

95. Chen, R., et al. (2011). Up-regulation of VEGF by small activator RNA in human corpus cavernosum smooth muscle cells. *J. Sex Med.*, **8**, 2773–2780.
96. Matilainen, J.M., et al. (2010). Primary effect of 1α,25(OH)2D3 on IL-10 expression in monocytes is short-term down-regulation. *Biochim. Biophys. Acta*, **1803**, 1276–1286.
97. Majid, S., et al. (2010). MicroRNA-205-directed transcriptional activation of tumor suppressor genes in prostate cancer. *Cancer*, **116**, 5637–5649.
98. Place, R.F., et al. (2008). MicroRNA-373 induces expression of genes with complementary promoter sequences. *Proc. Natl. Acad. Sci. U S A*, **105**, 1608–1613.
99. Hawkins, P.G., and Morris, K.V. (2010). Transcriptional regulation of Oct4 by a long non-coding RNA antisense to Oct4-pseudogene 5. *Transcription*, **1**, 165–175.
100. Huang, V., et al. (2012). Upregulation of cyclin B1 by miRNA and its implications in cancer. *Nucleic Acids Res.*, **40**, 1695–1707.
101. Wang, J., et al. (2010). Prognostic value and function of KLF4 in prostate cancer: RNAa and vector-mediated overexpression identify KLF4 as an inhibitor of tumor cell growth and migration. *Cancer Res.*, **70**, 10182–10191.
102. Wang, X., et al. (2012). Induction of NANOG expression by targeting promoter sequence with small activating RNA antagonizes retinoic acid-induced differentiation. *Biochem. J.*, **443**, 821–828.
103. Qin, Q., et al. (2012). RNAa-mediated overexpression of WT1 induces apoptosis in HepG2 cells. *World J. Surg. Oncol.*, **10**, 11.
104. Panning, B., and Jaenisch, R. (1998). RNA and the epigenetic regulation of X chromosome inactivation. *Cell*, **93**, 305–308.
105. Rougeulle, C., and Heard, E. (2002). Antisense RNA in imprinting: spreading silence through air. *Trends Genet.*, **18**, 434–437.
106. Qureshi, I.A., Mattick, J.S., and Mehler, M.F. (2010). Long non-coding RNAs in nervous system function and disease. *Brain Res.*, **1338**, 20–35.
107. Guttman, M., et al. (2011). lincRNAs act in the circuitry controlling pluripotency and differentiation. *Nature*, **477**, 295–300.
108. Huarte, M., et al. (2010). A large intergenic noncoding RNA induced by p53 mediates global gene repression in the p53 response. *Cell*, **142**, 409–419.

109. Wang, K.C., and Chang, H.Y. (2011). Molecular mechanisms of long noncoding RNAs. *Mol. Cell*, **43**, 904–914.

110. Yu, W., et al. (2008). Epigenetic silencing of tumour suppressor gene p15 by its antisense RNA. *Nature*, **451**, 202–206.

111. Tufarelli, C., et al. (2003). Transcription of antisense RNA leading to gene silencing and methylation as a novel cause of human genetic disease. *Nat. Genet.*, **34**, 157–165.

112. Turner, A.-M.W., and Morris, K.V. (2010). Controlling transcription with noncoding RNAs in mammalian cells. *BioTechniques*, **48**, ix–xvi.

113. Janowski, B.A., Hu, J., and Corey, D.R. (2006). Silencing gene expression by targeting chromosomal DNA with antigene peptide nucleic acids and duplex RNAs. *Nat. Prot.*, **1**, 436–443.

114. Reynolds, A., et al. (2004). Rational siRNA design for RNA interference. *Nat. Biotechnol.*, **22**, 326–330.

115. Sherman, M.P., and Greene, W.C. (2002). Slipping through the door: HIV entry into the nucleus. *Microbes Infect.*, **4**, 67–73.

116. Morris, K.V. (2005). siRNA-mediated transcriptional gene silencing: the potential mechanism and a possible role in the histone code. *Cell. Mol. Life Sci.*, **62**, 3057–3066.

117. Hwang, H.-W., Wentzel, E.A., and Mendell, J.T. (2007). A hexanucleotide element directs microRNA nuclear import. *Science*, **315**, 97–100.

118. Liao, J.-Y., et al. (2010). Deep sequencing of human nuclear and cytoplasmic small RNAs reveals an unexpectedly complex subcellular distribution of miRNAs and tRNA 3′ trailers. *PLoS ONE*, **5**, e10563.

119. Braasch, D.A., et al. (2003). RNA interference in mammalian cells by chemically-modified RNA. *Biochemistry*, **42**, 7967–7975.

120. Moreno, P.M.D., et al. (2009). A synthetic snRNA m3G-CAP enhances nuclear delivery of exogenous proteins and nucleic acids. *Nucleic Acids Res.*, **37**, 1925–1935.

121. Jackson, A.L., et al. (2006). Widespread siRNA "off-target" transcript silencing mediated by seed region sequence complementarity. *RNA*, **12**, 1179–1187.

122. Bridge, A.J., et al. (2003). Induction of an interferon response by RNAi vectors in mammalian cells. *Nat. Genet.*, **34**, 263–264.

123. Sledz, C.A., et al. (2003). Activation of the interferon system by short-interfering RNAs. *Nat. Cell Biol.*, **5**, 834–839.

124. Elbashir, S.M., et al. (2001). Duplexes of 21-nucleotide RNAs mediate RNA interference in cultured mammalian cells. *Nature*, **411**, 494–498.

125. Hornung, V., et al. (2005). Sequence-specific potent induction of IFN-alpha by short interfering RNA in plasmacytoid dendritic cells through TLR7. *Nat. Med.*, **11**, 263–270.

126. Judge, A.D., et al. (2005). Sequence-dependent stimulation of the mammalian innate immune response by synthetic siRNA. *Nat. Biotechnol.*, **23**, 457–462.

127. Kim, D.-H., et al. (2004). Interferon induction by siRNAs and ssRNAs synthesized by phage polymerase. *Nat. Biotechnol.*, **22**, 321–325.

128. Weinberg, M.S., et al. (2007). An RNA targeted to the HIV-1 LTR promoter modulates indiscriminate off-target gene activation. *Nucleic Acids Res.*, **35**, 7303–7312.

129. Moses, J., Goodchild, A., and Rivory, L.P. (2010). Intended transcriptional silencing with siRNA results in gene repression through sequence-specific off-targeting. *RNA*, **16**, 430–441.

130. Suzuki, K., et al. (2011). Transcriptional gene silencing of HIV-1 through promoter targeted RNA is highly specific. *RNA Biol.*, **8**, 1035–1046.

131. Das, A.T., et al. (2004). Human immunodeficiency virus type 1 escapes from RNA interference-mediated inhibition. *J. Virol.*, **78**, 2601–2605.

132. Boden, D., et al. (2003). Human immunodeficiency virus type 1 escape from RNA interference. *J. Virol.*, **77**, 11531–11535.

133. Westerhout, E.M., et al. (2005). HIV-1 can escape from RNA interference by evolving an alternative structure in its RNA genome. *Nucleic Acids Res.*, **33**, 796–804.

134. von Eije, K.J., ter Brake, O., and Berkhout, B. (2008). Human immunodeficiency virus type 1 escape is restricted when conserved genome sequences are targeted by RNA interference. *J. Virol.*, **82**, 2895–2903.

135. Li, M.-J., et al. (2005). Long-term inhibition of HIV-1 infection in primary hematopoietic cells by lentiviral vector delivery of a triple combination of anti-HIV shRNA, anti-CCR5 ribozyme, and a nucleolar-localizing TAR decoy. *Mol. Ther.*, **12**, 900–909.

136. Aguilar-Cordova, E., et al. (1994). A sensitive reporter cell line for HIV-1 tat activity, HIV-1 inhibitors, and T cell activation effects. *AIDS Res. Hum. Retroviruses*, **10**, 295–301.

137. Wu, X., and Pandolfi, P.P. (2001). Mouse models for multistep tumorigenesis. *Trends Cell Biol.*, **11**, S2–S9.

138. Pulukuri, S.M.K., Estes, N., Patel, J., and Rao, J.S. (2007). Demethylation-linked activation of urokinase plasminogen activator is involved in progression of prostate cancer. *Cancer Res.*, **67**, 930–939.

139. Ylä-Herttuala, S., and Alitalo, K. (2003). Gene transfer as a tool to induce therapeutic vascular growth. *Nat. Med.*, **9**, 694–701.

140. Grimm D, et al. (2009). Fatality in mice due to oversaturation of cellular microRNA. *Nature*, **441**, 537–41.

Chapter 14

Small RNA Molecules in Antiviral Therapy

Ye Qiu, Xin Ye, Maged Hemida, Mary Zhang, Paul Hanson, and Decheng Yang
Department of Pathology and Laboratory Medicine, University of British Columbia, The Institute for Heart and Lung Health, St. Paul's Hospital, 1081 Burrard Street, Vancouver, BC, V6Z 1Y6, Canada
decheng.yang@hli.ubc.ca

14.1 Introduction

Infectious viral diseases have existed throughout history and continue to be global threats to public health worldwide. In contrast to continually evolving vaccines to prevent viral infection, few direct antiviral therapies have been developed for treatment. The only drugs inhibiting viruses directly are interferon [1] and nucleoside analogs [2]. For patients with viral diseases, drugs against bacteria and parasites will be applied to alleviate the syndromes, while the viruses are actually wiped out by the immune system. That's the reason viruses targeting the immune system, such as human immunodeficiency virus (HIV), are so difficult to kill. It seems that prevention is more effective and important than treatment for viral

RNA Nanotechnology
Edited by Bin Wang

ISBN 978-981-4411-64-6 (Hardcover), 978-981-4411-65-3 (eBook)
www.panstanford.com

diseases. Unfortunately, long-term vaccines are unavailable for most kinds of viruses due to the high mutation rate of viruses [3, 4].

The life cycle of a virus generally includes steps of attachment, internalization, replication, assembly, and release. Theoretically, drugs stopping any step in the viral life cycle can be used in antiviral therapy. However, as other intracellular parasites dependent on host cellular machinery, viruses exploit many host components and processes for the sake of their own replication. Hence, ideal antiviral drugs should target viruses with high specificity so that these drugs can inhibit viruses with little harm to the host as well. As ribonucleic acid (RNA) technology is being rapidly developed, some small RNA molecules are found to specifically bind to certain nucleotide sequences or proteins and inhibit the function of targeted components. These small molecules may be promising candidates for antiviral drugs.

In this chapter, we will introduce several types of small RNA molecules that are potential candidates to be developed into antiviral therapies, including small interfering RNAs (siRNAs), microRNAs (miRNAs), RNA aptamers, and ribozymes (Rzs). Our discussion will focus on the key steps for antiviral small RNA design, which include target selection (Fig. 14.1), RNA molecule design, candidate screening, functional modification, and drug delivery (Table 14.1).

14.2 RNAi-Based Small RNAs

14.2.1 *Introduction to RNAi*

RNA interference (RNAi) was originally recognized as a post-transcription regulation process utilized by living cells to modulate gene expression [5]. Though some exceptions have been found now, in most cases, RNAi down-regulates expression levels of targeted genes via specific degradation or translation inhibition of messenger RNAs (mRNAs) by complementary double-stranded RNA (dsRNA) [6, 7]. Therefore, the process was named post-transcriptional gene silencing (PTGS) when it was firstly applied in plants [8].

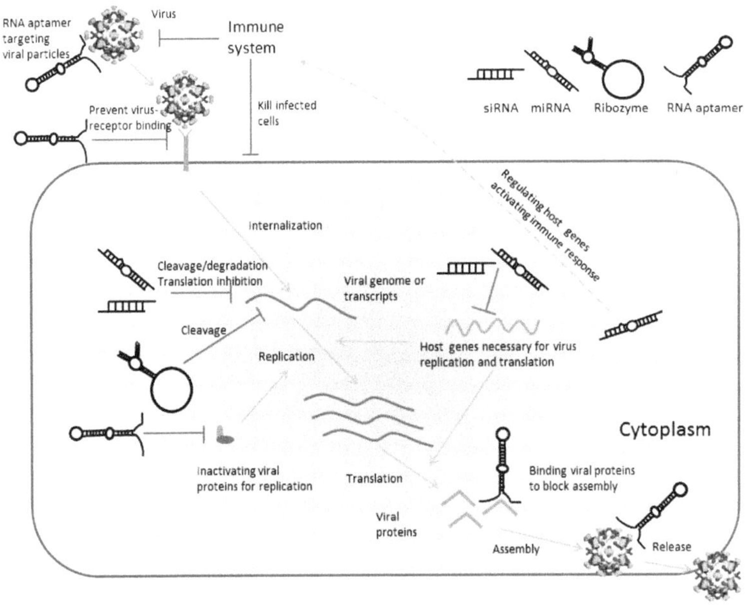

Figure 14.1 Antiviral strategies of small RNA molecules. Different components involved in viral infection and replication can be utilized as targets of antiviral small RNAs. Viral particles or individual proteins can be inactivated by RNA aptamers. The aptamer binding may also block viral attachment to viral receptors. These small RNA molecules can be transfected into cells or expressed intracellularly. siRNAs and Rzs can cleave or degrade genomic RNAs or transcripts of the virus to inhibit viral replication. miRNAs inhibit viral replication by degradation of viral RNA and/or suppression of translation. All these small RNAs can also target host cellular RNAs regulating the viral life cycle; particularly, certain miRNAs are able to inhibit virus replication by targeting host genes to enhance the immune response. The dashed line indicates the regulation is through multiple mediators.

Expression of both viral genes and some host genes is quite critical for amplification and survival of viruses. Accumulating evidence has indicated that RNAi is closely related to the protective effect against viruses in mammalian cells [9–12]. Thus, great attention is being drawn to the application of RNAi in antiviral therapy development [12, 13].

Table 14.1 Design of antiviral small RNAs

Small RNA	Target choice	Candidate screening	Functional modification	Delivery
siRNA	Viral genome (RNA viruses) Viral transcripts (DNA viruses) Host genes	Screen from database In vitro screening with proper reporters	Base modifications Addition of photocleavable protecting groups Conjugation with Rzs	Direct transfection Vector-mediated delivery Virus-mediated delivery Directed delivery with ligands
miRNA/AmiRNA	Viral genome (RNA viruses) Viral transcripts (DNA viruses) Host genes Immune response	Screen from database Prediction In vitro screening with proper reporters	Base modifications Addition of photocleavable protecting groups	Direct transfection shRNA-mediated delivery Directed delivery with ligands
RNA aptamer	Viral proteins Whole viruses Receptors	SELEX	Conjugation with siRNAs or Rzs	Direct addition Injection Direct transfection
Rz	Viral genome (RNA viruses) Viral transcripts (DNA viruses)	Design and in vitro screening with proper reporters	Multimerization with cleavable linkers DNA–RNA chimeric Rzs Conjugation with siRNAs or RNA aptamers	Direct transfection Vector-mediated delivery

Abbreviations: DNA, deoxyribonucleic acid; AmiRNA, artificial miRNA; SELEX, systematic evolution of ligands by exponential enrichment; shRNA, short hairpin RNA.

14.2.2 *Target Choice for RNAi-Based Antiviral Therapy*

To use RNAi as antiviral therapy, the first step is to choose the optimum target for such an RNAi molecule. Up to now, three kinds of targets are always considered:

(1) Viral RNA. RNA is critical for viruses to process their life cycle, since it is the chemical base of the viral genome (RNA viruses), intermediates of viral replication (some DNA viruses, such as hepatitis B virus (HBV) [14, 15]), or mRNAs for viral protein expression. Hence, silencing viral RNAs by RNAi is aimed to inhibit viruses directly.
(2) Host genes related to virus life cycles. Some host components are extremely important for viral infection and survival, such as receptors, transcription factors, and translation machineries. Though RNAi targeting these auxiliary factors will not eliminate the virus infection directly, it may result in dramatic inactivation of viruses by interfering with their entry and replication.
(3) Host genes related to inflammation and immune responses can also serve as targets of therapeutic RNAi, since these genes are critical in the antiviral process or syndromes related to viral infection.

14.2.3 *RNAi Mediators for Antiviral Therapy*

Selecting the type of molecules mediating RNAi is also a critical issue in developing antiviral therapy based on the RNAi process. There are two major groups of RNAi mediators that can be used as antiviral therapy, siRNA and miRNA. The mechanism of how these two groups of small RNAs work has already been introduced in other chapters, and for a better understanding of their antiviral effects, here we just mention some differences between these two kinds of small RNAs. Both groups of these small RNAs are eventually processed to an oligo RNA duplex in cells. One strand of the duplex (termed the "guide strand") is loaded in the RNA-induced silencing complex (RISC), which defines the specificity of RNAi [16], while the other strand (termed the "passenger stand") is degraded [17, 18]. siRNAs, originally generated from cleavage of a long dsRNA or a transgene-expressed shRNA, work with complementary regions ideally paired with their targets. The target site of siRNA can be any region on targeted RNAs, though those targeting on the coding region (especially the nonstructural coding region) will have higher antiviral activity [19–22]. Hence, siRNAs can achieve high specificity in antiviral activity, but they have low tolerance to mutations.

miRNAs, which are endogenous gene regulators, have their target specificity in the 7nt seed region of their 5′ end region, while being partially base paired in the rest of their complementary region is allowed. This feature of miRNAs provides higher tolerance to mutations than siRNAs [23] and also allows a certain miRNA to target multiple sites. Therefore, a variety of natural miRNAs from mammalian species can be used directly to target viral RNAs [24–26]. In addition, AmiRNAs, mimics of endogenous miRNA [27] with specific complementary sequences, can also be designed for an aimed target. An miRNA usually binds to the 3′ untranslated region (3′ UTR) of its target gene, and this target site may be in multicopy [28–31]. Now in most cases of antiviral therapy development, siRNAs are preferred, perhaps because their simple structure and various choices of target sites. While miRNAs are always investigated as drug targets, considering their roles in viral infection naturally, they're still promising candidates for antiviral drugs due to natural multifunctional features.

14.2.4 *Screening for Antiviral RNAi Mediators*

The big challenge faced by RNAi-based antiviral therapy is how to select the right siRNAs or miRNAs that can effectively inactivate the targets. Both experimental screening in vitro and bioinformatic prediction are currently used to search for potential RNAi mediators. A candidate pool of siRNAs can always be synthesized directly or expressed by shRNA plasmids based on target sequences. However, siRNAs in the pool do not always work well. Actually, only one of five siRNAs selected for targeting a region of a gene induces efficient gene silencing [32, 33]. Even though computer-based siRNA designing tools are available, it is still necessary to conduct in vitro screening to select the most efficient ones. Zhang et al. implemented a dual fluorescence reporter system to screen their candidate siRNAs for the HBV X gene (HBx) in an siRNA library [34]. The reporter plasmid expressed two fluorescence groups separately. One was enhanced green fluorescence protein (EGFP), which served as an internal control, and the other one was Ds-Red conjugated with HBx. They cotransfected the reporter plasmid and siRNAs into LinX cells and evaluated the efficiency of siRNAs by the intensity of Ds-Red. Finally, they got one with the highest silencing efficiency and confirmed its antiviral effect later [34]. Nevertheless, currently, libraries of commercial siRNAs with high efficiency are available from some biotechnological companies. Thus, researchers do not need to establish an siRNA library themselves.

Currently there are different databases for endogenous miRNAs of different species available online (a recommended database is Open Biosystems: http://www.openbiosystems.com/). Meanwhile, target prediction tools have been developed based on these databases for searching potential effective miRNAs. These miRNA prediction programs available online (Table 14.2) can give an index of efficiency of certain miRNA–target hybrids. Though different algorithms are used by different programs, generally they are all based on several parameters: (1) base-pairing pattern, Watson–Crick pair, or other pairing patterns, in which the Watson–Crick pair always enhances the silencing efficiency; (2) thermodynamic stability of the miRNA-targeted RNA hybrid, which can be measured by the melting temperature, T_m; (3) conservation level of the target sequences; and (4) number of target sites; generally the more the better [35–38]. Simply by referring to the results from different programs, the ones given high scores by most programs are probably the most effective miRNAs. In this way, the troublesome pool screening can be circumvented. However, the targeting effect still has to be experimentally validated by the luciferase assay or other methods.

14.2.5 *Application of RNAi-Based Small RNAs in Antiviral Therapy*

RNAi-based small RNAs are now widely used to develop therapy against different viruses. In our laboratory we study the effect of different RNAi molecules on the coxsackievirus B3 (CVB3) infection and replication. Here, we'll look at CVB3 as an example. CVB3 is a

Table 14.2 Online programs for miRNA target prediction

Name	URL	Input parameters	Application	Reference
DIANA- microT	http://diana.pcbi.upenn.edu/cgi-bin/micro_t.cgi	miRNA sequence & supposed target RNA sequence	Evaluate the RNAi efficiency of a known miRNA on a known target RNA	[102]
rna22	http://cbcsrv.watson.ibm.com/rna22.html			[103]
miRanda	http://www.microrna.org	1. miRNA sequence & target species 2. Target mRNA & species for miRNA	Get potential targets when the miRNA sequence is known or get candidate miRNAs when the targeted mRNA sequence is known	[104]
MirTarget2	http://mirdb.org	1. miRNA name & species 2. Target gene name & species	Get potential targets of a natural miRNA or get miRNAs for a certain gene	[105, 106]
PicTar	http://pictar.mdc-berlin.de	1. miRNA name & species 2. Target gene ID	Get potential targets of an miRNA or get miRNAs for a certain gene when the gene ID is known	[107]
RNAhybrid	http://bibiserv.techfak.uni-bielefeld.de/rnahybrid	1. miRNA sequence 2. Target RNA sequence & parameters for miRNA screening	Get potential targets when the miRNA sequence is known or get candidate miRNAs with special requirements when the targeted mRNA sequence is known	[108]
TargetScan	http://www.targetscan.org/	1. Species & gene symbol 2. Species & miRNA name	Get potential targets for a certain miRNA or get potential miRNAs targeting a certain gene when the conservation level in different species is considered	[109]

member of enteroviruses in the *Picornaviridae* family. Its genome is ~7.4 kb long, containing a single, long, open reading frame (encoding 11 proteins) flanked by 5′ and 3′ UTRs [39], which provide candidate regions for antiviral siRNA and miRNA.

The coding region of the CVB3 genome encodes some essential enzymes for viral replication and structure proteins necessary for viral assemblage, which are already targeted by a large amount of siRNAs designed in different scientific groups. For instance, our group designed an siRNA targeting the viral protease 2A of CVB3, achieving 92% inhibition of CVB3 replication [40]. Jin et al. designed 11 siRNAs targeting 7 distinct regions of the CVB3 genome, including structure proteins VP1, VP2, and VP3; proteases 2A, 2C, and 3C; and RNA-dependent RNA polymerase 3D. All the siRNAs were individually transfected into HeLa cells, and then the cells were challenged by CVB3. After evaluation of viral replication by different measures, five siRNAs were proved to be highly efficient in inhibiting viral replication, though they vary in efficiency, even as some of them target the same gene [41]. Park et al. designed six candidate siRNAs targeting 3D and VP1. According to the result of the tests, three prevented cell death from CVB3 infection in Cos-7 cells, among which two also exerted strong antiviral effects in viral replication in vivo [42].

As mentioned above, siRNAs have low tolerance to mutations. However, the mutation rate of viral genomes, especially genes encoding structure proteins, is usually high [121], which will help the virus escape from siRNA. To solve this problem, two approaches are being attempted: (1) using multiple distinct siRNAs or an siRNA pool to target more than one target genes of the virus [43, 44] and (2) using siRNA targeting conserved *cis*-acting replication elements (CREs) [45].

Besides the CVB3 genome, host cellular genes critical for CVB3 infection can also serve as the targets of anti-CVB3 siRNAs. One of the advantages of using host genes as siRNA targets is that low rates of mutation will occur. The viral receptor coxsackievirus and adenovirus receptor (CAR) is the most commonly used target for anti-CVB3 siRNAs. It was reported that silencing CAR led to ~60% reduction of CVB3 particle formation in HeLa cells [46].

Fewer cases of anti-CVB3 miRNAs have been reported. Our laboratory constructed three short hairpin AmiRNAs (AmiR-1, AmiR-2, and AmiR-3) targeting the stem loop of the 3′ UTR of CVB3 with mismatches at the middle region of the target. Two of them (AmiR-1 and AmiR-2) reduced the viral titers ~hundredfold in CVB3-infected HeLa cells [47]. In research of other viruses, more cases have been reported. hsa-miR-125a-5p was shown to interfere with HBV translation and down-regulate the expression of the surface antigen in cultured hepatic cells [24]. Ectopic expression of hsa-miR-29a, which targets the *nef* gene of HIV-1, was also reported to result in repression of Nef protein levels and a reduction in viral levels [48].

Although miRNAs are not quite as frequently used as antiviral therapy themselves, they can also be used as the target of antiviral drugs. For example, miRNA-122, a liver-specific miRNA, is required for hepatitis C virus (HCV) replication in hepatocytes by its binding to the 5′ UTR of HCV [49–51], which may serve as a potential target for anti-HCV therapy. It has been already reported that miR-122 antagonism had a potent antiviral effect against HCV genotypes 1–6 [26]. In addition, many other cases have been reported that miRNAs play critical roles in infection of various viruses, including Epstein-Barr virus (EBV), HIV-1, and West Nile virus [52–54], indicating their potential for developing antiviral therapies via these miRNAs.

14.2.6 *Immune Response–Related Antiviral siRNAs and miRNAs*

Immune response is a critical antiviral process. Both miRNAs and siRNAs are reported to have an influence on immune response [55, 56]. Hence, the two small RNAs can also present an antiviral effect by targeting immune components. For instance, hsa-miR-155 is reported to promote expression of inflammatory cytokines and the interferon response in primary macrophages and dendritic cells via repression of its direct targets, the suppressor of cytokine signaling 1 (SOCS1), and SH2 domain-containing inositol phosphatase (SHIP) [57, 58]. Treating with hsa-miR-155 in the early stage of viral infection may strengthen the innate immune system against viruses. Another miRNA has-miR-146a targeting tumor necrosis factor

(TNF) receptor–associated factor 6 (TRAF6) and interleukin-1 receptor-associated kinase 1 (IRAK1) [59, 60] serves as a negative feedback regulator of the TLR/NF-kB pathway [55] and can prevent an exaggerated immune response in the late stage of viral infection to protect healthy tissues.

In addition to specifically targeting immune components, some siRNAs are also able to induce immune response by a nonspecific effect. Intracellularly expressed siRNAs targeting the major capsid protein (MCP) gene of red sea bream iridovirus (RSIV), the hirame rhabdovirus (HIRRV) genome, and even the beta-galactosidase gene (miR-LacZ) induce an antiviral response to RSIV and HIRRV by activating interferon (IFN)-related pathways [61]. William's group introduced an miRNA-like nonpairing uridine bulge in the passenger strand of an siRNA targeting the semliki forest virus (SFV) genome to increase its immunostimulatory activity. Hence, this siRNA shows dual antiviral effects, blocking SFV replication by RNAi and inducing immune response to kill infected cells [56].

14.3 RNA Aptamers

14.3.1 *Introduction of Antiviral RNA Aptamers*

RNA aptamers are small RNA molecules that can bind to a specific target molecule [62]. The strategy to use RNA aptamers in antiviral therapy is simply based on a common thought—blocking functional components of the virus, just like antibodies, which are already widely used in antiviral research and treatment. However, antibodies have several disadvantages when served as antiviral drugs: (1) production of antibodies for a certain antigen takes a long time, while most viruses are highly mutable, and hence antibody development can't always compete against viruses, and (2) as exogenous proteins, manufactured antibodies always induce an immune response, which will inactivate the drugs or even cause harmful side effects. Generally, aptamers can be peptides, small RNAs or DNAs. Perhaps RNA aptamers are the best choice, considering the following reasons: (1) compared with peptide aptamers, RNA aptamers have even lower immunogenicity and will hardly activate immune response; (2) RNA aptamers are easier to be synthesized than peptides; and (3) like peptides, RNA can be expressed and released in cells, while DNA can't, and RNA is thought to be more biochemically active than DNA. Despite the advantages, RNA's instability in vivo may be the biggest problem for the application of RNA aptamers in antiviral therapy, and some attempts have been made to overcome it, which will be mentioned later. Like antibodies, the target of an antiviral RNA aptamer can be any of the viral or host components necessary for viral infection, such as viral proteins, whole viruses, and receptors.

14.3.2 *Screening for Antiviral RNA Aptamers*

Unlike other therapeutic RNAs, RNA aptamers' function depends on their secondary structure rather than the sequence [63]. Thus, it is more difficult to design an RNA aptamer than to design an siRNA. The RNA aptamers can be screened from a library of randomized molecules through the so-called "SELEX," a high-flux screening technique involving repeated rounds of partitioning and amplification [64, 65].

As a generally used technique, SELEX screening may be conducted in different processes by different laboratories, but the core principle remains similar. Here we take the process of Kumar's group as an example. In this SELEX screening, an RNA random pool was established, in which the RNAs contained a central domain of a randomized sequence of tens of nucleotides flanked by the 5′ and 3′ regions of invariable sequences [66]. The RNA pool was obtained from a corresponding DNA library via in vitro transcription. Then the RNA pool was sent into several cycles of selection. In each cycle, the RNAs were coincubated with the target (viral proteins or the whole virus) and nonspecific competitors (usually total tRNAs of *Escherichia coli* (*E. coli*)). The RNAs binding to the target were selected by filtration or the affinity process, and some laboratories also used target-conjugated beads to increase the specificity [67]. The selected RNAs were reverse-transcribed to complementary DNAs (cDNAs) and amplified by polymerase chain reaction (PCR). Then the cDNAs were again transcribed into RNAs for the next cycle of selection. During these selections, concentration of the

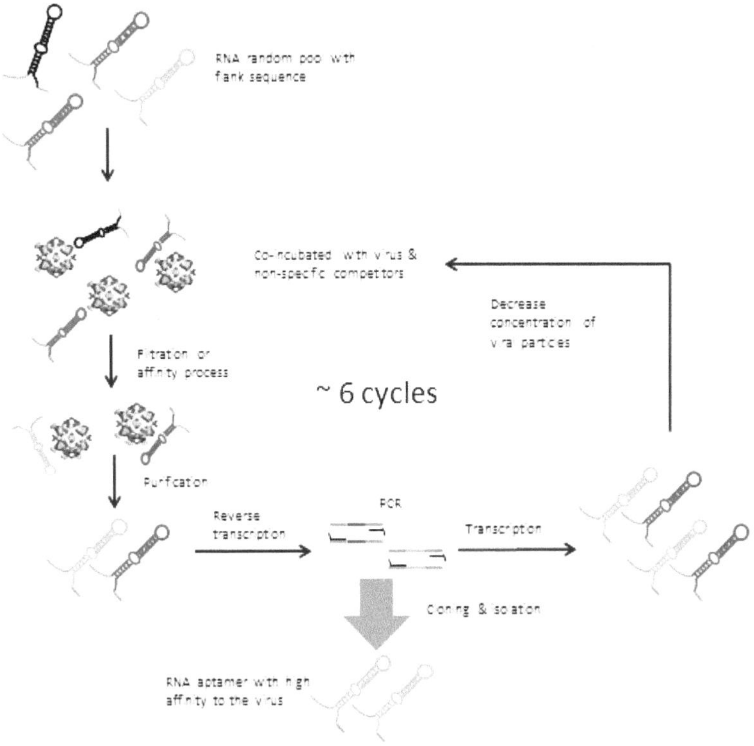

Figure 14.2 The process of SELEX to screen aptamers targeting a specific virus [63].

targets was decreased cycle by cycle in order to get high-affinity RNA aptamers. After six cycles of selection, the final products were reverse-transcribed into cDNAs and then cloned into vectors to isolate individual plasmids expressing the RNA aptamers via bacterial transformation [66, 68, 69] (Fig. 14.2).

The RNA aptamers finally acquired from SELEX screening always display very high specificities (molecular discrimination as much as 10,000-fold) and target affinities (K_d at subnanomolar levels), which are comparable to the affinities achieved by antibodies for antigens [66]. These RNA aptamers are ready for antiviral effect evaluation.

14.3.3 *Application of RNA Aptamers in Antiviral Research*

Via the process of SELEX screening, many RNA aptamers were developed against viral proteins or whole viruses, such as the Tat protein of HIV [70], the NS3 protein of HCV [68], and whole human cytomegalovirus [66].

Since RNA aptamers directly target viral proteins, they can be applied for almost all kinds of viruses. Some RNA aptamers have been designed that specifically bind to a conserved region of the HIV-1 envelope glycoprotein, gp120, and thereby neutralized a broad spectrum of clinical HIV-1 isolates [63, 71–74]. Actually, to enhance the antiviral effect, the aptamer was usually merged with anti-HIV siRNA [75], and these designs worked well in several different HIV-sensitive cell lines. However, as drugs are aimed to prevent acquiring HIV infection through unprotected vaginal or rectal sex [76], the aptamer should keep its stability to nucleases in those environments. By chemical modification of the aptamer with 2-O-Me-modified nucleotides, this issue has been successfully resolved [77].

Interestingly, it is not only viral proteins that can serve as targets of RNA aptamers, but also viral genomes. A viral genome, especially an RNA-based genome, often forms a secondary structure for certain functions, which makes it possible for aptamers to target. Internal ribosome entry sites (IRESs) are a good example of viral RNA with a secondary structure. It has been used as a target for screening aptamers for anti-HCV studies. Among the selected candidates one aptamer targeting the IRES domain IIId dramatically inhibited translation of HCV both in vitro and in vivo [78].

14.4 Ribozymes

14.4.1 *Introduction of Antiviral Ribozymes*

Rzs are small RNA molecules (~30–100 nt) that can act as enzymes to cleave and inactivate RNAs. At least 12 groups of Rzs have been defined, which harbor various Rz catalytic sequences (Rz core sequences). Fused with RNA sequences complementary to the target sequences at the 5′ and 3′ ends of the Rz core sequence, Rzs can recognize the target sequences with high specificity [79]. The specifically catalytic feature of modified Rzs makes them possible to serve as antiviral drugs.

Rzs share several critical advantages over other small RNA molecules as therapies: (1) Rzs do not require any cellular factors for their activity, resulting in minimal cellular toxicity and no interferon response [80, 81]; (2) since the target of an Rz is decided by the complementary sequences added artificially, ideally an Rz targeting any sequence close to the cleavage site can be designed and produced; (3) like protein enzymes, Rzs have high catalytic ability and will not be consumed in cleavage reactions, which means that a low dose of therapeutic Rzs is needed in treatment, which will minimize the cost and side effects. However, Rzs also face some limitations. Firstly, though Rzs can be designed to target any sequences, the cleavage sites are limited and should be located near targeted sequences. Secondly, in virtual application, selections of target sequences for highly efficient Rzs are still challenging due to structural constraints that interfere with Rz function [82]. Finally, as with other RNAs, Rzs are susceptible to degradation in cells, and hence it will be difficult for them to be applied against chronic disease [83].

The routine to get a functional antiviral Rz is comparatively easier than that of siRNAs, miRNAs, or aptamers. The first and also most important step is to find the cleavage sites on the targeting sequence. Then a recognition sequence of the Rz can be designed based on the adjacent sequence. To obtain better therapeutic effects, several optimizations are applied in using an Rz for treatment. Some modifications are introduced in the following sections.

14.4.2 *Hammerhead Ribozymes and Multimeric Hammerhead Ribozymes*

Among the 12 groups of Rzs, the hammerhead Rzs draw much attention in therapeutic application [84], and now most Rz-related therapeutics are based on the hammerhead Rz. Hammerhead Rzs are widely used due to their comparatively simple structure (about 30 nt) and their minimal requirements for the cleavage site. Any motif with the dinucleotide sequence UU, UC, or UA can be targeted [85].

Monomeric hammerhead Rzs already show good effects in several antiviral studies. For instance, a monomeric hammerhead Rz targeting position 23 within the C–C chemokine receptor type 5 (CCR5) ORF can inhibit R5-tropic HIV-1 replication up to 70% on day 7 postinfection (p. i.) [86]. It has been evaluated as part of a triple combination gene therapy strategy in two clinical trials [87, 88]. However, in most cases, monomeric hammerhead Rzs are often used to target host genes rather than viral RNAs. This is due to the high rate of mutation of viral RNAs that can help viruses escape from Rzs targeting a single site on viral RNAs. Therefore, the strategy of delivering several Rzs at the same time may be tried, but the delivery efficiency and lasting time of every individual Rz are hardly guaranteed and unified [89], which may weaken the synergistic action. Thus, multimeric hammerhead Rzs (MHRs), in which different Rzs are merged into the same molecule, are considered to target different sites on viral genome [90].

Since the catalytic domain of a hammerhead Rz is small and can establish its active motif independently, hammerhead Rzs with different recognition sequences can be merged together directly and sequentially, even without any linkers between monomers [90]. The merging of the coding DNA can be achieved by both overlapping PCR [91] and normal enzyme digestion and ligation. In most studies, the whole fragment containing different Rz monomers will be inserted into moloney murine leukemia virus-based vectors [92], such as the MGIN vector [93], or lentivirus-based vectors [90]. The delivery will then be conducted by the pseudo-viral particles.

MHRs have been reported to bring a dramatic inhibition to viruses on either the entry or the replication level. Joshi et al. used MHRs targeting the HIV genome to inhibit replication of HIV and reported that even challenged with HIV-1 for as long as 60 days after treatment with their MHRs, a human $CD4^{+}$ T lymphoid cell line MT4, which is supposed to be highly sensitive to HIV-1, showed very little or undetectable virus production [94]. The same group also used the MHR targeting different sites of CCR5, the receptor for HIV-1, to prevent the virus from entry into another human $CD4^{+}$ T lymphoid cell line PM1 and no reverse transcription DNA was detected even during 90 days p.i. [90]. These findings suggest that plasmid-expressed MHRs can keep their function in both high efficiency and long lasting time.

However, up to now, very less in vivo data has been obtained to support MHRs' function, perhaps because no appropriate ways can be utilized to deliver MHRs to target cells in vivo. But MHRs still have quite a potential as candidates for antiviral therapy.

14.4.3 *Chimeric Ribozymes*

As mentioned above, despite many advantages, Rzs still have several shortcomings when used as antiviral therapy, such as short function time and weakened cleavage effect on some targeting sequences. To circumvent these problems, basic Rzs are modified with other components to create chimeric Rzs.

The short function time is a major barrier for clinical application of Rzs. Though plasmids can be used to stably express Rzs in cells, this way seems not so applicable in vivo. To solve this problem, Fukushima et al. designed a chimeric DNA–RNA hammerhead Rz based on the fact that DNAs are much more stable in cells than RNAs. In their design, the catalytic domain and recognition sequence of the modified Rz are kept as ribonucleotides but the noncatalytic residues are replaced with deoxyribonucleotides [95]. The DNA–RNA chimeric Rz targeting the genome of severe acute respiratory syndrome coronavirus (SARS-CoV) was proved to inhibit SARS-CoV dramatically [95]. Though no long-term data about the stability was shown in this study, other reports claimed that the chimeric DNA–RNA hammerhead Rz could enhance catalytic turnover and stability [96]. Considering the great advantage that the chimeric DNA–RNA Rz may benefit its in vivo application, this method is worth further studying in therapeutic researches.

In some cases, the catalytic activity of Rzs may not be strong enough to inactivate viruses for several reasons, which have been described above. Moreover, some RNAs cleaved by Rzs may still retain some activity. To improve the antiviral effect, other components are merged to Rzs to help cleavage or degrade target RNAs thoroughly. siRNAs can be chosen as the assistant components, considering their specificity and that they can degrade the target

RNAs. For example, siRNA-chimeric-Rz constructs targeting the influenza virus matrix protein M1 have been tested to inhibit viral gene expression of influenza A. In this testing, Rz-163, an Rz targeting the 161-CUC-163 site of M1 RNA, was linked to an siRNA targeting the adjacent sequence with a cleavable spacer. The cleavable spacer can be cut by nuclease in the cytoplasm; thus the two components will not influence the function of each other. Both results of gene expression and the cell survival test showed that the siRNA chimeric Rz could inhibit virus amplification and viral-induced cell death more efficiently compared to individual treatment with the siRNA or Rz [97].

14.5 Drug Delivery

Effective drug delivery to the target organ or cell population is one of the major challenges in drug development. The way to deliver a certain small RNA mainly depends on its antiviral mechanism. Some small RNAs that block viral attaching to cell surface receptors or inactivate viruses before cellular entry, such as the majority of RNA aptamers binding to viral proteins or the whole virus, can be introduced without the help of any assistant components. For cultured cells, the small RNAs can be added directly to the culture medium. For animals, the drugs can be injected into the veins directly or with some inert protective components.

For most antiviral small RNAs targeting intracellular components, methods should be found to deliver them across the plasma membrane. For those that can function without any cellular modification, including siRNAs, RNA aptamers and some monomeric ribozymes, they may be transfected into cells by using certain chemical carriers such as liposomes [115], and polyethylene glycol (PEG) [116]. Meanwhile, these small RNAs, together with those that require intracellular modification or stable expression, can also be delivered using DNA vectors. In these cases, the coding sequences of the small RNAs are cloned into certain DNA vectors and then tranfected into target cells for endogenous expression. However, if high efficiency of the delivery is required, especially for in vivo applications, recombinant viral vectors, which can be delivered by pseudo-viral-particle transduction, will be a better choice. Viral vectors are among the most efficient gene transfer vehicles because of the high efficiency of viral infection and the great capacity of the viral genome for carrying exogenous DNA fragments. Numerous different viral vector systems have been derived from murine and human DNA- and RNA-viruses. These include retroviruses (or lentiviruses), adenoviruses, adeno-associated viruses (AAVs), herpes viruses and poxviruses [117].

The most widely used RNA-virus-derived vectors are retroviral vectors. Retroviral vectors are the predominant system in gene therapy, which were utilized in about 40% of clinical gene therapy experimental protocols [117]. The advantages of retroviral vectors include 1) stable expression via integration into the host genome, 2) generation of viral titres sufficient for efficient gene transfer, and 3) infectivity to a broad variety of target cell types. The retroviral vectors derived from the Moloney murine leukaemia virus (MLV) are commonly used for small RNA therapy. For instance, MLV-based vectors MGIN were used in a number of studies to deliver ribozymes targeting various sites within the HIV-1 RNA via pseudo-viral infection [92]. In the construction of MLV-based vectors, the viral RNA is reverse transcribed and inserted into proviral DNA plasmids without viral structural and packaging genes for safe and efficient production of recombinant virus particles. Another popular series of retroviral vectors are lentiviral vectors derived from the human immunodeficiency virus (HIV). The key advantage of lentiviral vectors is that HIV-derived viral particles are able to infect and integrate into nondividing cells which cannot be achieved by MLV-based vectors [122, 123]. siRNAs delivered via shRNA on lentiviral vectors is a commonly used method to get long-term effect [111–114]. Kim et al. constructed a lentiviral plasmid expressing shRNAs targeting the CRE within CVB3 2C and produced recombinant pseudo-lentiviruses by the co-transfection of 293FT cells with lentiviral plasmids, pMDLg/pRRE, pRSV-Rev and pVSV-G. Intraperitoneal injection of mice with these viruses clearly showed a protective effect against viral myocarditis by elimination of CVB3 infection and reducing pro-inflammatory cytokines, such as IL6 and INF-α [118]. Now the third generation of lentiviral system [119] has showed quite favorable efficiency [120] and is commercial available.

In addition to recombinant RNA viruses, some DNA viruses have been also modified for therapy applications. One critical advantage of DNA-virus-derived vectors is their great packaging capacities, which is suitable for delivering long coding sequences such as RNA aptamers and ribozymes. AAV-related transducing system has been proved to be an efficient tool for the transferring of small RNAs. Fechner et al. constructed a plasmid expressing siRNAs targeting CVB3 3D based on pseudotyped AAV2.9 vector and produced the recombinant AAV by the co-transfection of HEK 293T cells with the viral plasmids, p5E18-VD2/9 and pHelper. Finally they successfully reduced CVB3 replication both in vitro and in vivo with this recombinant virus [121].

In some cases, directed delivery for antiviral small RNAs is needed. This kind of delivery can be achieved by conjugating small RNAs with a ligand, which interacts with receptors on certain cells. The ligands can be an antibody, vitamin, short peptide, RNA aptamer, folic acid, etc. [98]. Our laboratory has successfully delivered siRNAs targeting CVB3 2A to HeLa (cancer) cells specifically via interactions between folate, carried by the bacterial phage-29 packaging RNA (pRNA), and the folate receptor almost exclusively expressed on the surface of cancer cells [99]. Another trial for targeted delivery of AmiRNAs was also successful [47]. The detailed mechanism of pRNA-based, targeted delivery was reviewed in other chapters.

Besides, during delivery, the therapeutic small RNAs should be prevented from degradation and binding to nonspecific targets. Ideally, the small RNAs should be completely inactive in their delivery form but fully active after reaching the target site. One commonly used way to achieve this is adding a photocleavable protecting group onto the proper site of small RNA molecules [100]. For example, Mikat and Heckel introduced the 2-(2-nitrophenyl)propyl group (NPP group) on the nucleobases guanine or thymine near the miRNA cleavage site and successfully blocked RNAi activity [100]. When the miRNAs reach the target site, the normal level of RNAi activity can be recovered with irradiation. However, considering the reverse effects of irradiation, the prospect of this strategy on clinical application is still obscure. Meanwhile, some researchers also make use of the special conformation of siRNAs to protect siRNAs from nonspecific binding to off-target proteins such as PKR [101]. In this case, N2-Alkyl-8-oxo-2'-deoxyguanosine was introduced in the guide strand of siRNA. This modification reduced the nonspecific binding between siRNA and off-target proteins but did not interfere with the target mRNA cleavage process [101].

14.6 Conclusion and Prospective

The development of RNA technology makes it possible to utilize small RNA molecules for new antiviral strategies. To be applied as antiviral therapy, small RNAs show some promising advantages, including (1) high specificity based on the nucleotide base-paired rule, (2) direct effect on the viral life cycle, (3) a comparatively simple structure with low immunogenicity, and (4) ease in synthesis and modification. Besides, for each kind of small RNAs, routine techniques and processes are available in antiviral application because of the chemical similarity of RNA molecules, which greatly speed up the pace of drug development.

However, challenges always accompany the progress. Importantly, most research regarding the therapeutic effect of small RNAs remains still in the stage of in vitro experiments and tests in animal models, while few proceed with clinical trials. One great challenge may be to deliver RNA molecules in the human body with high stability and low toxicity. In addition, more databases of small RNAs are still in need, considering the importance of bioinformatics analysis in therapeutic development of small RNAs, such as target prediction, high-order structure analysis, etc. Moreover, many therapeutic companies are still not devoting much into small RNA-related therapy, since it is quite time and money consuming.

Nevertheless, as a comparatively new field of therapeutics, the research on small RNA-based antiviral therapy has already achieved promising progress. With the help of rapid development of biotechnology, bioinformatics, and bionanotechnology, small RNA molecules may have a bright prospect in the field of antiviral therapy.

Acknowledgments

This work was supported by grants from the Canadian Institutes of Health Research and the Heart and Stroke Foundation of BC and Yukon (D. Yang). Dr. Maged Gomaa Hemida is a recipient of the CIHR-IMPACT postdoctoral training fellowship. Xin Ye is supported by an UGF Award from the University of British Columbia.

References

1. Samuel, C.E. (2001). Antiviral actions of interferons. *Clin. Microbiol. Rev.*, **14**(4), 778–809.
2. Ruth, J.L., and Cheng, Y.C. (1981). Nucleoside analogues with clinical potential in antivirus chemotherapy. The effect of several thymidine and 2′-deoxycytidine analogue 5′-triphosphates on purified human (alpha, beta) and herpes simplex virus (types 1, 2) DNA polymerases. *Mol. Pharmacol.*, **20**(2), 415–422.
3. Xin, K.Q., et al. (2001). A novel recombinant adeno-associated virus vaccine induces a long-term humoral immune response to human immunodeficiency virus. *Hum. Gene Ther.*, **12**(9), 1047–1061.
4. Kim, H., et al. (2004). Study on persistent infection of Japanese encephalitis virus Beijing-1 strain in serum-free Sf9 cell cultures. *J. Microbiol.*, **42**(1), 25–31.
5. Sachidanandam, R. (2005). RNAi as a bioinformatics consumer. *Brief Bioinf.*, **6**(2), 146–162.
6. Bushati, N., and Cohen, S.M. (2007). microRNA functions. *Annu. Rev. Cell Dev. Biol.*, **23**, 175–205.
7. Morris, K.V., and Rossi, J.J. (2006). Antiviral applications of RNAi. *Curr. Opin. Mol. Ther.*, **8**(2), 115–121.
8. Napoli, C., Lemieux, C., and Jorgensen, R. (1990). Introduction of a chimeric chalcone synthase gene into petunia results in reversible co-suppression of homologous genes in trans. *Plant Cell*, **2**(4), 279–289.
9. Bennasser, Y., et al. (2005). Evidence that HIV-1 encodes an siRNA and a suppressor of RNA silencing. *Immunity*, **22**(5), 607–619.
10. Berkhout, B., and Jeang, K.T. (2007). RISCy business: MicroRNAs, pathogenesis, and viruses. *J. Biol. Chem.*, **282**(37), 26641–26645.
11. Cullen, B.R. (2006). Is RNA interference involved in intrinsic antiviral immunity in mammals? *Nat. Immunol.*, **7**(6), 563–567.

12. Lecellier, C.H., et al. (2005). A cellular microRNA mediates antiviral defense in human cells. *Science*, **308**(5721), 557–560.

13. Otsuka, M., et al. (2007). Hypersusceptibility to vesicular stomatitis virus infection in Dicer1-deficient mice is due to impaired miR24 and miR93 expression. *Immunity*, **27**(1), 123–134.

14. Beck, J., and Nassal, M. (2007). Hepatitis B virus replication. *World J. Gastroenterol.*, **13**(1), 48–64.

15. Bruss, V. (2007). Hepatitis B virus morphogenesis. *World J. Gastroenterol.*, **13**(1), 65–73.

16. MacRae, I.J., et al. (2008). In vitro reconstitution of the human RISC-loading complex. *Proc. Natl. Acad. Sci. U S A*, **105**(2), 512–517.

17. Matranga, C., et al. (2005). Passenger-strand cleavage facilitates assembly of siRNA into Ago2-containing RNAi enzyme complexes. *Cell*, **123**(4), 607–620.

18. Rand, T.A., et al. (2005). Argonaute2 cleaves the anti-guide strand of siRNA during RISC activation. *Cell*, **123**(4), 621–629.

19. Gitlin, L., Stone, J.K., and Andino, R. (2005). Poliovirus escape from RNA interference: short interfering RNA-target recognition and implications for therapeutic approaches. *J. Virol.*, **79**(2), 1027–1035.

20. Merl, S., Michaelis, C., Jaschke, B., Vorpahl, M., Seidl, S. and Wessely, R. (2005). Targeting 2A protease by RNA interference attenuates coxsackieviral cytopathogenicity and promotes survival in highly susceptible mice. *Circulation*, **111**(13), 1583–1592.

21. Saleh, M.C., Van Rij, R.P. and Andino, R. (2004). RNA silencing in viral infections: insights from poliovirus. *Virus Res.*, **102**(1), 11–17.

22. Yuan, J., et al. (2005). Inhibition of coxsackievirus B3 replication by small interfering RNAs requires perfect sequence match in the central region of the viral positive strand. *J. Virol.*, **79**(4), 2151–2159.

23. Brodersen, P., and Voinnet, O. (2009). Revisiting the principles of microRNA target recognition and mode of action. *Nat. Rev. Mol. Cell Biol.*, **10**(2), 141–148.

24. Potenza, N., et al. (2011). Human microRNA hsa-miR-125a-5p interferes with expression of hepatitis B virus surface antigen. *Nucleic Acids Res.*, **39**(12), 5157-5163.

25. Wilson, J.A., et al. (2011). Human Ago2 is required for efficient microRNA 122 regulation of hepatitis C virus RNA accumulation and translation. *J. Virol.*, **85**(5), 2342–2350.

26. Li, Y.P., et al. (2011). MicroRNA-122 antagonism against hepatitis C virus genotypes 1-6 and reduced efficacy by host RNA insertion or

mutations in the HCV 5′ UTR. *Proc. Natl. Acad. Sci. U S A*, **108**(12), 4991–4996.

27. Hemida, M.G., et al. (2010). Exploiting the therapeutic potential of microRNAs in viral diseases: expectations and limitations. *Mol. Diagn. Ther.*, **14**(5), 271–282.
28. Brennecke, J., Stark, A., Russell, R.B., and Cohen, S.M. (2005). Principles of microRNA-target recognition. *PLoS Biol.*, **3**(3), e85.
29. Grimson, A., et al. (2007). MicroRNA targeting specificity in mammals: determinants beyond seed pairing. *Mol. Cell*, **27**(1), 91–105.
30. Krek, A., et al. (2005). Combinatorial microRNA target predictions. *Nat. Genet.*, **37**(5), 495–500.
31. Lewis, B.P., et al. (2003). Prediction of mammalian microRNA targets. *Cell*, **115**(7), 787–798.
32. McManus, M.T., and Sharp, P.A. (2002). Gene silencing in mammals by small interfering RNAs. *Nat. Rev. Genet.*, **3**(10), 737–747.
33. Kumar, R., et al. (2003). High-throughput selection of effective RNAi probes for gene silencing. *Genome Res.*, **13**(10), 2333–2340.
34. Tong, W.P., et al. (2008). An accurate quantitative method for screening effective siRNA probes targeting a Hepatitis B virus transcript in single living cells. *Biochem. Biophys. Res. Commun.*, **367**(4), 866–873.
35. Min, H., and Yoon, S. (2010). Got target? Computational methods for microRNA target prediction and their extension. *Exp. Mol. Med.*, **42**(4), 233–244.
36. Takane, K., et al. (2010). Computational prediction and experimental validation of evolutionarily conserved microRNA target genes in bilaterian animals. *BMC Genomics.*, **11**, 101.
37. Witkos, T.M., et al. (2011). Practical Aspects of microRNA Target Prediction. *Curr. Mol. Med.*, **11**(2), 93–109.
38. Peacock, H., et al. (2011). Nucleobase and ribose modifications control immunostimulation by a microRNA-122-mimetic RNA. *J. Am. Chem. Soc.*, **133**(24), 9200–9203.
39. Klump, W.M., et al. (1990). Complete nucleotide sequence of infectious Coxsackievirus B3 cDNA: two initial 5′ uridine residues are regained during plus-strand RNA synthesis. *J. Virol.*, **64**(4), 1573–1583.
40. Yuan, J., et al. (2005). Inhibition of coxsackievirus B3 replication by small interfering RNAs requires perfect sequence match in the central region of the viral positive strand. *J. Virol.*, **79**(4), 2151–2159.

41. Luan, Y., et al. (2012). Small interfering RNA against the 2C genomic region of coxsackievirus B3 exerts potential antiviral effects in permissive HeLa cells. *Virus Res.*, **163**(1), 183–189.
42. Kim, J.Y., et al. (2007). Expression of short hairpin RNAs against the coxsackievirus B3 exerts potential antiviral effects in Cos-7 cells and in mice. *Virus Res.*, **125**(1), 9–13.
43. Merl, S., and Wessely, R. (2007). Anti-coxsackieviral efficacy of RNA interference is highly dependent on genomic target selection and emergence of escape mutants. *Oligonucleotides*, **17**(1), 44–53.
44. Nygardas, M., et al. (2009). Inhibition of coxsackievirus B3 and related enteroviruses by antiviral short interfering RNA pools produced using phi6 RNA-dependent RNA polymerase. *J. Gen. Virol.*, **90**(Pt 10), 2468–2473.
45. van Ooij, M.J., et al. (2006). Structural and functional characterization of the coxsackievirus B3 CRE(2C): role of CRE(2C) in negative- and positive-strand RNA synthesis. *J. Gen. Virol.*, **87**(Pt 1), 103–113.
46. Werk, D., et al. (2005). Developing an effective RNA interference strategy against a plus-strand RNA virus: silencing of coxsackievirus B3 and its cognate coxsackievirus-adenovirus receptor. *Biol. Chem.*, **386**(9), 857–863.
47. Ye, X., et al. (2011). Targeted delivery of mutant tolerant anti-coxsackievirus artificial microRNAs using folate conjugated bacteriophage Phi29 pRNA. *PLoS ONE*, **6**(6), e21215.
48. Ahluwalia, J.K., et al. (2008). Human cellular microRNA hsa-miR-29a interferes with viral nef protein expression and HIV-1 replication. *Retrovirology*, **5**, 117.
49. Nasheri, N., et al. (2011). Competing roles of microRNA-122 recognition elements in hepatitis C virus RNA. *Virology*, **410**(2), 336–344.
50. Nunnari, G., and Schnell, M.J. (2011). MicroRNA-122: a therapeutic target for hepatitis C virus (HCV) infection. *Front. Biosci. (Schol. Ed.)*, **3**, 1032–1037.
51. Jopling, C.L., et al. (2008). Position-dependent function for a tandem microRNA miR-122-binding site located in the hepatitis C virus RNA genome. *Cell Host Microbe*, **4**(1), 77–85.
52. Tang, Y., et al. (2011). Human-derived IgG level as an indicator for EBV-associated lymphoma model in Hu-PBL/SCID chimeras. *Virol. J.*, **8**, 213.
53. Triboulet, R., et al. (2007). Suppression of microRNA-silencing pathway by HIV-1 during virus replication. *Science*, **315**(5818), 1579–1582.

54. Hussain, M., et al. (2012). West Nile virus encodes a microRNA-like small RNA in the 3′ untranslated region which up-regulates GATA4 mRNA and facilitates virus replication in mosquito cells. *Nucleic Acids Res.*, **40**(5), 2210–2223.
55. Contreras, J., and Rao, D.S. (2012). MicroRNAs in inflammation and immune responses. *Leukemia*, **26**(3), 404–413.
56. Gantier, M.P., and Williams, B.R. (2010). Monitoring innate immune recruitment by siRNAs in mammalian cells. *Methods Mol. Biol.*, **623**, 21–33.
57. Androulidaki, A., et al. (2009). The kinase Akt1 controls macrophage response to lipopolysaccharide by regulating microRNAs. *Immunity*, **31**(2), 220–231.
58. O'Connell, R.M., et al. (2010). MicroRNA-155 promotes autoimmune inflammation by enhancing inflammatory T cell development. *Immunity*, **33**(4), 607–619.
59. Nahid, M.A., et al. (2009). miR-146a is critical for endotoxin-induced tolerance: Implication in innate immunity. *J. Biol. Chem.*, **284**(50), 34590–34599.
60. Nahid, M.A., et al. (2011). MicroRNA in TLR signaling and endotoxin tolerance. *Cell Mol. Immunol.*, **8**(5), 388–403.
61. Dang, L.T., et al. (2008). Inhibition of red seabream iridovirus (RSIV) replication by small interfering RNA (siRNA) in a cell culture system. *Antiviral Res.*, **77**(2), 142–149.
62. Proske, D., et al. (2005). Aptamers—basic research, drug development, and clinical applications. *Appl. Microbiol. Biotechnol.*, **69**(4), 367–374.
63. Khati, M. (2010). The future of aptamers in medicine. *J. Clin. Pathol.*, **63**(6), 480–487.
64. Tuerk, C., and Gold, L. (1990). Systematic evolution of ligands by exponential enrichment: RNA ligands to bacteriophage T4 DNA polymerase. *Science*, **249**(4968), 505–510.
65. Ellington, A.D., and Szostak, J.W. (1990). In vitro selection of RNA molecules that bind specific ligands. *Nature*, **346**(6287), 818–822.
66. Gopinath, S.C., et al. (2006). An efficient RNA aptamer against human influenza B virus hemagglutinin. *J. Biochem.*, **139**(5), 837–846.
67. Kobayashi, K., et al. (2005). Epidemiological and outcome characteristics of major burns in Tokyo. *Burns*, **31**(Suppl 1): S3–S11.
68. Osborne, S.E., et al. (1997). Aptamers as therapeutic and diagnostic reagents: problems and prospects. *Curr. Opin. Chem. Biol.*, **1**(1), 5–9.

69. Misono, T.S., and Kumar, P.K. (2005). Selection of RNA aptamers against human influenza virus hemagglutinin using surface plasmon resonance. *Anal. Biochem.*, **342**(2), 312–317.
70. Yamamoto, R., et al. (2000). Molecular beacon aptamer fluoresces in the presence of Tat protein of HIV-1. *Genes Cells*, **5**(5), 389–396.
71. Cohen, C., et al. (2008). An aptamer that neutralizes R5 strains of HIV-1 binds to core residues of gp120 in the CCR5 binding site. *Virology*, **381**(1), 46–54.
72. Khati, M., et al. (2003). Neutralization of infectivity of diverse R5 clinical isolates of human immunodeficiency virus type 1 by gp120-binding 2′F-RNA aptamers. *J. Virol.*, **77**(23), 12692–12698.
73. Sayer, N., et al. (2002). Structural characterization of a 2′F-RNA aptamer that binds a HIV-1 SU glycoprotein, gp120. *Biochem. Biophys. Res. Commun.*, **293**(3), 924–931.
74. Joubert, M.K., et al. (2010). A modeled structure of an aptamer-gp120 complex provides insight into the mechanism of HIV-1 neutralization. *Biochemistry*, **49**(28) 5880–5890.
75. Zhou, J., et al. (2008). Novel dual inhibitory function aptamer-siRNA delivery system for HIV-1 therapy. *Mol. Ther.*, **16**(8), 1481–1489.
76. McGowan, I., and Taylor, D.J. (2010). Heterosexual anal intercourse has the potential to cause a significant loss of power in vaginal microbicide effectiveness studies. *Sex Transm. Dis.*, **37**(6), 361–364.
77. Moore, M.D., et al. (2011). Protection of HIV neutralizing aptamers against rectal and vaginal nucleases: implications for RNA-based therapeutics. *J. Biol. Chem.*, **286**(4), 2526–2535.
78. Kikuchi, K., et al. (2005). A hepatitis C virus (HCV) internal ribosome entry site (IRES) domain III-IV-targeted aptamer inhibits translation by binding to an apical loop of domain IIId. *Nucleic Acids Res.*, **33**(2), 683–692.
79. Peracchi, A. (2004). Prospects for antiviral ribozymes and deoxyribozymes. *Rev. Med. Virol.*, **14**(1), 47–64.
80. Rossi, J.J. (1999). Ribozymes, genomics and therapeutics. *Chem. Biol.*, **6**(2), R33–R37.
81. Shiota, M., et al. (2004). Ribozymes: applications to functional analysis and gene discovery. *J. Biochem.*, **136**(2), 133–147.
82. Frese, M., and Bartenschlager, R. (2006). Inhibition of hepatitis C virus by nucleic acid-based antiviral approaches, in *New Concept of Antiviral Therapy*, E. a. H. Bogner, A. (ed.), Springer, ISBN: 0387310460, 47–86.

83. Fukushima, A., et al. (2009). Development of a chimeric DNA-RNA hammerhead ribozyme targeting SARS virus. *Intervirology*, **52**(2), 92–99.
84. Scherer, L.J., and Rossi, J.J. (2003). Approaches for the sequence-specific knockdown of mRNA. *Nat. Biotechnol.*, **21**(12), 1457–1465.
85. Haseloff, J., and Gerlach, W.L. (1992). Simple RNA enzymes with new and highly specific endoribonuclease activities. 1988. *Biotechnology*, **24**, 264–269.
86. Cagnon, L., and Rossi, J.J. (2000). Downregulation of the CCR5 beta-chemokine receptor and inhibition of HIV-1 infection by stable VA1-ribozyme chimeric transcripts. *Antisense Nucleic Acid Drug Dev.*, **10**(4), 251–261.
87. Li, M.J., et al. (2005). Long-term inhibition of HIV-1 infection in primary hematopoietic cells by lentiviral vector delivery of a triple combination of anti-HIV shRNA, anti-CCR5 ribozyme, and a nucleolar-localizing TAR decoy. *Mol. Ther.*, **12**(5), 900–909.
88. Rossi, J.J. (2007). Targeted cleavage: tuneable cis-cleaving ribozymes. *Proc. Natl. Acad. Sci. U S A*, **104**(38), 14881–14882.
89. Zhang, Y.C., et al. (2005). Antisense inhibition: oligonucleotides, ribozymes, and siRNAs. *Methods Mol. Med.*, **106**, 11–34.
90. Nazari, R., et al. (2008). Inhibition of human immunodeficiency virus-1 entry using vectors expressing a multimeric hammerhead ribozyme targeting the CCR5 mRNA. *J. Gen. Virol.*, **89**(Pt 9), 2252–2261.
91. Medina, M.F., and Joshi, S. (1999). Design, characterization and testing of tRNA3Lys-based hammerhead ribozymes. *Nucleic Acids Res.*, **27**(7), 1698–1708.
92. Ramezani, A., et al. (2002). Development and testing of retroviral vectors expressing multimeric hammerhead ribozymes targeted against all major clades of HIV-1. *Front. Biosci.*, **7**, a29–a36.
93. Cheng, L., et al. (1997). A GFP reporter system to assess gene transfer and expression in human hematopoietic progenitor cells. *Gene Ther.*, **4**(10), 1013–1022.
94. Ramezani, A., et al. (2002). Development and testing of retroviral vectors expressing multimeric hammerhead ribozymes targeted against all major clades of HIV-1. *Front. Biosci.*, **7**, a29–a36.
95. Fukushima, A., et al. (2009). Development of a chimeric DNA-RNA hammerhead ribozyme targeting SARS virus. *Intervirology*, **52**(2), 92–99.
96. Kotani, M., et al. (2003). Chimeric DNA-RNA hammerhead ribozyme targeting PDGF A-chain mRNA specifically inhibits neointima formation in rat carotid artery after balloon injury. *Cardiovasc Res.*, **57**(1), 265–276.

97. Kumar, P., et al. (2010). Potent inhibition of influenza virus replication with novel siRNA-chimeric-ribozyme constructs. *Antiviral Res.*, **87**(2), 204–212.

98. Ye, X., and Yang, D. (2009). Recent advances in biological strategies for targeted drug delivery. *Cardiovasc. Hematol. Disord. Drug Targets.*, **9**(3), 206–221.

99. Zhang, H.M., et al. (2009). Targeted delivery of anti-coxsackievirus siRNAs using ligand-conjugated packaging RNAs. *Antiviral Res.*, **83**(3), 307–316.

100. Mikat, V., and Heckel, A. (2007). Light-dependent RNA interference with nucleobase-caged siRNAs. *RNA*, **13**(12), 2341–2347.

101. Peacock, H., et al. (2011). Chemical modification of siRNA bases to probe and enhance RNA interference. *J. Org. Chem.*, **76**(18), 7295–7300.

102. Kiriakidou, M., et al. (2004). A combined computational-experimental approach predicts human microRNA targets. *Genes Dev.*, **18**(10), 1165–1178.

103. Miranda, K.C., et al. (2006). A pattern-based method for the identification of MicroRNA binding sites and their corresponding heteroduplexes. *Cell*, **126**(6), 1203–1217.

104. Enright, A.J., et al. (2003). MicroRNA targets in Drosophila. *Genome Biol.*, **5**(1), R1.

105. Xiaowei, Wang and Issam, M. El Naqa (2008) Prediction of both conserved and nonconserved microRNA targets in animals. *Bioinformatics*, **24**(3), 325–332.

106. Xiaowei, Wang (2008). miRDB: a microRNA target prediction and functional annotation database with a wiki interface. *RNA*, **14**(6), 1012–1017.

107. Grün, D., Wang, Y.L., Langenberger, D., Gunsalus, K.C., Rajewsky, N. (2005). MicroRNA target predictions across seven Drosophila species and comparison to mammalian targets. *PLoS Comput. Biol.*, **1**, e13.

108. Rehmsmeier, M., et al. (2004). Fast and effective prediction of microRNA/target duplexes. *RNA*, **10**(10), 1507–1517.

109. Lewis, B.P., Shih, I.H., Jones-Rhoades, M.W., Bartel, D.P., Burge, C.B. (2003). Prediction of mammalian microRNA targets. *Cell*, **115**, 787–798.

110. Hsu, P.W., et al. (2007). ViTa: prediction of host microRNAs targets on viruses. *Nucleic Acids Res.*, **35**(Database issue), D381–D385.

111. Li, M., and Rossi, J.J. (2005). Lentiviral vector delivery of siRNA and shRNA encoding genes into cultured and primary hematopoietic cells. *Methods Mol. Biol.*, **309**, 261–272.
112. Barnor, J.S., et al. (2005). Lentiviral-mediated delivery of combined HIV-1 decoy TAR and Vif siRNA as a single RNA molecule that cleaves to inhibit HIV-1 in transduced cells. *Nucleosides Nucleotides Nucleic Acids*, **24**(5-7), 431–434.
113. Zhang, L., et al. (2008). Lentiviral vector-mediated siRNA knockdown of SR-PSOX inhibits foam cell formation in vitro. *Acta Pharmacol. Sin.*, **29**(7), 847–852.
114. Li, M., and Rossi, J.J. (2008). Lentiviral vector delivery of siRNA and shRNA encoding genes into cultured and primary hematopoietic cells. *Methods Mol. Biol.*, **433**, 287–299.
115. Ewert, K.K., et al. (2010). Cationic liposome-nucleic acid complexes for gene delivery and silencing: pathways and mechanisms for plasmid DNA and siRNA. *Top. Curr. Chem.*, **296**, 191–226.
116. Ryan, S.M., et al. (2008). Advances in PEGylation of important biotech molecules: delivery aspects. *Expert Opin. Drug Delivery*, **5**(4), 371–383.
117. Walther, W., and Stein, U. (2000). Viral vectors for gene transfer: a review of their use in the treatment of human diseases. *Drugs*, **60**(2), 249–271.
118. Kim, Y.J., et al. (2008). Recombinant lentivirus-delivered short hairpin RNAs targeted to conserved coxsackievirus sequences protect against viral myocarditis and improve survival rate in an animal model. *Virus Genes*, **36**(1), 141–146.
119. Dull, T., et al. (1998). A third-generation lentivirus vector with a conditional packaging system. *J. Virol.*, **72**(11), 8463–8471.
120. Dishart, K.L., et al. (2003). Third-generation lentivirus vectors efficiently transduce and phenotypically modify vascular cells: implications for gene therapy. *J. Mol. Cell Cardiol.*, **35**(7), 739–748.
121. Coyne, K.P., et al. (2007). Evolutionary mechanisms of persistence and diversification of a calicivirus within endemically infected natural host populations. *J. Virol.*, **81**(4), 1961–1971.
122. Weinberg, J.B., et al. (1991). Productive human immunodeficiency virus type 1 (HIV-1) infection of nonproliferating human monocytes. *J. Exp. Med.*, **174**(6), 1477–1482.
123. Lewis, P., et al. (1992). Human immunodeficiency virus infection of cells arrested in the cell cycle. *EMBO J.*, **11**(8), 3053–3058.

Chapter 15

Cell Specificity of siRNA as a Key for Therapeutic Applications

Tobias G. Poehlmann, Juliane Reiche, Mirko Ludwig, and Rolf Guenther

Friedrich-Schiller-University of Jena, Jena Center for Soft Matter, Intelligent-siRNA Group, Ernt-Ruska-Ring 1, 07745 Jena, Germany

Tobias.Poehlmann@Intelligent-siRNA.com, Juliane.Reiche@Intelligent-siRNA.com, Mirko.Ludwig@Intelligent-siRNA.com, rolf.guenther@bianoscience.com

15.1 Introduction

The use of short interfering ribonucleic acid (siRNA) to induce RNA interference (RNAi) and to silence gene expression is well established in basic research and widely used in biological research. Its high potential for new therapeutic strategies stimulated many groups and companies worldwide to develop siRNA therapies for diseases like cancer, viral infections, and chronic diseases. These efforts revealed, however, that siRNA could only be applied to patients once several technical and systemic problems are solved: (1) siRNA stability and the prevention of its degradation have to be improved, (2) efficient siRNA delivery and gene silencing are still a major hurdle in vivo, (3) unspecific stimulation of immune

RNA Nanotechnology
Edited by Bin Wang

ISBN 978-981-4411-64-6 (Hardcover), 978-981-4411-65-3 (eBook)
www.panstanford.com

response by siRNA has to be reduced, (4) the balance between target sequence specificity and the risk of escape mutations has to be addressed, (5) siRNA production for therapeutic applications has to be up-scaled, and (6) the cell specificity of RNAi for high-impact therapeutic strategies has to be addressed.

By simply applying siRNA to cells of the human body using optimized delivery methods it is possible to reduce the expression of a relevant gene. But the genetic network in the human body is complex and often even more complex in the case of a disease. So it seems not likely that silencing of a single gene associated with a disease would give rise to an effective therapeutic strategy.

In the following sections the authors describe general strategies of siRNA delivery and RNAi induction in cells and tissues related to specific diseases. Furthermore novel therapeutic strategies based on cell-specific RNAi will be discussed.

15.2 siRNA Delivery Strategies

siRNA can be applied to animals and humans via several routes, including local application (e.g., to the eye, skin, or muscle) and systemic application (e.g., intravenously). In many cases local injection shows advantages over systemic application because lower amounts of siRNA and less cell specificity are required for an effect, thereby leading to lower siRNA degradation and systemic side effects. However, for some therapeutic siRNA applications, for example, cancer treatment, local injections are often not suitable since tumor cells may have already spread throughout the entire body. On the other hand, molecules delivered systemically have to cross several barriers until they reach their target area of action, for example, tumor tissue. Furthermore the siRNA has to remain active when circulating in the bloodstream and passing through organs comprising different tissues, like liver, kidney, or spleen. Additionally the delivery systems have to be biocompatible and biodegradable and have to be able to penetrate the endothelium to transport the siRNA to its target tissue. In most cases, however, the transport of macromolecular structures across the endothelium is poor, whereas in tumor endothelium, the "enhanced permeation and retention" (EPR) effect allows the penetration of high-molecular-mass macromolecules. Subsequently, molecules need to diffuse through the extracellular matrix (Zámecník et al., 2003) to reach the target cell, and finally the siRNA has to pass the plasma membrane to reach the cytoplasm of the target cell. The delivery system has to ensure the liberation of the cargo into the cytoplasm without being trapped in the endosome. In that case the delivery system should invoke an endosomal escape. To cope with the acidic environment in endosomes, delivery systems are often pH sensitive, so they are able to destroy or exit the endosome. Subsequently, the siRNA has to detach from the delivery system and be released into the cytosol (Shim and Kwon, 2010). There are mainly two strategies to deliver siRNA to the cell, viral and nonviral (chemical) delivery systems. While **viral delivery systems** achieve high efficiency and reproducibility using retro-, lenti-, or adenoviral vectors to induce RNAi via small hairpin RNA (shRNA) sequences, their therapeutic use for RNAi is limited due to concerns toward side effects induced by integration into the host genome, functional gene disruption, immunogenicity, and difficult production at a large scale (Lares et al., 2010). Today, viral systems are usually not preferred over nonviral strategies to deliver therapeutic siRNA, and meanwhile nonviral siRNA delivery strategies comprise numerous, quite different classes of molecular carriers, such as liposomes, cationic polymers, protein-based vectors, and metallic core nanoparticles.

Liposomes were one of the first vectors developed showing siRNA protection regarding nuclease degradation, low toxicity, and low immunogenicity. They are biocompatible and suitable for adding conjugates to enable targeted delivery. To stabilize complexes of liposomes and their cargo, neutral lipids, cationic lipids, and polyethylene glycol (PEG) are often added to form stable nucleic acid–lipid particles (SNALPs). The conjugation of a pH-sensitive histidine-lysine (HK) peptide improves endosomal release of the siRNA (Guo et al., 2010).

Polyethyleneimines (PEIs) are the most important class of **cationic polymers** for siRNA delivery. Branched or linear versions of PEI are applicable. Less cytotoxic but more expensive to manufacture are linear compared to branched PEIs. siRNA complexed with PEI is protected from nuclease degradation. Additionally PEIs exhibit the proton sponge effect, leading to water intake and thereby to swelling of endosomes and finally to increased release of the siRNA into the cytosol (Günther et al., 2011). **Protein-based vectors** are positively charged proteins, like protamine, polyarginine, penetratin, poly(L-lysine), or atelocollagen, that complex negatively charged siRNA via electrostatic interaction and are able to penetrate cells. A subgroup of protein-based vectors is formed by short cell-penetrating peptides (CPPs) (Richard et al., 2003). CPPs with a maximum length of 30 amino acids bind to the negatively charged cell surface and induce cellular internalization via several ways (e.g., endocytosis, binding to heparan sulfate proteoglycans) (Richard et al., 2003; Lecifert et al., 2002).

Another class of molecules used for siRNA delivery are **metallic core nanoparticles**. These nanoparticles consist of a stable metal core (iron oxide, iron cobalt, iron gold, or iron nickel) and a coating layer (polymers, sugars, or other compounds) to form a core-shell structure. The siRNA is conjugated to the metal core via thiol bindings (Moore et al., 2004), dextran (Medarova et al., 2007), biotin-streptavidin, or cationic polymers (Huth et al., 2004).

For all delivery strategies it is important to be biodegradable and to exhibit low cytotoxicity.

15.3 Strategies to Induce Cell-Specific RNAi and Their Therapeutic Relevance

The specific gene silencing in disease-related target cells is the most challenging aspect in the development of delivery strategies. The better the cellular targeting is, the less siRNA will be needed and unwanted side effects of the siRNA will be reduced. Cell-specific RNAi can be achieved via four different approaches: selective distribution to certain tissues only, targeted cellular uptake, a pro-drug approach, and target sequences with restricted expression in the target tissue.

So far, none of the known delivery strategies for systemic injection of siRNA will result in homogeneous **biodistribution** (see Fig. 15.1). SNALPs, for example, deliver siRNA particularly to the

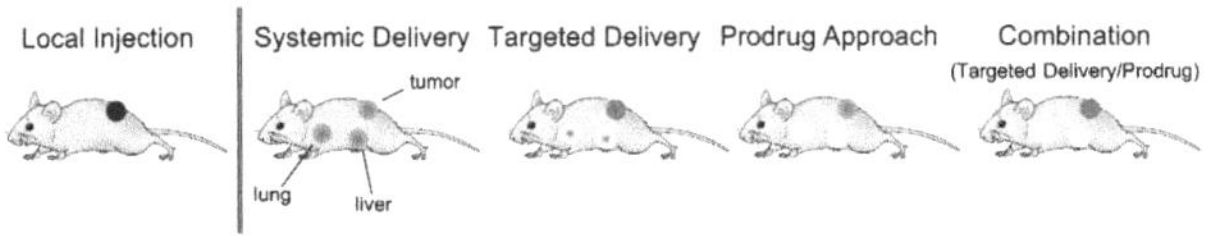

	Local Injection	Systemic Delivery	Targeted Delivery	Prodrug Approach	Combination
Target organ/cell specificity	++	--/+	-/++	+	++
Delivery efficacy (injected siRNA versus cellular active siRNA)	++	-/+	+	--/+	+
Risk of cellular escape from delivery	low	low	elevated	elevated	high
Accessability to target cells/tissue (e.g. metastases)	-	+	+	+	+

Figure 15.1 Overview of delivery systems. High organ specificity and delivery efficacy can be reached using local injection; the risk of cellular escape is low, but cellular specificity cannot be performed using this kind of application except for immune-stimulatory RNAs (van den Boorn and Hartmann, 2013). Systemic treatment of diseases like cancer is hardly possible. Using chemical delivery mechanisms, systemic injection induces some kind of organ-specific RNAi—typical organs with good siRNA delivery are the liver, spleen, and tumor tissue. Using targeted delivery mechanisms the amount of siRNA reaching the desired target cells can be increased, but still some siRNA will induce RNAi in other nontarget cells by unspecific uptake (e.g., in the liver). Furthermore, any targeted delivery strategy (e.g., by receptor–ligand systems) increases the risk of cellular escape by reduced expression of the specific receptor. A pro-drug approach targets cell populations with high specificity, but most of the injected siRNA will be delivery unspecificly in the liver. A knowledge-based combination of using biodistribution, targeted delivery, and the pro-dug approach could lead to effective siRNA delivery and cell specificity at the same time but might again increase the risk of cellular escape during long-term disease treatment.

liver, spleen, and tumors (Guo et al., 2010, Morrissey DV et al., 2005), whereas PEI-complexed siRNA accumulates in the lung and tumors (Höbel et al., 2010). In general, organs that are highly vasculated will be exposed to higher siRNA concentrations than other organs. Although delivery tools provide a certain mechanistic selectivity for specific organs, siRNA uptake into other cells or organs is not prevented and RNAi is still induced albeit at lower levels. **Targeted cellular uptake** of siRNA can be reached by conjugating/complexing aptamers or receptor-specific ligand systems, peptides, carbohydrates, or antibodies to the siRNA and/or to the delivery system. These molecules have high specificity for cellular target recognition, but large molecules are particularly prone to stimulate immune responses. Several smaller peptides display a distinct specificity for well-known surface targets, for example, the Arg-Gly-Asp (RGD) peptide targets tumor vascularization. RGD binds to transmembrane integrins, which are expressed at high levels in the process of vascularization of cancer tissue (Shim and Kwon, 2010; Guo et al., 2010).

To target specific cells, siRNA can be linked to antibodies. For example, a transferrin receptor–specific single-chain antibody fragment bound to cationic liposomes was designed to specifically target tumor xenografts in nude mice (Pirollo et al., 2007). Approaches like this lead to a tremendous accumulation of siRNA molecules in target tissue so that the required siRNA concentrations in vivo could be reduced, resulting in less side effects. Furthermore these approaches reduce the accumulation of siRNA in typical delivery sites, for example, the liver (Viel et al., 2008). It should be noted, however, that the expression of specific receptors varies among cancer cells even within the same tumor tissue. A reduced expression of the target receptor in only a small subpopulation of tumor cells would, for example, provide a survival advantage of that cancer cell population and would end up in a therapeutic failure.

To minimize the effects of unspecific siRNA uptake and unwanted delivery into nontarget cells, much effort has to be put into designing the optimal siRNA target sequence. For therapeutic approaches aiming at cancer treatment, cytokine receptors (e.g., colony stimulating factor (CSF-1)) (Morandi et al., 2011) and proteins involved in cell-signaling pathways, such as signal transducers of cytokine signaling (STATs) (Du et al., 2011), *Ras* (Sunaga et al., 2011), or mitogen-activated protein (MAP) kinases (Guo et al., 2012) are often used as siRNA targets. As these genes play an important role in noncancer cells as well, strong side effects have to be avoided for these therapeutic strategies. Obviously fewer side effects are expected with therapeutic approaches against viral infections, because the siRNA targets are often **restricted to viral genes**. By specifically targeting viral genes essential for viral replication, for production of the virus capsule, or for viral enzymes, it might be possible to control or even to cure viral infections (Kneidinger et al., 2012). But the flexibility of viruses to develop escape mutations may hamper the therapeutic success. Any mutation within the mRNA sequence might lead to a direct mismatch, or the accessibility of the siRNA can be affected by structural changes of the messenger RNA (mRNA). In either case an escape mutant could emerge and result in a therapeutic failure similar to the viral adaptation to protease inhibitor drugs (Doyon et al., 1996). In this case one siRNA targeting several mRNAs at once might be beneficial (see Fig. 15.2). This approach reduces the risk of escape mutants and at the same time might increase the potential of the siRNA to induce toxicity in the target cells. If the siRNA is aimed to induce cytotoxicity, it should exert its function exclusively in the target cell population. However, to date, no strategies mentioned before exert this level of cellular specificity. **An siRNA pro-drug approach** may help to advance this strategy. By using peptides bound to the siRNA backbone it could be possible to regulate the siRNA activity. The strategy is to use cancer cell specific proteases to cleave the bound peptides and thereby promote siRNA activity (Koehn et al., 2010). Although still in development, this promising approach may help reduce siRNA side effects leading to a powerful siRNA therapy.

15.4 Specific Characteristics of Therapeutic Cell-Specific siRNA

When an siRNA sequence becomes a lead candidate for any therapeutic applications and is optimized to act in a cell-specific manner, its transfer to clinics is all but straightforward: The proof of concept for a therapeutic approach in animals is just the first step in candidate development. First, the siRNA sequence needs to be optimized in most cases for target specificity to increase the silencing effect and to reduce side effects. Second, the sequence might be immune stimulatory, requiring further optimization steps. Third, the siRNA sequence could be optimized for nuclease resistance and stability. One way to stabilize nucleotides is to introduce chemical modifications such as locked nucleic acids (LNAs) (Elmén et al., 2005). Finally the sequence should be optimized for the

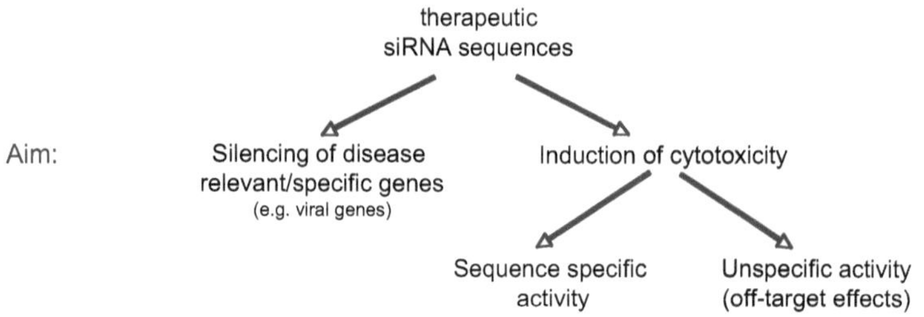

Cell specific Delivery useful:	yes	yes	yes
Cell specificity needed:	no	yes	yes
Risk of cellular escape from RNAi:	high	high	low

Figure 15.2 Strategies of therapeutic siRNA sequences. Whereas the silencing of specific disease-relevant genes (e.g., viral genes or specific *Ras* mutations) reduces the need of cell specificity, the risk of cellular escape from the gene silencing is very high, especially for long-term therapeutic applications. Selection of conserved target regions may help to reduce this risk but at the same time makes the task to identify relevant therapeutic sequences even more challenging. Whenever the aim of the siRNA is to kill its cellular targets (e.g., cancer cells), then the therapeutic nucleotide should induce RNAi exclusively in its target cells. If the siRNA does exert its cytotoxic function by silencing of specific mRNA sequences, then the risk of cellular escape is higher than for siRNA sequences having more than one target mRNA and/or inducing off-target or unspecific effects.

production process since the particular oligonucleotide sequence strongly influences the production yield of the chemical process, potentially leading to an immense cost reduction for the final drug. Similar optimization steps have to be performed for the part of the molecule bearing the functionality for cell specificity—for antibodies and antibody fragments or synthetically produced peptides in a comparable manner.

Nevertheless the development of a therapeutic nucleic acid candidate can be much faster and less cost intensive compared to other pharmaceutical compounds due to the highly predictable properties of the nucleic acid, so nucleic acids are still attractive to pharmaceutical companies and investors.

15.5 Recent Therapeutic Developments

After the melt of the therapeutic RNAi market during the past years most pharmaceutical companies do not afford their own RNAi centers anymore. Most of the recent clinical RNAi developments were performed by small to midsize companies benefiting from strong support given by their pharmaceutical partners to reduce the risk of developmental failure to a minimum. Although the overall investments in the therapeutic RNAi market were reduced during the last two years, they are again rapidly growing now, and the number of new companies focusing on therapeutic RNAi developments is growing again.

So the question arises, Why are companies and investors still interested in therapeutic nucleotide developments despite the fact that the US$10 billion spent during the last years led to only two Food and Drug Administration (FDA)-approved and marketed drugs (Arthur Krieg, 2011)? From our point of view therapeutic RNAi approaches do have major advantages compared to other techniques (e.g., small molecules or antibodies): (1) the potential to develop therapies for "undrugable targets" and (2) the finding of a candidate being faster and less cost intensive. While lead candidate development for a specific small molecule took US$32 million and six years, development of a specific antibody therapy required just US$8 million and took three years only, and just US$5 million were spent for a nucleotide drug candidate within one year (Arthur Krieg, 2011). This makes nucleotide therapies attractive to investors, even though a major-breakthrough and high-potential drug for diseases is still under development.

During the last few years a total of 25 clinical trials with siRNA as the active substance were launched according to the National Institutes of Health (NIH) (July 2012, www.clinicaltrials.gov), 8 of which are completed so far and 3 terminated. Major disease targets were ocular diseases (e.g., age-related macular degeneration (AMD), diabetic macular edema, intraocular pressure) and cancer (chronic myeloid leukemia (CML), advanced solid cancers, liver cancer, adenocarcinoma of the pancreas). Furthermore, different delivery strategies have been the focus of clinical trials. Besides these major studies and therapeutic targets, siRNA has recently been tested in clinical trials for the prophylaxis of delayed graft function in kidney transplantation and to treat pachyonychia congenita and hypercholesterolemia. Besides therapeutic siRNA developments, shRNA is the focus of clinical trials as well: according to the NIH database five clinical trials have been started recently, targeting several types of cancer. In these studies already, strategies for cell-specific gene silencing are used. One example is from Calando Pharmaceuticals. It delivers its siRNA candidate (CALAA-01) via nanoparticles specifically targeting the transferrin receptor (TfR) highly expressed on the cell membrane of tumor cells. Upon internalization of the siRNA–nanoparticle complex via TfR, the siRNA is released from endosomes and activates gene silencing of the M2 subunit of ribonucleotide reductase (RRM2), resulting in tumour reduction.

This overview gives only a small impression toward therapeutic siRNA developments because most recent activity is still in the preclinical part. The authors assume that in the near future many more clinical trials will be started where therapeutic nucleic acid candidates will display better cell specificity and its function will be more than just gene silencing.

15.6 Conclusion

Even though siRNA is a great tool for silencing gene expression, a powerful siRNA therapy against diseases like cancer or viral infections is still far from being developed. Delivery, the selection of an appropriate siRNA sequence, and cell specificity remain major obstacles. While siRNA in combination with nontargeted delivery systems may exert some therapeutic effect, this approach is burdened with a severe risk of unacceptable side effects. Specific delivery strategies and pro-drug approaches, however, show high therapeutic potential despite their early developmental stage.

Even though less money was spent for the pharmaceutical siRNA development during the last two years, a persisting commercial interest in new developments and therapeutical strategies can be observed. The scientific community looks forward seeing the first siRNA candidates passing successfully the clinical trials soon, so new funding to this awesome technology of *siRNA* becomes very likely.

Competing Interests

All authors are employees of BianoScience GmbH and the Friedrich-Schiller-University of Jena, Germany, which are engaged in the discovery and development of siRNA therapeutics.

References

1. Doyon, L., et al. (1996). Second locus involved in human immunodeficiency virus type-1 resistance to protease inhibitors. *J. Virol.*, **70**(6), 3763–3769.
2. Du, W., et al. (2012). Inhibition of JAK2/STAT3 signaling induces colorectal cancer cell apoptosis via mitochondrial pathway. *J. Cell Mol. Med.*, **16**(8), 1878–1888.
3. Elmén, J., et al. (2005). Locked nucleic acid (LNA) mediated improvements in siRNA stability and functionality. *Nucleic Acids Res.*, **33**(1), 439–447.
4. Günther, M., et al. (2011). Polyethylenimines for RNAi-mediated gene targeting in vivo and siRNA delivery to the lung. *Eur. J. Pharm. Biopharm.*, **77**, 438–449.
5. Guo, P., et al. (2010). Engineering RNA for targeted siRNA delivery and medical application. *Adv. Drug Delivery*, **62**(6), 650–666.
6. Guo, S.Y., et al. (2012) RNAi silencing of the MEKK3 gene promotes TRAIL-induced apoptosis in MCF-7 cells and suppresses the transcriptional activity of NF-κB. *Oncol. Rep.*, **27**(2), 441–446.
7. Höbel, S., et al. (2010). Polyethylenimine/small interfering RNA-mediated knockdown of vascular endothelial growth factor in vivo exerts anti-tumor effects synergistically with Bevacizumab. *J. Gene Med.*, **12**(3), 287–300.

8. Huth, S., et al. (2004). Insights into the mechanism of magnetofection using PEI-based magnetofectins for gene transfer. *J. Gene Med.*, **6**, 923–936.
9. Kneidinger, D., Ibrišimović, M., Lion, T., and Klein, R. (2012) Inhibition of adenovirus multiplication by short interfering RNAs directly or indirectly targeting the viral DNA replication machinery. *Antiviral Res.*, **94**(3), 195–207.
10. Koehn, S., et al. (2010). Cell-specific RNA interference by peptide-inhibited-peptidase-activated siRNAs. *J. RNAi Gene Silencing*, **6**(2), 422–430.
11. Krieg A. (2011) Is RNAi dead? *Mol. Ther.*, **19**(6), 1001–1002.
12. Lares, M.R., Rossi, J.J., and Oullet, D.L. (2010). RNAi and small interfering RNAs in human diseases therapeutic applications. *Trends Biotechnol.*, **28**(11), 570–579.
13. Lecifert, J.A., Harkins, S. and Whitton, J.L. (2002). Full-length proteins attached to the HIV tat protein transduction domain are neither transduced between cells, nor exhibit enhanced immunogenicity. *Gene Ther.*, **9**, 1422–1428.
14. Medarova, Z., et al. (2007). In vivo imaging of siRNA delivery and silencing in tumors. *Nat. Med.*, **13**, 372–377.
15. Morandi, A., et al. (2011). The colony-stimulating factor-1 (CSF-1) receptor sustains ERK1/2 activation and proliferation in breast cancer cell lines. *PLoS ONE*, **6**(11), e27450.
16. Morrissey, D.V., et al. (2005). Potent and persistent in vivo anti-HBV activity of chemically modified siRNAs. *Nat. Biotechnol.*, **23**, 1002–1007.
17. Moore, A., Medarova, Z., Potthast, A., and Dai, G. (2004). In vivo targeting of underglycosylated MUC-1 tumor antigen using a multimodal imaging probe. *Cancer Res.*, **64**, 1821–1827.
18. Pirollo, K.F., et al. (2007). Materializing the potential of small interfering RNA via a tumor-targeting nanodelivery system. *Cancer Res.*, **67**, 2938–2943.
19. Richard, J.P., et al. (2003). Cell-penetrating peptides: a reevaluation of the mechanism of cellular uptake. *J. Biol. Chem.*, **278**, 585–590.
20. Shim, M.S., and Kwon, Y.S. (2010). Efficient and targeted delivery of siRNA in vivo. *FEBS J.*, **277**, 4814–4827.
21. Sunaga, N., et al. (2011). Knockdown of oncogenic KRAS in non-small cell lung cancers suppresses tumor growth and sensitizes tumor cells to targeted therapy. *Mol. Cancer Ther.*, **10**(2) 336–46.

22. van den Boorn, J.G. and Hartmann, G. (2013). Turning tumors into vaccines: co-opting the innate immune system. *Immunity*, **39**(1), 27–37.
23. Viel, T., et al. (2008). Molecular imaging study on in vivo distribution and pharmacokinetics of modified small interfering (siRNAs). *Oligonucleotides*, **18**, 201–212.
24. Zámecník. J., et al. (2003). Extracellular matrix glycoproteins and diffusion barriers in human astrocytic tumours. *Neuropathol. Appl. Neurobiol.*, **30**, 338–350.

Chapter 16

Targeting Alternative Splicing in Human Genetic Disease

Michael R. Ladomery
Faculty of Health and Applied Sciences, University of the West of England, Coldharbour Lane, Bristol, BS16 1QY, UK
michael.ladomery@uwe.ac.uk

16.1 Introduction: The Discovery of Splicing

Following the discovery of the structure of deoxyribonucleic acid (DNA) in the 1950s it nonetheless took several years to work out the physical structure of a gene. A clear understanding of the structure of genes had been achieved initially in relation to prokaryotes; however, the structure of eukaryotic genes turned out to be more complex. While working on eukaryotic genes researchers in the 1960s and 1970s were confronted with a number of puzzling observations. The first related to eukaryotic heterogenous ribonucleic acid (RNA) (hnRNA). hnRNA is now better known as pre-messenger RNA (mRNA), in other words the primary transcript and nuclear precursor of cytoplasmic mRNA. hnRNA was observed to sediment in rather large ribonucleoprotein (RNP) particles—much larger when compared to corresponding

RNA Nanotechnology
Edited by Bin Wang

ISBN 978-981-4411-64-6 (Hardcover), 978-981-4411-65-3 (eBook)
www.panstanford.com

cytoplasmic messenger RNP particles. Like mRNA, hnRNA contained a methylated cap at the 5′ end, and a processed 3′ end—in other words a poly(A) tail. It therefore appeared to be somehow related to mRNA but was much larger. Another puzzling observation came from experiments that measured the kinetics of hybridization of nuclear hnRNA to cytoplasmic mRNA. These experiments suggested that hnRNA was, on average, kinetically five times more complex than mRNA—in other words it corresponded to and contained mRNA sequences within it, but it also contained a lot of additional extra sequences. The identity and functions of these extra sequences were not understood.

These apparently bizarre observations remained essentially unexplained until the discovery of pre-mRNA splicing: in fact hnRNA turned out to be the *unspliced nuclear precursor of mRNA*. What happens is that to generate mature mRNA, extra nucleotides need to be removed, or *spliced*, from the pre-mRNA. There are other types of splicing in nature (when RNA sequences are excised from pre-existing RNA sequences). For the purposes of this chapter we will focus on nuclear **pre-mRNA splicing** only. The other forms of splicing include group I and group II self-splicing introns and the splicing of transfer RNA (tRNA) precursors.

The story of the discovery of pre-mRNA splicing began with studies on adenovirus. Viruses are very good models for studying the many steps in the regulation of gene expression, including the processing of pre-mRNA transcripts into mRNA. In a classic experiment, Sharp et al. first used gel electrophoresis to purify the most abundant adenoviral mRNA; it encodes the viral hexon protein. They then proceeded to hybridize this mRNA to the adenoviral genome using a technique known as "R-looping," in which RNA hybridizes with DNA. The results were surprising in that several loops were observed, indicating that there are intervening sequences in the *hexon* gene that are not represented in the mature mRNA (Fig. 16.1). This was, in effect, the direct physical observation of what became known as **introns** (intervening sequences) as opposed to **exons** (which are incorporated, or spliced together, in mature mRNAs). Sharp and Roberts both received the Nobel Prize in Physiology or Medicine for their discovery of "split genes" in 1993.

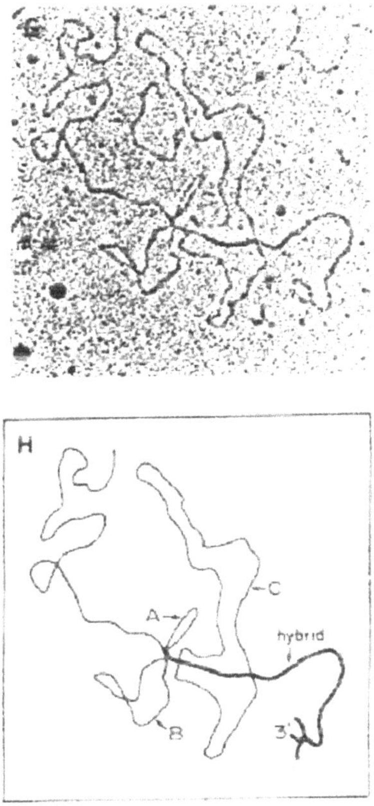

Figure 16.1 The discovery of splicing in the adenoviral *hexon* gene. Observation of R-loops following hybridization of hexon mRNA to corresponding genomic DNA. The cartoon is an interpretation of the electron micrograph. The thicker line represents the DNA:RNA hybrid. The loops labeled A, B and C represent DNA that is not hybridized. These "intervening sequences" are now known as introns, whereas exons are spliced together in the mature mRNA. Figure taken from *PNAS USA* 74: 3173, Fig4GH (2007).

Soon after the discovery of splicing in viral pre-mRNAs, it became apparent that the same process occurs in cellular genes. In other words, pre-mRNA splicing is a general process that occurs in animals, plants, and unicellular eukaryotes. It is now considered to be a vital aspect of gene regulation and a key driver of proteomic complexity. The energy cost of splicing to the cell is offset by its evolutionary benefits. Exons can be duplicated, shortened, or lengthened, and cryptic exons can arise within introns—all of which can potentially help generate new or modify existing protein domains.

16.2 The Splicing Reaction

The key problem in splicing is the following: how can exons be joined together precisely? This is not a trivial problem; after all an error in exon definition of one nucleotide can cause a frame-shift that has deleterious consequences. The answer lies in the **spliceosome**: a complex molecular machine made up of several RNP complexes that work together to define exon boundaries and catalyze the splicing reaction. The spliceosome's components are called small nuclear RNPs (snRNPs; pronounced as "snurps"). These are made up of small nuclear RNAs (snRNAs), several core proteins (common to all snRNPs), and several snRNP-specific proteins. The main snRNPs are the U1, U2, U4, U5, and U6 snRNPs, and their snRNA components are highly conserved in evolution.

The boundaries of introns and exons are defined by **splice sites**. Thus the 3′ end of an exon (and the 5′ end of the following intron) is known as the 5′ splice site. The 3′ end of the intron, and the 5′ end of the next exon, is known as the 3′ splice site. How are splice sites recognized by the spliceosome? They conform to a consensus sequence that is highly conserved in evolution. Introns contain an additional critical sequence—the **branchpoint A** (adenosine)—itself surrounded by a consensus sequence. Following the branchpoint A, a pyrimidine-rich tract precedes the 3′ splice site. Together these sequences define where splicing can occur. There are splice site sequences present in pre-mRNAs that are sometimes weaker or less well defined; in some conditions, or following mutation, these can be used and are referred to as **cryptic splice sites**.

The splicing reaction involves two **transesterification reactions** (Fig. 16.2). Exon definition occur very early in spliceosome assembly. The U1 snRNP base-pairs across the 5′ splice site. The splice factor U2AF65 (U2AF: U2 auxiliary factor) binds to the pyrimidine tract and helps to recruit the U2 snRNP, which

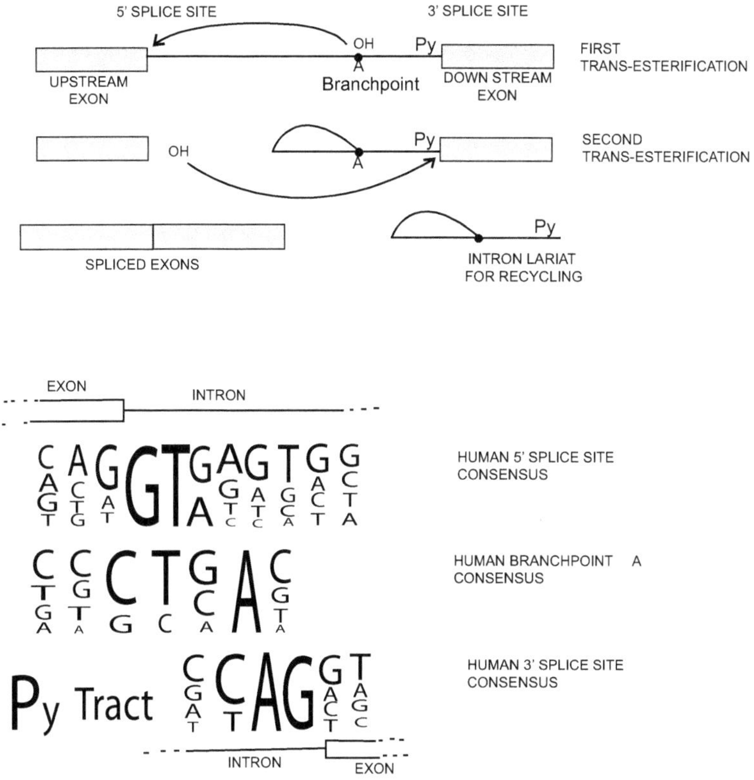

Figure 16.2 The splicing reaction and splice sites. The pre-mRNA splicing reaction consists of two transesterification reactions. The first begins with a nucleophilic attack by a hydroxyl group provided by the intronic branchpoint A, linking it to the 5′ splice site. In the second reaction, the upstream exon's free hydroxyl group attacks the 3′ splice site, joining the exons precisely together. The human consensus splice sites and branchpoint A context are shown below. *Abbreviation*: *Py*, pyrimidine-rich tract.

base-pairs across branchpoint A, allowing branchpoint A to bulge out. This is important because branchpoint A initiates the first transesterification reaction. Next the remaining three snRNPs enter the picture—initially combined together in the U4/U5.U6 tri-snurposome. The U1 and U4 snRNPs are then released. The U6 snRNP interacts with both the 5′ splice site and the U2 snRNP, while the U5 snRNP basepairs with both exons, keeping them in place in what is now the catalytically active spliceosome. The U5 snRNP contains specific **RNA helicases** that facilitate the splicing reaction. At this point the first transesterification reaction occurs, and the 2′-OH of branchpoint A attacks the phosphodiester bond between the last base of the upstream exon and the first base of the intron, cleaving the two and circularizing the intron into an **intron lariat structure**. In the second transesterification reaction, the upstream exon now attacks the downstream exon and the second transesterification reaction takes place. The intron lariat is thus released and recycled in the nucleus, and the exons are joined precisely together. This sequence of events is known as the

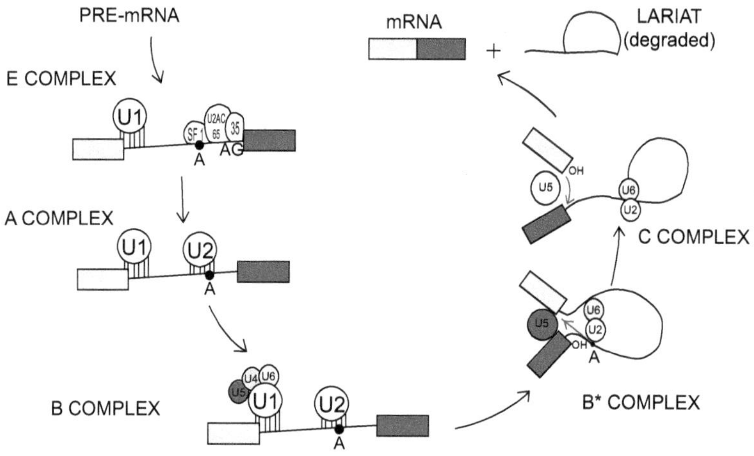

Figure 16.3 The spliceosome cycle. A molecular machine called the spliceosome has evolved to align the nucleotides involved and catalyze the splicing reaction. Splice factors first bind to nascent pre-mRNA (H-complex). The first step is to define the splice sites in order to commit to splice. The U1 snRNP base-pairs across the 5′ splice site. U2AF65 (U2 auxiliary factor) binds to the pyrimidine tract, facilitating binding of SF1 to branchpoint A; this is the E complex. Then the U2 snRNP hybridizes across branchpoint A, making it bulge out (the A complex). Now the U4/U5.U6 "trisnurposome" associates with the U1 snRNP, and the spliceosome is fully formed (B complex). Molecular rearrangements allow branchpoint A to be brought close to the upstream exon, ready for the first transesterification reaction (B* complex). Next the two exons are joined together (C complex), and splicing is completed.

spliceosome cycle (Fig. 16.3). Other than the core snRNP proteins, several, if not hundreds of, additional factors help to assemble the spliceosome in order to carry out the splicing reaction.

Most eukaryotic spliceosomal introns begin with GU and end with AG (this is known as the "GT–AG rule"). However, a small proportion of introns, present both in metazoans and in plants, follow the "AT–AC rule." It turns out that to be spliced out these introns require a different, minor snRNP complex called U12 snRNP. They are spliced more slowly than the GT–AG introns, despite following the same biochemical steps (two transesterifications starting with the branchpoint adenosine attacking the upstream 5′ splice site). AT–AC introns are handled by the "minor spliceosome" (minor because it is the least abundant). In the minor spliceosome a different set of snRNPs is used (U11, U12, U4atac, and U6atac). Only the U5 snRNP is shared between major and minor spliceosomes.

A detailed explanation of the complex interactions between snRNPs and pre-mRNA is not the purpose of this chapter. However, a basic understanding of the splicing reaction is required in order to understand the process known as **alternative splicing**, in which exons are spliced together in many different ways, as we shall see in the next section.

To understand how the spliceosome might have evolved, it is useful to consider the existence of other forms of splicing that predate the spliceosome. Group I and group II self-splicing introns are essentially parasitic sequences found in bacteria, Archaea, mitochondria, and chloroplasts (both also of bacterial origin). They catalyze their own excision from transcripts in which they are found as RNA molecules, without the aid of accessory proteins. Group II introns are particularly interesting in that their structure is quite reminiscent of the pre-mRNA spliceosome, as follows.

Group II introns are a single RNA molecule that has the ability, through a complex RNA structure that arises from basepairing interactions, to catalyze two transesterification reactions. The first reaction involves a nucleophilic attack by a 2′-OH group in an adenosine residue. This results in cleavage of the RNA and the formation of a loop structure. In the second transesterification reaction, the cleaved free 3′ end attacks another residue, causing a second cleavage and the release of the looped structure generally known as a lariat. Remarkably, the same process occurs in eukaryotic spliceosomal splicing. However, in the spliceosome the critical components of the splicing machinery have been separated during evolution into the five snRNAs. The reaction is essentially the same, in the sense that splicing is initiated by a nucleophilic attack by an A residue (known as branchpoint A in spliceosomeal splicing). However, the catalytically active spliceosome is, interestingly, still characterized by a high density of RNA performing critical catalytic functions, in the same way as in the group II self-splicing introns.

16.3 Modes and Consequences of Alternative Splicing

Following the discovery of splicing, the next question was, are exons always joined together in a precise, predictable, and consistent manner? Now over 30 years after the discovery of splicing it is overwhelmingly clear that many transcripts are in fact **alternatively spliced**. One of the first examples of alternative splicing was described in 1980. In B lymphocyte tumor cell lines there are two isoforms of the immunoglobulin mu chain, of 64 kDa and 67 kDa [2]. The 67 kDa isoform is membrane bound, whereas the 64 kDa isoform is secreted—in other words they are functionally distinct. The isoforms arise due to differences in the 3′ end of the mRNA generated through differential splicing (Calame et al. 1980). The immunoglobulin mu example illustrated an important point: splice isoforms can encode functionally distinct proteins. At first, alternative splicing was thought to be unusual; instead, we now know that alternative splicing is a widespread mechanism of gene regulation.

It is now in fact abundantly clear that alternative splicing is widespread across eukaryotes. The advent of next-generation sequencing now provides an excellent opportunity to survey alternative splicing systematically in different tissues or disease contexts. In humans, it is estimated that the vast majority, perhaps as many as 95%, of genes are subjected to alternative splicing (Pan et al. 2008).

What then are the modes of alternative splicing? These are shown in Fig. 16.4. The most prominent modes are **exon skipping**,

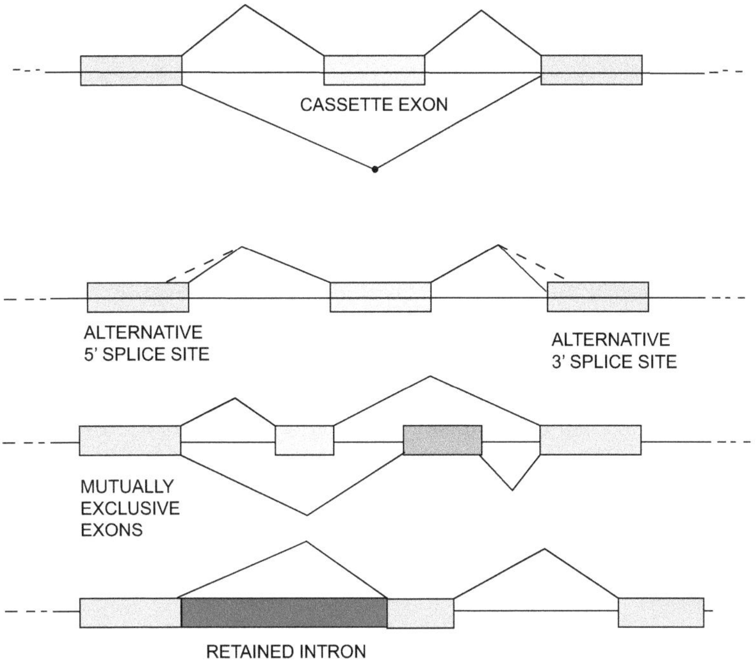

Figure 16.4 The main modes of alternative splicing. Exons can sometimes be skipped; these are known as cassette exons. The lengths of exons can differ due to either alternative 5′ or alternative 3′ splice sites. Mutually exclusive exons arise when either one or the next exon (but not both) can be spliced. Occasionally, introns are not removed; these are known as retained introns.

when individual exons are not spliced out. These exons are also referred to as **cassette exons**. Another mode is the **alternative exon size**; this occurs when there are **alternative 5′ or 3′ splice sites**, which can effectively lengthen or shorten an exon. Alternative splice sites can sometimes occur when mutations, or changes in levels of regulatory factors, can activate **cryptic splice sites**—that is, splice sites that are otherwise less prone to being selected by the splicing machinery. Another form of alternative splicing is **intron retention**, when introns are simply not removed from the pre-mRNA. Exon skipping is the most common form of alternative splicing in metazoans (multicellular animals), followed by alternative exon size. Intron retention is less common in metazoans but the most common in yeasts and plants. The consequences of alternative splicing can be profound.

There are some additional subtle variations of these modes of alternative splicing. **Mutually exclusive exons** are two exons in succession where only one can be included in the mature mRNA. Alternative promoters can result in different exon inclusion at the 5′ end of mRNAs; and lastly, alternative polyadenylation signals can truncate mRNAs at different points, thus affecting exon inclusion at the 3′ end. It is important to note that individual exons can provide both alternative 5′ and 3′ splice sites, while also being cassette exons. It is also important to realize that all parts of an mRNA can be affected by alternative splicing—not only the open reading frame (ORF) but also the 5′- and 3′-untranslated regions (UTRs), which regulate mRNA stability, translation, and localization (Fig. 16.5). In other words, alternative splicing can affect both protein coding information and gene expression.

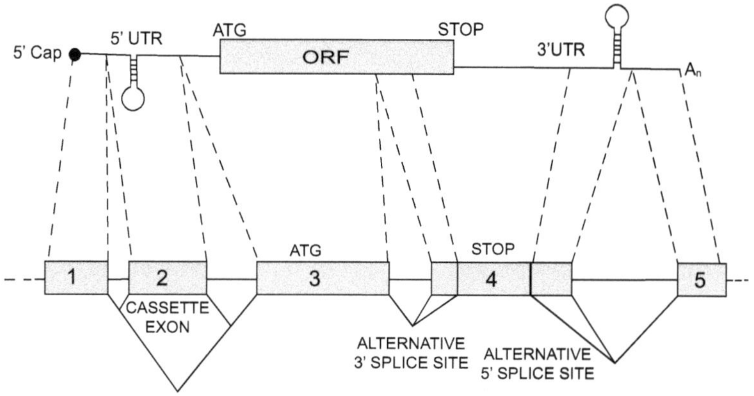

Figure 16.5 The functional consequences of alternative splicing. Alternative splicing can affect all aspects of an mRNA. In this theoretical mRNA, cassette exon 2 encodes part of the 5′-UTR with a strong stem loop that could be involved in repressing mRNA translation. An alternative 3′ splice site in exon 4 could add, in the frame, an extra protein domain within the ORF. An alternative 5′ splice site in exon 4 could result in an extra sequence in the 3′-UTR, perhaps adding another stem–loop structure, involved in the regulation of mRNA stability or localization.

Taking into account the versatility of alternative splicing, it is easy to see how even a simple multiexon gene can express, potentially, a significant number of splice variants. This principle is powerfully illustrated by the notorious example of the Down syndrome cell adhesion molecule (*DSCAM*) gene in the fruit fly *Drosophila melanogaster*. The *DSCAM* gene is a member of the immunoglobulin superfamily and is involved in the establishment of neural connections. It encodes an axon guidance receptor. The gene essentially consists of 95 exons that, however, include 4 clusters of mutually exclusive exons: 12 alternative versions of exon 4, 48 alternatives of exon 6, 33 alternatives of exon 9, and 2 versions of exon 17 (Celotto and Graveley, 2001). Taken together, the *DSCAM* gene can therefore express, potentially, 38016 splice variants! This number exceeds, by a large margin, the total number of genes in the fruit fly genome (around 15,000).

Why is *DSCAM* alternative splicing so extreme? The consequences of its alternative splicing are to alter the biochemical properties of the protein isoforms to facilitate substantial plasticity in the way that neurons can connect to each other. The take-home message is that alternative splicing can, and does, hugely increase proteomic complexity. The emergence of exons and introns in evolution contributed very significantly to an expansion of the genome's coding potential.

16.4 Alternative Splicing in Disease

The emergence of alternative splicing in eukaryotes therefore provides an excellent mechanism to increase proteomic complexity. Not surprisingly, alternative splicing is involved in normal development. This is illustrated by the interaction between the epithelium and the mesenchyme—an essential process in the development of several organ systems. Mesenchymal cells are relatively unstructured and interact with the extracellular matrix so that they can migrate toward their intended destinations. Epithelial cells normally form part of established structures involved in selective absorption secretion, protection, and the detection of sensation. Epithelial cells are polarized, so their apical surface faces a lumen, whereas the basolateral surfaces interact with the basal lamina. These two cell types can be converted reciprocally in development—what is known as epithelial–mesenchymal transition (EMT) and mesenchymal–epithelial transition (MET). Both types of transitions are required for normal development; but in disease EMT has been associated with the development of metastasis due to the properties of mesenchymal cells. There is therefore considerable interest in the genetic programs that control EMT.

Recent work has shown that there are complex changes in alternative splicing during EMT (Warzecha and Carstens, 2012). Genes whose alternative splicing changes during EMT include fibroblast growth factor receptor 2 (*FGFR2*), p120-*catenin* (a regulator of cadherin activity), *CD44* (involved in cell adhesion), and *Mena* (a regulator of cell motility). The splice isoforms that are expressed in mesenchymal cells are consistent with the properties of these cells.

So if alternative splicing plays a critical role in normal development, then it stands to reason to expect that aberrant splicing might be associated with disease. This is indeed the case. It has been estimated that at least 15% of all disease-causing mutations in humans affect alternative splicing. Several of these mutations affect, directly, 5′ or 3′ splice sites; however, mutations can also affect the branchpoint A, the pyrimidine tract, and splice enhancer or silencer sequences. For that reason it is important, when looking for novel mutations, ideally, to look at intronic sequences and not just exonic sequences. The breast cancer 1 (*BRCA1*) gene illustrates this principle. *BRCA1* is involved in the DNA repair process, and mutations are strongly associated with a predisposition to breast cancer. A point mutation in exon 18 was shown to lead to increased exon skipping (and thus dysfunctional *BRCA1*). The point mutation affects the binding site of a well-known splice factor called SRSF1 (previously known as ASF/SF2; as we shall see in section 16.5, splice factors are proteins that can regulate alternative splicing). Interestingly, SRSF1 was the first splice factor described as a proto-oncogene, as its overexpression is linked to cancer (Karni et al. 2007).

Individual splice factors such as SRSF1 have multiple targets; in other words they affect the alternative splicing of several genes. One of SRSF1's best-known targets is *RON*. The *RON* gene encodes a receptor tyrosine kinase with a cassette exon 11. Skipping of exon 11 results in an in-frame loss of 49 amino acids, which makes the receptor constitutively active (this property is oncogenic). SRSF1 binds to exon 12, the consequence being that the splicing machinery is drawn preferentially towards exon 12, so exon 11 tends to be skipped (Ghigna et al. 2010). The ability of SRSF1 to promote the oncogenic isoform of *RON* is consistent with its having the properties of a proto-oncogene.

A quick survey of the literature will reveal that changes in alternative splicing are associated with all types of diseases, from cancer to diabetes, vascular disease, aging, neurodegenerative diseases, etc. Cancer is an increasing problem due to the aging human population; not surprisingly a lot of resources go into cancer research. Due to the heterogeneity and complexity of cancer, it is important to tackle it from all possible angles. Alternative splicing has emerged as a very important new area in cancer research. This is illustrated by systematic research into alternative splicing markers, for example in breast cancer. The new concept is that it is not necessarily the overall expression of biomarker genes that matters, but rather their splicing patterns. Alternative splicing patterns could even be used as a diagnostic or prognostic tool. One high-throughput study looked at 600 cancer-associated genes; 41 alternative splicing events differed significantly in breast tumors relative to normal tissue. Most of the alternative splicing changes were associated with changes to proteins linked to increased proliferation and survival (Venables et al. 2008). It should be noted, however, that alternative splicing patterns can change within tumors and in metastases—but this heterogeneity potentially applies to all cancer biomarkers.

So in summary, mutations can affect splice sites, splice regulatory sequences, or the expression and activity of splice factors (regulators of splicing), and these mutations have been associated with a wide range of diseases. However, it should be noted that mutations can also affect the core splicing machinery. For example, mutations in the *PRP8* gene are associated with the disease retinitis pigmentosa, a disorder in which the loss of photoreceptor function leads to visual loss. PRP8 is an essential component of the U4/U5.U6 tri-snurposome. The disease is caused by dominant mutations in the gene; the wild-type copy ensures that the core splicing machinery is not entirely compromised.

Another classical example of a mutation that affects the core splicing machinery relates to the case of spinal muscular atrophy (SMA). SMA is a neurodegenerative disease that causes muscular wasting and impaired mobility. SMA occurs in 1 in 6,000 live births. The disease is caused by an autosomal recessive mutation in the *SMN1* gene. *SMN1* encodes a multifunctional protein involved in snRNP assembly, nuclear snRNP recycling, and even cytoplasmic transport of mRNAs. Adjacent to *SMN1*, in humans there is a second, closely related gene, *SMN2*. Although less efficiently expressed, wild-type *SMN2* is thought to render the *SMN1* mutation sublethal. However, mutations that affect *SMN2* have also been found, including mutations that cause the skipping of exon 7, leading to a nonfunctional protein (Cartegni et al. 2006).

16.5 The Regulation of Alternative Splicing

The need to determine, in pre-mRNA, what is an exon and what is an intron is a difficult problem. There are two principal models that explain how this is achieved, the exon definition model and the intron definition model. In the exon definition model, snRNPs and **splice factors** (proteins that facilitate snRNP binding) bind to the pre-mRNA. Of particular importance are the U1 snRNP (in defining 5′ splice sites) and the U2 snRNA (in defining the branchpoint A). In the exon definition model, snRNPs and splice factors interact across exons. For example, the U2 snRNP and U2AF and other splice factors interact with the U1 snRNP at the end of the following exon, thus helping to define the exon. In contrast in the intron definition model, early components of the spliceosome assemble across introns and interact across introns. Whichever model is correct, the key concept is that splice factors and snRNPs interact and "talk" to each other to bring about the definition of exons.

Splice factors are proteins that can recognize specific RNA sequences—that is, they are RNA-binding proteins. Several protein domains have evolved that can bind RNA; the most frequently encountered RNA-binding domain is the RNA recognition motif

(RRM) which generally recognizes single-stranded RNA (ssRNA) target sequences. Auxiliary domains in the splice factors facilitate protein–protein interactions; the best known is the serine-arginine (SR)rich domain. Among the splice factors, the **SR protein** family is widely studied; U2AF65 is in fact an SR protein that binds the pyrimidine-rich tract that lies upstream of 3′ splice sites. Splice factors can bind throughout exons and introns to sequences known as exonic splice enhancers (ESEs), exonic splice silencers (ESSs), intronic splice enhancers (ISEs), and intronic splice silencers (ISSs). These sequences can help define where splicing occurs. Thus **splice enhancers** can promote what might otherwise be a weak splice site; alternatively **splice silencers** could repress what is otherwise a strong splice site. Among the splice factors, the SR proteins are generally known to work as splicing activators whereas, the hnRNP proteins (another class of abundant RNA-binding proteins) are generally splicing repressors.

Thus the interaction of specific, regulatory RNA sequences, and splice factors that bind them, can help to regulate alternative splicing in a tissue, and developmental-specific way (Fig. 16.6). The information required to regulate alternative splicing is therefore embedded in the pre-mRNA itself and is referred to as the **splicing code**.

The activity of splice factors can be regulated by reversible protein phosphorylation (Stamm, 2008). Phosphorylation can affect splice factors in two ways, by altering their biochemical properties (their ability to bind RNA or to interact with other proteins) or by altering their intracellular localization. Several splice factors are known to shuttle between the nucleus and the cytoplasm. This is a means of regulating their intranuclear concentration; some splice factors (such as SRSF1) have cytoplasmic functions in the regulation of mRNA translation, stability, or localization. The dephosphorylation of SRSF1 promotes its nuclear export, whereas its phosphorylation by SR protein kinase (SRPK1) promotes import into the nucleus (Giannakouros et al. 2011). SRPK1's substrate is the SR domain itself, rich in serine residues. In the nucleus, SRSF1 accumulates in the **nuclear speckles**, sites that are involved in the storage of splice factors. SRSF1 can be further hyperphosphorylated by the Clk kinases (CDC2-like or Lammer

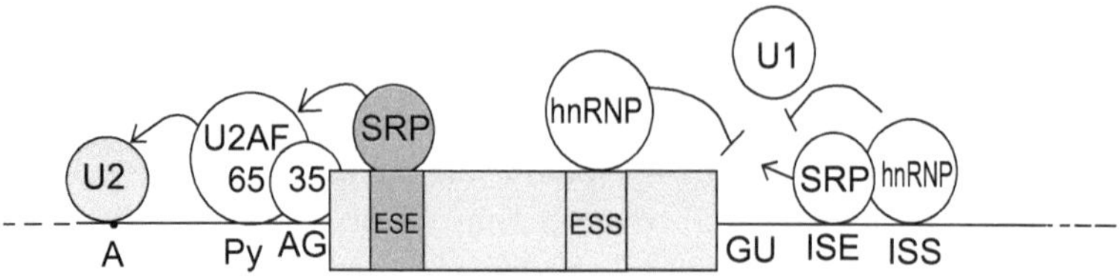

Figure 16.6 The regulation of alternative splicing. Splicing can be regulated in several ways. In this theoretical exon, there is an ESE at its 5′ end. It is bound by SR proteins (an abundant family of splice factors) that through protein–protein interactions facilitate recruitment of the U2AF (specifically the U2AF65:U2AF35 complex). U2AF65 is bound to the intronic pyrimidine tract, just upstream of the 3′ splice site following the AG dinucleotide that defines the end of an intron. The purpose of U2AF is to then help recruit the U2 snRNP, which hybridizes across branchpoint A. Toward the 3′ end of the exon, an ESS could be bound by hnRNP proteins that sterically compete the association of a U1 snRNP with the 5′ splice site. Downstream, an ISE could also be bound by SR proteins that help recruit the U1snRNP. The ISE could overlap with an ISS also bound by hnRNP proteins that compete for binding with SR proteins. In this scenario, the relative levels of various splice factors can determine the extent to which particular splice sites are used.

kinases); its hyperphosphorylation promotes the release of SRSF1 from the nuclear speckles to areas where the splicing reactions occur (Colwill et al. 1996). Thus splice factor kinases can affect the compartmentalization and availability of splice factors.

An example of the biological importance of the phosphorylation of SRSF1 is in the regulation of alternative splicing of *VEGF-A* which encodes vascular endothelial growth factor. *VEGF-A* is alternatively spliced producing pro- or antiangiogenic growth factor isoforms. SRSF1 promotes the pro-angiogenic isoform by facilitating the use of a proximal 3′ splice site in exon 8, the terminal exon in the *VEGF-A* gene. When SRPK1 levels are high, SRSF1 accumulates in the nucleus, and this causes high pro-angiogenic VEGF-A expression. Knockdown or chemical inhibition of the SRPK1 results in cytoplasmic accumulation of SRSF1, so the antiangiogenic isoform of VEGF-A is preferentially expressed (Amin et al. 2011).

Protein phosphatases are also involved in modulating splice factor function. Cell signaling through ceramide, a sphingolipid second messenger, activates the protein phosphatase PP1. In response to tumor necrosis factor α(TNFα), PP1 dephosphorylates SRSF5 (SRp40), causing exon skipping of PKCβ (Ghosh et al. 2007). The result is an isoform of PKCβ that does not promote glucose uptake—a potential mechanism of insulin resistance.

The involvement of protein kinases and phosphatases suggests that cell signaling can play an important role in regulating alternative splicing. This is indeed the case; a classic example is the *CD44* gene and its cassette exon v5. *CD44* encodes a cell membrane glycoprotein that plays a role in cell adhesion. *CD44* exons 6–14 are known as the 10 alternative exons v1–v10; these exons are generally included in proliferating cells and tumors. Inclusion of exon v5 is facilitated by a splice factor called Sam68. In nonproliferating cells, exon v5 is bound and repressed by the abundant RNA-binding protein hnRNP A1. After activation, the well-known mitogenic RAS-signaling pathway is then followed by the mitogen-activated protein (MAP) kinase cascade. As a result of this signaling cascade, Sam68 is phosphorylated, changing its properties so that it now binds to the v5 exon, promoting its inclusion (Matter et al. 2002).

Sometimes RNA sequences that form stable stem–loop structures can also, in themselves, and without the aid of splice factors, repress splicing. This can occur when they cover a splice site, as illustrated by *Tau* exon 10. *Tau* encodes a microtubule-associated protein that is involved in neurodegenerative disease. Mutations that affect the stability of this stem loop are associated with differential exon 10 inclusion and tauopathy (Ray et al. 2011). Another factor is the structure of the gene itself. Very long introns can mean that upstream exons are spliced together before a competing (strong) splice site has been transcribed. This point illustrates the fact that splicing is thought to be generally cotranscriptional and in fact that RNA polymerase II itself can even help recruit splice factors when its C-terminal domain (CTD) is hyperphosphorylated (Du and Warren, 1997).

However, it should be noted that there are other ways in which alternative splicing can be regulated—through the use of different promoters. That transcription can affect alternative splicing became apparent when researchers noticed that the type of promoter used in a minigene construct influences the ratio of splice isoforms expressed. (A minigene is a synthetic gene in which a subset of exons and introns is transcribed in order to study a specific alternative splicing event). A mechanism through which promoters can affect alternative splicing is simply the speed of transcriptional elongation by RNA polymerase II. When transcription is fast, the spliceosomes will tend to prefer stronger splice sites, whereas if transcription is slow, weaker splice sites have more time to be recognized. A classic example is the EDI cassette exon in the *fibronectin* gene. The cotransfection of a construct that encoded an RNA polymerase II that was mutated so as to be slower increased EDI exon inclusion (de la Mata et al. 2003). The speed of RNA polymerase II elongation can also change in different parts of the gene, affecting the alternative splicing of specific exons.

A further layer of complexity is due to the fact that the chromatin-remodeling machinery can also affect alternative splicing. The SWI/SNF complex is best known for its ability to remodel chromatin by altering nucleosome positioning. Brahma, an adenosine triphosphatase (ATPase), is one of its components. Unexpectedly, Brahma has also been shown to affect splice isoform levels and is detected on transcripts (Tyagi et al. 2009). Details of the mechanism are not yet fully understood, but there are very strong indications that nucleosomes are preferentially enriched in exons compared to introns, suggesting that chromatin structure could contribute to the process of exon definition and therefore to alternative splicing (Luco et al. 2011). Table 16.1 summarizes the several factors that are thought to affect alternative splicing.

Finally, it is also worth noting that there are interactions between the splicing machinery and post-transcriptional processes. A complex of RNA-binding proteins known as the **exon junction complex** (EJC) is deposited just upstream of exon–exon splice junctions. The EJC is complex in composition; it contains factors that affect several aspects of an mRNA's life cycle. So the EJC promotes export from the nucleus; this makes sense because correctly spliced mRNAs need to be exported. The EJC also contains factors that, perhaps surprisingly, promote mRNA degradation in the cytoplasm. However, when an mRNA is correctly translated, the EJC is removed in the so-called pioneer round of translation. When mutations create premature termination codons (PTCs), EJCs remain bound

Table 16.1 Factors that affect alternative splicing. RNA sequence, RNA structure, RNA-binding proteins, and chromatin work together to regulate alternative splicing

Factors that affect alternative splicing	General principles involved
Intrinsic strength of splice sites	Some splice sites are easily recognized by the splicing machinery; weaker splice sites require auxiliary factors.
Pre-mRNA structure	RNA has a tendency to form structures, e.g., stem loops, that can mask a splice site or a regulatory sequence.
Splice enhancer sequences (ESE, ISE)	Sequences in pre-mRNA bound by splice factors (typically members of the SR protein family) facilitate spliceosome assembly.
Splice silencer sequences (ISE, ISS)	Sequences in pre-mRNA bound by splice factors (typically members of the hnRNP protein family) prevent spliceosome assembly.
Transcription rates	Rates of transcription can affect alternative splicing; slower transcription can favor weaker splice sites.
Gene structure	The sizes of exons and introns can affect alternative splicing.
Chromatin modification	The chromatin structure can influence alternative splicing; nucleosomes are preferentially concentrated in exons.
Splice factor kinases and phosphatases	The reversible phosphorylation of splice factors modulates their intracellular localization and activity. Therefore cell signaling can influence alternative splicing.

to mRNAs, thus promoting the destruction of mutant mRNAs—this process is known as nonsense-mediated decay (NMD). Exons that contain PTCs can therefore be used to target mRNAs for degradation—these are known as **poison exons**. It is estimated that up to 30% of alternative splicing introduces poison exons in humans—a significant proportion indicating that they have an important role in gene regulation.

16.6 The Manipulation of Alternative Splicing in Therapy

Given its increased importance in disease it is then appropriate to think of ways of manipulating alternative splicing in terms of potential therapies (Wilton and Fletcher, 2011). In the main, there are at least two possible approaches (Fig. 16.7). The first is to

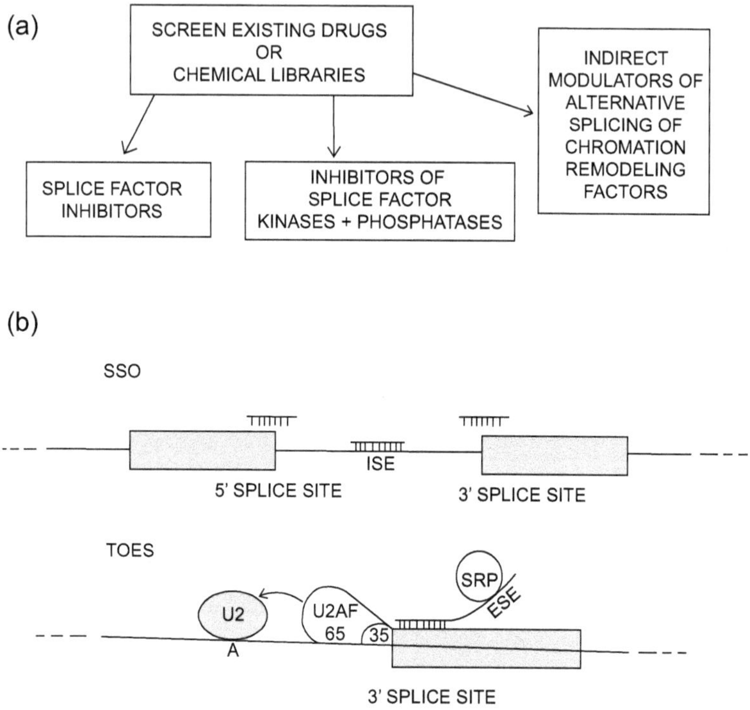

Figure 16.7 Therapeutic strategies aimed at alternative splicing. (A) The alternative splicing of disease-associated splice isoforms can be targeted directly using compounds that target splice factors or protein kinases and phosphatases that modify splice factor function. It can also be targeted at factors that modify alternative splicing indirectly; for example, chromatin-remodeling factors. (B) Antisense oligonucleotides can be used to target specific pre-mRNAs directly. SSOs can sterically block splice sites or regulatory sequences (e.g., an ISE). TOES can be used to recruit splice factors that promote the inclusion of an otherwise skipped exon, effectively acting as ESEs, *Abbreviations*: SSO, splice-switching oligonucleotide; TOES, targeted oligonucleotide enhancers of splicing.

use compounds to target the machinery that regulates alternative splicing. Thus splice factors themselves could be targeted, or the protein kinases and phosphatases that modulate their function, or even the chromatin-remodeling factors that participate in exon definition. However, targeting these factors could have side effects due to the fact that they affect the expression of many genes. This is a common problem that does not necessarily invalidate the approach a priori—several drugs are not necessarily particularly specific, and side effects are often a problem. However, in the appropriate context, and with appropriate doses and delivery methods, this type of approach can be effective.

Several groups around the world are attempting to isolate compounds that can modify alternative splicing. There are two approaches: one is to determine if existing drugs (that target other pathways) might affect alternative splicing. The second is to develop new compounds that specifically target the splicing machinery. Amiloride is an example of the former approach. Amiloride is an antihypertensive drug; it works as a potassium-sparing diuretic. A recent study has shown that amiloride can modulate the alternative splicing of genes involved in human leukemia (Chang et al. 2011). Specifically, amiloride can induce apoptosis. Several genes that are involved in apoptosis are alternatively spliced into pro- or antiapoptotic isoforms. Amiloride appears to alter the alternative splicing of genes such as *apaf-1*, *survivin*, and *bcl-x*. The mechanism of action of amiloride is not yet clear, but there appear to be changes in the expression of several splice factors and in the phosphorylation status of SR protein splice factors.

The other approach is to develop novel compounds. For example, the Hagiwara Laboratory at the University of Kyoto screens chemical libraries to identify compounds that target splice factor kinases (Ogawa and Hagiwara, 2012). One of the compounds they developed is SRPIN340, an isonicotinamide that targets SRPKs. Initially SRPIN340 was shown to have potential in the treatment of viral infection (several viruses need to splice their own pre-mRNAs in a controlled fashion). SRPIN340 can, for example, inhibit hepatitis C virus replication (Karakama et al. 2010). It is now clear that SRPIN340 can also reduce angiogenesis in tumors and in a model of retinal neovascularization relevant to eye disease (Amin et al.

2011). By inactivating SRPKs, SRPIN340 alters the intracellular distribution of key splice factors such as SRSF1—the effect of this is to alter the alternative splicing of vascular endothelial growth factor (VEGF-A) to favor the antiangiogenic isoform. However, the consequences of targeting SRPK1 on the other hallmarks of cancer need to be determined, and the half-life of the compound in serum needs to be improved before research can move on to clinical trials. Nonetheless, the SRPIN340 example shows how the alternative splicing machinery can be targeted relatively generically to achieve a highly significant reduction in angiogenesis in vivo (Fig. 16.8).

Another approach is to target specific RNA sequences directly using antisense technology. In theory the antisense approach is more specific, but in practice there are several practical problems with antisense oligonucleotides that need to be overcome, not least the delivery, toxicity, and stability of the molecules. Toxicity and stability can be addressed by chemically modifying oligonucleotides. Phosphorodiamidate morpholino oligomers (PMOs), DNA phosphorothioates, 2′-*O*-methyl (2′OMe) oligomers, or peptide nucleic acid (PNA) have all been used when attempting to modify splicing in vivo (Bauman et al. 2009).

Apoptosis is one of the hallmarks of cancer, and so therapies are targeted at blocking the tendency of tumor cells to evade apoptosis. Apoptosis is a complex process that is controlled by several genes. Among these the B cell lymphoma 2 (*Bcl-2*) family is known to have a significant role in apoptosis by altering mitochondrial function. *Bcl-X* is a member of this family; an alternative 5′ splice site in exon 2 results in the generation of two isoforms, an antiapoptotic isoform (Bcl-X_L) and a pro-apoptotic isoform (Bcl-X_S). SSOs are a class of antisense oligonucleotide designed to block, sterically, specific RNA sequences. They work in the nucleus specifically to modify alternative splicing outcomes (Bauman and Kole 2011). Thus SSOs are not designed to cause mRNA translation repression or mRNA degradation. Several antisense approaches work well in vitro, but it is often difficult to then demonstrate their in vivo effectiveness due to difficulties delivering the compound to the intended target. However, Bauman et al. were able to use lipid nanoparticles to deliver the SSOs, thereby altering *Bcl-X* splicing not only in cell line

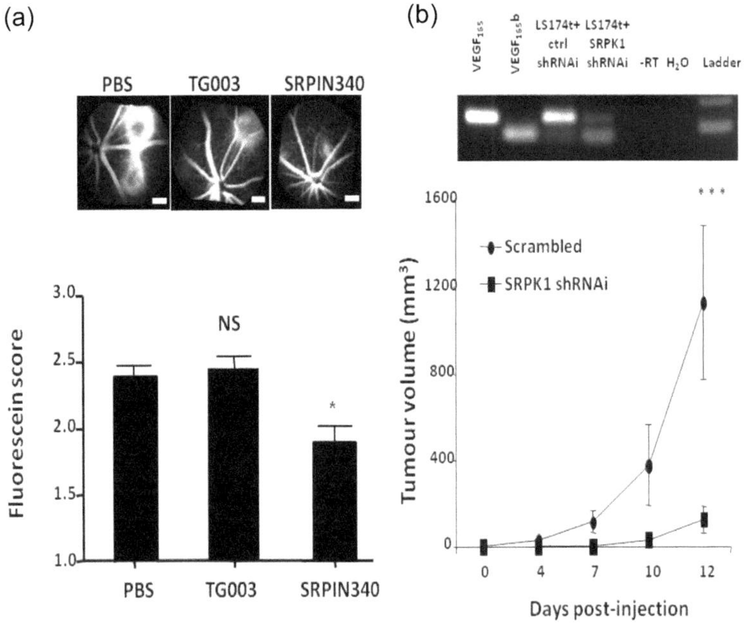

Figure 16.8 The inhibition of SRPK1 is antiangiogenic. (a) In a mouse model of choroidal neovascularization, the administration of the SRPK1 inhibitor SRPIN340 greatly reduced angiogenesis (measured via fluorescein angiography). Treatment with PBS alone or the Clk1/4 kinase inhibitor TG003 did not have the same effect. (b) shRNA-mediated knockdown of SRPK1 in the LS174t colorectal carcinoma cell line shifted VEGF splicing toward the antiangiogenic isoform ($VEGF_{165}$ b). When these tumor cells were implanted into nude mice, SRPK1 knockdown severely restricted tumor growth in vivo. Adapted from Amin et al. (2011). *Abbreviations*: PBS, phosphate buffer saline; shRNA, small hairpin RNA.

models (Fig. 16.9) but also in vivo, reducing lung metastases and tumor load in mouse models (Baumann et al. 201).

There is an alternative approach when using antisense oligonucleotides. The idea is that rather than attempting to mask a splice site or splice enhancer sequence, antisense oligonucleotides could be used to deliver splicing enhancer sequences to particular sites—for example, to exons that are being inappropriately skipped. This approach was developed in the Eperon Laboratory at the University of Leicester and is known as TOES. The idea is to design an

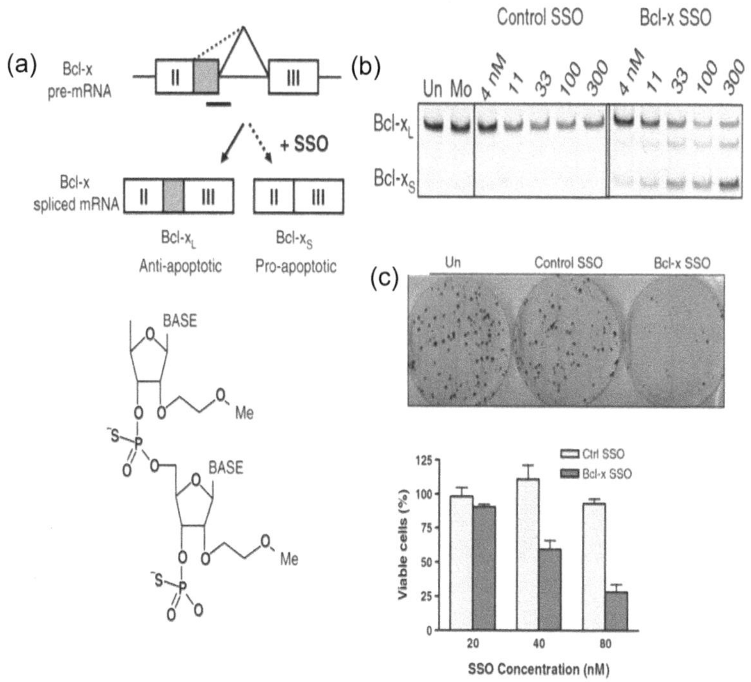

Figure 16.9 SSOs increase pro-apoptotic Bcl-X_S in vivo. (a) The downstream 5′ splice site in exon II produces antiapoptotic Bcl-X. The idea is to use an SSO to block this splice site, favoring the expression of the pro-apoptotic isoform. (b) As little as 4nM of the SSO causes a discernible shift in Bcl-X splicing. (c) By promoting pro-apoptotic Bcl-X_S, the net effect is significantly increased cell death as demonstrated by using a clonogenic assay. Adapted from Bauman et al. (2010).

oligonucleotide with two domains, an annealing domain and a tail domain. The annealing domain is complementary to any convenient sequence in the target exon; the tail domain is engineered to carry high-efficiency splice enhancer sequences that attract an appropriate SR protein that can then promote the inclusion of the intended exon. This approach has been applied to the case of the skipped exon 7 in the *SMN2* gene, which was briefly discussed in section 16.4 in the context of Spinal Muscular Atrophy (SMA). TOES oligonucleotides were specifically targeted to *SMN2* exon 7, resulting in a significant increase in exon 7 inclusion in patient-derived fibroblasts (Owen et al. 2011). The notion that manipulating *SMN2* exon 7 inclusion could lead to a therapy had been demonstrated before; this was achieved using an approach similar to TOES. The expression of a modified U7 snRNA (a special snRNA involved in histone mRNA processing) that contains a targeting sequence and splice enhancer domain rescued the SMA phenotype in a mouse model of the disease (Meyer et al. 2009).

Lastly, we turn our attention to muscular dystrophy, a disease in which muscles weaken progressively. The most common is Duchenne's muscular dystrophy (DMD; incidence of 1 in 3,500 males). The gene involved is called the *dystrophin* gene, present on the X chromosome. Dystrophin connects the actin cytoskeleton to the extracellular matrix and is involved in muscle contraction. Mutations have been found to cause PTCs in several exons, causing the expression of a truncated, nonfunctional protein. The antisense strategy applied to DMD is to design oligonucleotides that promote the skipping of exons that contain PTCs (reviewed in Aartsma-Rus 2010). Research was begun in cell lines and then taken to mouse models; it then progressed to clinical trials involving direct injections into muscle tissue. This caused the desired exon skipping resulting in the expression of normal dystrophin. A phase II study has confirmed good tolerance and effectiveness of an antisense oligonucleotide aimed at achieving *dystrophin* exon skipping (Cirak et al. 2011). Thus although more work is still required to turn this approach into an effective and reliable treatment, again the validity of the principle of targeting alternative splicing in disease is confirmed.

16.7 Summary and Key Points

- Nuclear pre-mRNA splicing was discovered in the late 1970s in the adenoviral *hexon* gene and subsequently observed in cellular genes. It consists of the precise removal of introns from pre-mRNA and the precise joining together of exons. The splicing reaction consists of two transesterification reactions catalyzed by the spliceosome, a complex of the snRNP complexes consisting of the U1, U2, U4, U5, and U6 snRNPs. Splice site and exon definition requires the assistance of RNA-binding splice factors.

- The vast majority of human genes (>95%) is alternatively spliced, and alternative splicing has emerged as a key generator of proteomic complexity across eukaryotes. The main modes of alternative splicing include skipped (cassette) exons, mutually exclusive exons, intron retention, alternative 5′ splice sites, and alternative 3′ splice sites.
- Regulatory sequences in pre-mRNA, both in exons and in introns, can influence alternative splicing. These sequences are bound by splice factors that can either activate or repress specific splice isoforms. RNA sequences can also give rise to secondary structures that affect splice site availability. Splice factor kinases and phosphatases, through cell signaling, modulate splice factor function. Transcription rates and chromatin structure can also influence alternative splicing indirectly.
- Alternative splicing is very important in both development and disease. One in six disease-causing mutations in humans is thought to affect splicing. Studying the alternative splicing of key genes could in the future contribute to the diagnosis and prognosis of disease. The manipulation of alternative splicing has therapeutic potential; there are two approaches, namely, targeting the splicing machinery directly and targeting specific pre-mRNA sequences using antisense oligonucleotides.

References

1. Aartsma-Rus, A. (2010). Antisense-mediated modulation of splicing: therapeutic implications for Duchenne muscular dystrophy. *RNA Biol.*, **7**, 453–461.
2. Alt, F.W., et al. (1980). Synthesis of secreted and membrane-bound immunoglobulin mu heavy chains is directed by mRNAs that differ at their 3′ ends. *Cell*, **20**, 293–301.

3. Amin, E.M., et al. (2011). WT1 mutants reveal SRPK1 to be a downstream angiogenesis target by altering VEGF splicing. *Cancer Cell*, **20**, 768–780.
4. Bauman, J., Jearawiriyapaisarn, N., and Kole, R. (2009). Therapeutic potential of splice-switching oligonucleotides. *Oligonucleotides*, **19**, 1–13.
5. Bauman, J.A., et al. (2010). Anti-tumor activity of splice-switching oligonucleotides. *Nucleic Acids Res.*, **38**, 8348–8356.
6. Calame, K., et al. (1980). Mouse Cmu heavy chain immunoglobulin gene segment contains three intervening sequences separating domains. *Nature*, **284**, 452–455.
7. Cartegni, L., et al. (2006). Determinants of exon 7 splicing in the spinal muscular atrophy genes, *SMN1* and *SMN2*. *Am. J. Hum. Genet.*, **78**, 63–77.
8. Celotto, A.M., and Graveley, B.R. (2001). Alternative splicing of the Drosophila Dscam pre-mRNA is both temporally and spatially regulated. *Genetics*, **159**, 599–608.
9. Chang, W.H., et al. (2011). Amiloride modulates alternative splicing in leukemic cells and resensitizes Bcr-AblT315I mutant cells to imatinib. *Cancer Res.*, **71**, 383–392.
10. Cirak, S., et al. (2011). Exon skipping and dystrophin restoration in patients with Duchenne muscular dystrophy after systemic phosphorodiamidate morpholino oligomer treatment: an open-label, phase 2, dose-escalation study. *Lancet*, **378**, 595–605.
11. Colwill, K., et al. (1996). The Clk/Sty protein kinase phosphorylates SR splicing factors and regulates their intranuclear distribution. *EMBO J.* **15**, 265–275.
12. De la Mata, M., et al. (2003). A slow RNA polymerase II affects alternative splicing in vivo. *Mol. Cell*, **12**, 525–532.
13. Du, L., and Warren, S.L. (1997). A functional interaction between the carboxy-terminal domain of RNA polymerase II and pre-mRNA splicing. *J. Cell. Biol.*, **136**, 5–18.
14. Ghigna, C., et al. (2010). Pro-metastatic splicing of Ron proto-oncogene mRNA can be reversed: therapeutic potential of bifunctional oligonucleotides and indole derivatives. *RNA Biol.*, **7**, 495–503.
15. Ghosh, N., et al. (2007). Ceramide-activated protein phosphatase involvement in insulin resistance via Akt, serine/arginine-rich protein 40, and ribonucleic acid splicing in L6 skeletal muscle cells. *Endocrinology*, **148**, 1359–1366.

16. Giannakouros, T., Nikolakaki, E., Mylonis, I., and Georgatsou, E. (2011). Serine-arginine protein kinases: a small protein kinase family with a large cellular presence. *FEBS J.*, **278**, 570–586.
17. Karakama, Y., et al. (2010). Inhibition of hepatitis C virus replication by a specific inhibitor of serine-arginine-rich protein kinase. *Antimicrob. Agents Chemother*, **54**, 3179–3186.
18. Karni, R., et al. (2007). The gene encoding the splicing factor SF2/ASF is a proto-oncogene. *Nat. Struct. Mol. Biol.*, **14**, 185–193.
19. Luco, R.F., et al. (2011). Epigenetics in alternative pre-mRNA splicing. *Cell*, **144**, 16–26.
20. Matter, N., Herrlich, P., and König, H. (2002). Signal-dependent regulation of splicing via phosphorylation of Sam68. *Nature*, **420**, 691–695.
21. Meyer, K., et al. (2009). Rescue of a severe mouse model for spinal muscular atrophy by U7 snRNA-mediated splicing modulation. *Hum. Mol. Genet.*, **8**, 546–555.
22. Ogawa, Y., and Hagiwara, M. (2012). Challenges to congenital genetic disorders with "RNA-targeting" chemical compounds. *Pharmacol. Ther* **134**, 298–305.
23. Owen, N., et al. (2011). Design principles for bifunctional targeted oligonucleotide enhancers of splicing. *Nucleic Acids Res.*, **39**, 7194–7208.
24. Pan, Q., et al. (2008). Deep surveying of alternative splicing complexity in the human transcriptome by high-throughput sequencing. *Nat. Genet.*, **40**, 1413–1415.
25. Ray, P., et al. (2011). PSF suppresses tau exon 10 inclusion by interacting with a stem-loop structure downstream of exon 10. *J. Mol. Neurosci.*, **45**, 453–466.
26. Stamm, S. (2008). Regulation of alternative splicing by reversible protein phosphorylation. *J. Biol. Chem.*, **283**, 1223–1227.
27. Tyagi, A., et al. (2009). SWI/SNF associates with nascent pre-mRNPs and regulates alternative pre-mRNA processing. *PLoS Genet.*, **5**, e1000470.
28. Venables, J.P., et al. (2008). Identification of alternative splicing markers for breast cancer. *Cancer Res.*, **68**, 9525–9531.
29. Warzecha, C.C., and Carstens, R.P. (2012). Complex changes in alternative pre-mRNA splicing play a central role in the epithelial-to-mesenchymal transition (EMT). *Semin. Cancer Biol.*, **22**(5–6), 417–427.
30. Wilton, S.D., and Fletcher, S. (2011). RNA splicing manipulation: strategies to modify gene expression for a variety of therapeutic outcomes. *Curr. Gene Ther.*, **11**, 259–275.

Chapter 17

Messenger RNA Metal Sensing: Iron-Responsive Element (IRE)-mRNA Is a Metal-Sensitive Riboregulator

Mateen A. Khan
Department of Chemistry & Biochemistry, Hunter College, City University of New York, 695 Park Ave, New York, NY 10065, USA
makhan@hunter.cuny.edu

17.1 Introduction

Iron-responsive elements (IREs) are *cis*-acting messenger ribonucleic acid (mRNA) stem–loop or hairpin structures that specifically bind cytoplasmic iron regulatory proteins (IRP1, IRP2) [1–4]. An IRE-RNA is a ~30-nucleotide structure formed by two RNA helices that are separated by a bulged cytosine residue, and by a 6-nucleotide loop of the sequence 5′-CAGUG*N*-3′ (*N* is usually a pyrimidine). The sequence of the loop and the bulged nucleotide are highly conserved [2, 5–9] (Fig. 17.1). The two IRPs, which are highly conserved themselves, bind IRE-RNA structures in a variety of animal mRNAs that appeared at various times during evolution [10] and have been extensively characterized [11–15]. The term "iron

RNA Nanotechnology
Edited by Bin Wang

ISBN 978-981-4411-64-6 (Hardcover), 978-981-4411-65-3 (eBook)
www.panstanford.com

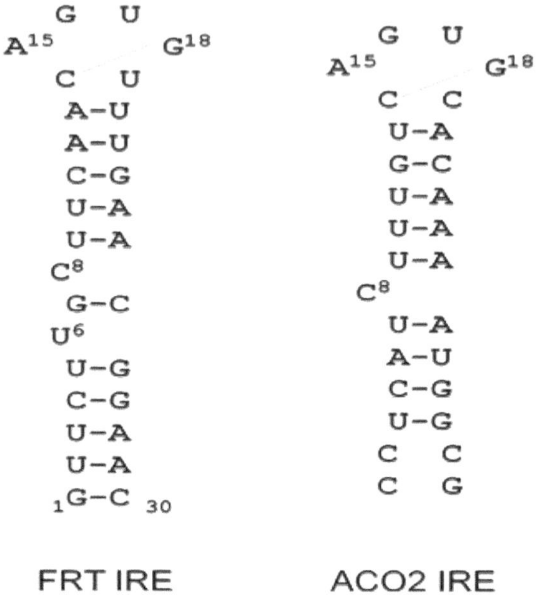

Figure 17.1 Structure of ferritin and mt-aconitase IRE-RNA. *Abbreviation*: mt-aconitase, mitochondrial aconitase.

responsive element" was developed based on effects of increasing iron in animals or in cultured cells of animals, and most models depended on increased IRP degradation [11, 12]. Reporters used to monitor iron effects were entry of IRE-RNA to polysomes, protein accumulation; and the changes in RNA mobility with cell extracts. Models that developed reflected iron-induced protein degradation of both IRP repressors and Fe–S cluster insertion to IRP1.

Recent studies showed [16] that metal ions (Fe^{2+} and Mn^{2+}) directly weaken IRP1/IRE stability. Metal ions directly modulate the function of many RNA classes, for example, ribosomal RNA (rRNA) [17];, transfer RNA (tRNA) [18, 19]; ribozymes [20–24]; riboswitches in bacterial mRNAs, where metals contribute both to RNA function and to metal sensing [21, 25–28]; and possibly hammerhead, mammalian mRNAs [29]. Ferritin IRE-RNA binds metal ions (Mg^{2+}) with a 1:2 stoichiometry [30, 31]; Mg^{2+} also binds to many rRNAs, tRNAs, riboswitches, and ribozymes. In addition, the IRE-RNA also binds shape-specific metal complexes

(1,10-phenanthrolene and Ru(tpy)bpy) and other small molecules at even more specific sites [32–34].

The effect of Fe^{2+} and Mn^{2+} was much larger for ferritin IRE-RNA binding than for mitochondrial aconitase (mt-aconitase) IRE-RNA [16], which has a different IRE-RNA sequence (Fig. 17.1). Biologically, manganese homeostasis may have crossover points with iron homeostasis, exemplified by divalent metal transporter 1 (DMT1) transport of both Mn^{2+} and Fe^{2+} in animals and yeast [35]. When IRP1 binds the ferritin IRE-RNA conformational changes occur, based on solution nuclear magnetic resonance (NMR) of the free RNA and the crystal structure of the ferritin RNA/protein complex [31, 36]. The triloop bases A15 and G16 and helix bulge base C8, present in all IRE-RNAs, are flipped out, and a large surface of the RNA remains exposed in the RNA protein complex (Fig. 17.2). Important as RNA/protein-binding equilibria are, cells must respond rapidly to changes in metabolism and the environment, which makes IRP/IRE-RNA turnover kinetics likely to be a more sensitive regulatory target. IRE-RNA/IRP1 binding is fast, and turnover of the complex is rapid, with quantitative differences among different IRE structures. The association of two IRE-RNA complexes with IRP1 decreased in the presence of Mn^{2+} [16], but the percent decrease is IRE-RNA specific, explaining why the response to iron in vivo [37] is larger for the translation of ferritin mRNA molecules than for mt-aconitase mRNA molecules.

Ferritin mRNA contains the oldest IRE-RNA in the IRE-mRNA family, based on evolutionary comparisons [10, 38], and has the most extensively characterized structure (solution NMR and X-ray crystallography) [31, 36, 39, 40]. The IRE-RNA hairpin in ferritin mRNA contains the conserved terminal loop and the midhelix bulge C, as well as a ferritin IRE-RNA-specific bulge U that is associated with tighter IRP binding and larger iron responses in vitro and in vivo [10, 38]. In the IRE-RNA terminal loop, the base pair between conserved C and G bases creates an AGU triloop [31, 38]. The three main protein/RNA contact sites in IRE-RNA complexed to IRP1 are in the terminal loop A15, terminal loop G16, and helix bulge C8 (Fig. 17.2) [31]. A large RNA surface remains exposed, inviting RNA interactions with other molecules and ions even while complexed to IRP1. Probing IRE-mRNA with metalloncleases such

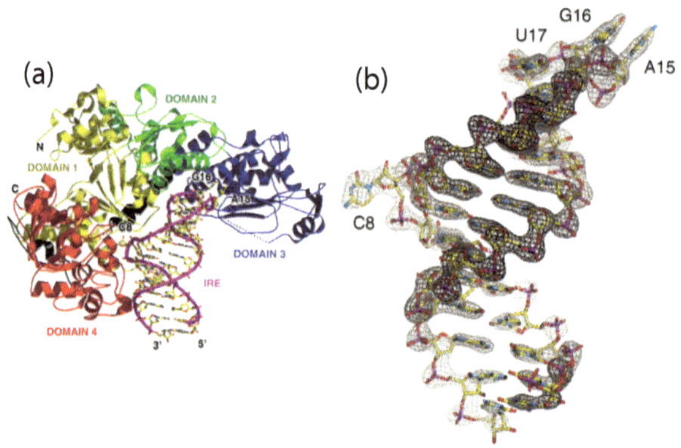

Figure 17.2 Structure of ferritin IRE-RNA binding to IRP1. The IRE-RNA helix bends, and contact bases C8 and triloop bases A15 and G16 are flipped out. (A) IRE-RNA:IRP complex showing IRP domains' (1, yellow; 2, green; 3, blue; 4, red, and linker, black) contact with IRE-RNA. (B) Electron density of IRE-RNA molecule in an IRE:IRP complex. (Modification of the figure originally published in Ref. 31.)

as Cu^{1+} –1,10–phenathroline or Mg^{2+} [30, 41] to locate metal sensitive sites, combined with the X-ray crystal structure of the IRE-RNA/IRP1 complex, shows that some metal sites are located on exposed surfaces that are available for additional RNA interactions even in the IRP1 complex [38] (Fig. 17.2). The selective role of Fe^{2+} in dissociating IRE-RNA from the protein synthesis inhibitor, IRP1, identified the chemical nature of the biological iron signal [16] and complemented regulatory models of IRP degradation by iron-sensitive E3 ligase [42, 43] or IRP inhibition of IRE-RNA binding by insertion of a [4Fe-4S] cluster at the RNA-binding site. Insertion of the [4Fe-4S] cluster converts IRP1 to c-aconitase [11, 44]. How the IRE-RNA enhances protein synthesis rates in the absence of IRP [45] has remained obscure.

IRE-mRNA and IRPs are key players in the control of iron metabolism in animal cells. Iron increases rates of ferritin protein synthesis in animals by facilitating mRNA binding; metabolic iron, that is, "free" iron in cells, is considered to be ferrous [46]. IRPs bind with different affinities to the IRE-RNAs family [15, 16], creating a hierarchal set of mRNA responses to iron in vivo. Deletion of the 30 nt IRE-RNA not only removes IRP regulation but also decreases the rate of IRP-independent protein synthesis [45]. A number of current models of IRE-RNA/IRP regulation feature iron-dependent degradation/modification of the IRPs as the main control point [14, 47–49]. Regulation by metabolic iron (Fe^{2+}) binding to IRE-RNA, to decrease inhibitor protein (IRP)-RNA interactions and increase activator protein (eIF4F)-RNA interactions, is a novel genetic regulatory mechanism in eukaryotes. The IRE riboregulator illustrates the potential of specific metal interactions with folded, noncoding mRNA structures for gene regulation.

Selective metal interactions with RNA structures are becoming increasingly well known, such as those in ribozymes and riboswitches, which are predominantly found in bacteria [50, 51]. Less is known about metal interactions with riboregulators in eukaryotic mRNAs, which function in cytoplasmic protein synthesis uncoupled from mRNA synthesis/deoxyribonucleic acid (DNA) transcription, a contrast with bacterial mRNA riboswitches. Recently, metal-sensitive quadruplexes have been shown to function as a protein synthesis rate regulator in a human mRNA (ADAM10) [52, 53]. Regulation of stable mRNAs permits rapid cellular responses. Eukaryotic protein synthesis can be divided into three main phases: initiation, elongation, and termination. Different sets of proteins ("factors") are required for each phase; energy (adenosine triphosphate (ATP)) is consumed in each phase [54, 55]. Most regulation of protein synthesis rates occurs during the initiation phase, when mRNAs are bound to and oriented on ribosomes by initiation factors (eIFs) and the mRNA AUG start codon is correctly aligned for recognition by the initiator $tRNA^{met}$ and ribosome. eI4F binding to eukaryotic mRNA is the first step in protein synthesis and thought to be rate limiting. The first committed step in protein synthesis is the binding of the 5′ mRNA cap (m7Gppp*N*, where *N*represent any nucleotide) to the eIF4E, the small subunit of eIF4F. eIF4G, the large subunit of eIF4F, recruits additional initiation factors including eIF4A, which unwinds the secondary structure within the 5′-leader sequence to allow 40 S ribosomal scanning—poly(A)-binding protein, which stabilizes the eIF4F complex, and eIF3, which promotes 40 S ribosomal subunit binding. After the 40 S subunit binds and scans to the initiation codons, it assembles with the 60 S subunit to form the 80 S ribosome and carry out translation of the coding region [54, 55]. This noncoding region between the cap and the coding region is called the 5′–untranslated region (5′-UTR). A noncoding mRNA structure can make a major contribution to protein synthesis rates, and gene expression in eukaryotic cells is illustrated by the ability of IRE-RNA to bind, competitively, two regulatory proteins. IRE-RNA also binds the protein synthesis enhancer, eIF4F, in addition to a protein synthesis inhibitor, IRP, a known binding. Phosphorylation of proteins factors (eIFs) [54] alters mRNA binding in various cellular conditions. The stabilization of eIF/RNA interactions by binding of a metabolic metal ion (Fe^{2+}) to a noncoding IRE-RNA is a novel mechanism. Fe^{2+} facilitates the binding of eIF4F/IRE-mRNA with higher stabilities than for eIF4F binding to β-globin mRNA or the m^7G cap structure [56, 57]. Metal ion–stabilized binding demonstrates enhanced in vitro translation of ferritin IRE-RNA.

Metal binding to selective sites on IRE-RNA by small, shapely molecules such as Cu^{1+}- phenathroline and/or Ru^{2+}-tris(bipyridyl) has been known for some time [41, 58]. In the presence of dioxygen and the RNA-bound complexes, reactive intermediates form, which cleave the folded RNA at the binding site of the metal complex; sequencing of the cleaved RNA (primer extension analysis) identifies the RNA-binding sites. Such studies showed specific binding of the metallo-complexes at sites near helix distortions or in the terminal loop of the IRE-RNA structure [41, 58]. The possibility of Fe^{2+}binding to IRE-RNA in ferritin mRNA is suggested by the results of IRE-RNA cleavage with Fe^{2+} and hydrogen peroxide [59]. The unusual pattern of hypersensitive cleavage sites was initially explained as solvent access prevented by RNA folding [59], a supposition that was not supported by NMR structures obtained later [36, 39]. Alternatively, the hypersensitive sites could indicate Fe^{2+}-/RNA-binding sites. Some of the IRE-NMR metal sites observed by metallonuclease probing are at bases that are on the exposed RNA surfaces in the IRE-RNA conformation observed in the IRP/IRE-RNA crystal structure (Figs. 17.2 and 17.3) and thus could be accessible to metal ions while the IRE-RNA is bound to IRP. Recent studies with 2AP-substituted IRE-RNA showed that Fe^{2+} altered the conformation of both the terminal IRE loops: A15 was less stacked, and the bulge C8 was more ordered. A15 and C8 are part of the two IRE-RNA/IRP1 contact sites and have multiple contacts [31].

17.2 Structure of Iron-Responsive Element–RNA

IRE-RNAs are a conserved 30-nucleotide structure formed by two RNA helices that are separated by a bulged cytosine residue and a six-nucleotide loop of the sequence 5′CAGUG*N*3′ (usually *N* is a pyrimidine). Functional IRE-RNAs have been identified in the 5′-UTR of ferritin H and L-chain, erythroid aminolevulinate synthase (eALAS), mt-aconitase, and succinate dehydrogenase (SDH) mRNAs, as well as the 3′-UTR of transferrin receptor (TfR) mRNA [60]. The conserved structure of IRE-RNA family are a double-stranded RNA helix of 9/10 base pairs with a terminal hexaloop and an unpaired C8 residue in the helix, creating a 5–base pair upper stem and a variable-length lower stem. The terminal loop is spanned by a G–C base pair, which effectively creates a triloop of AGU that is exposed to the solvent. In the terminal loop, C14 and G18 are base-paired and create a pseudotriloop of AGU [30, 31, 36, 39–41, 61]. G16 and U17 are disordered. Base pairing and sequencing around the unpaired C8 residue vary between different IRE-RNAs, separating IRE-RNAs into two groups, one with an internal loop constructed from the unpaired C8 and the other with an isolated unpaired C, an unpaired base at position 6, 5′-UTR to the unpaired C, and an additional base pair between them. Helix distortion around the conserved C8 residue plays an important role in selective repressor binding [30]. Five base pairs separate the conserved apical loop from a bulge, which is a conserved C residue (Figs. 17.1 and 17.2). Each IRE-RNA have three specific structural components: 1-helix bulge/loop; 2-flanking structure; 3-helix base pairs that are contribute to the signal responses observed in vivo [38]; that modulate and shared features of the short RNA helix and terminal loop structure. Differences among the helix sequence, the flanking structure, and the helix bulge/loop are associated with variations in IRP binding in vitro and changes in the protein synthesis in vivo [62]. The helix distortion, which surrounds the conserved C residue, influences the binding of IRP2 much more than IRP1. In a large group of IRE-RNAs, mt-aconitase, eALAS, ferrotin, and DMT1, and in isolated TfR1 IREs, the unpaired C is a bulge in the helix that makes a relatively weak IRP2 recognition site. When the unpaired C is separated from an unpaired U by a G–C pair, as in ferritin IREs or in a loop induced by an AU-rich context and linked to four other IRE structures by sequences with conserved motifs, IRP2 binding is enhanced [63]. Interactions of IRE-RNA/IRP are more stable than typical enzyme–substrate complexes. Sequence comparisons of different animal IRE-RNAs and footprinting reveal that the IRP-binding site is 30 oligonucleotides long. The importance of a helix structure around the unpaired C8 is illustrated by the effect of deleting a single, unpaired U6, characteristic of ferritin IRE-RNA, which decreases the stability of the IRP1 and IRP2 complexes significantly, with the consequence of decreasing translational repression [30, 64]. IRE-RNA flanking sequences serve an important role in ferritin mRNA. The length of the base-paired flanking region differs among different ferritin IRE-RNAs. Binding of ferritin IRE-RNA with IRP increases the helicity of the flanking regions, showing the conformational change between IRE-RNA and the flanking region helix [59]. Changes occurred on the IRE-RNA structure after binding with IRP. The residues A15 and G16 of the terminal loop were extruded from the IRE-RNA molecule and extended into a large protein cavity, with U17 completing the helix stack (Figs. 17.1 and 17.2). However, A15 was stacked over the C13–G18 base pair, and G16 with U17 was disordered in solution. The second RNA-binding site was the unpaired C8 in the helix, which was disordered in free RNA in solution [39]. RNA was extended completely away from the stem and inserted into a protein pocket; however, the unpaired U6, which was stacked in an RNA cavity of the free RNA, was flipped out toward the protein surface in the IRP1 complex (Fig. 17.2). The correct sequence and structure of the IRE-RNA stem loop is important in human genetic disorders such as hereditary hyperferritinemia-cataract syndrome, which arises from point mutations or deletions within the IRE or flanking regions in the ferritin 5′-UTR. Mutations clustered in the terminal loop and within the helix loop of ferritin IRE-RNA affect IRP binding. Also IRE-RNA regulation could be altered in patients with iron overload from transfusion therapies of hemoglobinapathies, such as sickle cell anemia, thalassemia; an increase in ferritin synthesis could be beneficial. The development of small molecules and/or drugs that selectively disrupt ferritin IRE-RNA/IRP interactions could have therapeutic potential.

17.3 IRE-RNA Binding to Iron-Regulatory Protein

IRPs were identified as RNA-binding proteins that regulate cellular iron metabolism. Interaction of IRPs with target mRNAs regulates key aspects of cellular iron metabolism. IRP1 is a unique protein. IRP1 not only binds to IRE-RNA but also binds to [4Fe-4S] cluster confers cytoplasmic aconitase (c-aconitase) activity. IRPs that selectively recognize IRE-mRNA structures of the aconitase form of IRP1 are homologues to mt-aconitase. The IRP effect is generally related to the position of the IRE in the mRNA, with IREs near the 5′ cap repressing ribosome binding and IREs after the mRNA-coding region and stop codon repressing mRNA degradation. All members of the aconitase family have a four-domain organization. The first three domains of typical mt-aconitase are in tight association to nestle the Fe–S cluster, while the fourth is bound through a long amino acid (~60) linker. Binding of IRE-RNA creates conformational changes on IRP1. (Fig. 17.3). The IRP1 molecule adopts an L-shaped conformation after binding to IRE-RNA, and RNA is bent. The IRP1/IRE-RNA complex is achieved with two widely separated contact sites [31].

In the complex, the RNA is inserted between protein domains 1–2 and 4 (Fig. 17.3), using the numbering of aconitases. Aconitases are globular structures with close interactions among all four domains. The IRE-RNA stem loop is tethered across the concave opening of IRP1, with direct contacts from the IRE-RNA terminal loop and the lower stem (Fig. 17.3). Exposed terminal loop residues A15 and G16 extend into an IRP cavity at the interface between domains 2 and 3. Stability of the IRE-RNA/IRP complex is provided by the bonds A15, G16, and U17. IRP tethering to the stem of the IRE-RNA is centered on the C8 oligonucleotide, which inserts into a pocket on the inner face of domain 4. Bending of 31° in the IRE-RNA redirects the path of the lower stem along the same face of domain 4, which binds the side of the IRE-RNA lower stem farthest from the

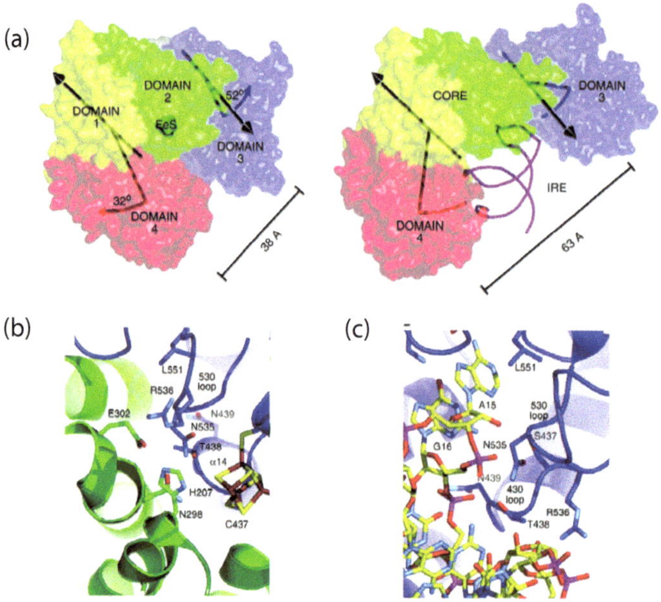

Figure 17.3 Crystal structure of two functional conformations of IRP1 as c-aconitase and an IRP1:IRE-RNA complex. (a) (left) c-aconitase, Fe-S cluster in the center of the molecule (FeS-apo-IRP); (right) IRP/IRE-RNA complex. Protein domains, red, yellow, green, and blue; RNA, magenta. (b) Close-up of the IRP–IRE-RNA contacts at the C8 bulge. (Modification of the figure originally published in Ref. 31.)

terminal loop (Fig. 17.3). The terminal loop and C8 of the IRE-RNA are separated by 30 Å, held in their orientations by the five–base pair helix of the upper stem. Twenty-two bonds hold the IRE-RNA and IRP together. Ten contacts are formed between A15 and G16 in the pseudotriloop at the IRE-RNA terminus (Fig. 17.3b) and amino acids, such as residues Ser371 and Lys379 in a pocket generated in domain 3 at a site that is blocked by domains 1 and 2 in the globular form (Fig. 17.3b). Eight bonds are formed between IRE-RNA C8 and amino acid residues Arg713 and Arg780 in domain 4 of IRP1; four additional bonds occur between amino acids Ser681, Pro682, Asp781, and Trp782 in domain 4 of IRP1 and the IRE stem below C8 (Fig. 17.3c). In addition to flipping out of terminal loop bases A15 and G16 and the helix bulge C8, [39] in the IRP/IRE-RNA complex, the IRE-RNA backbone is distorted by a sharp midhelix turn, the unpaired U19 in the hexaloop is flipped out of the IRE-RNA helix, and the unpaired U6 is tucked into the minor groove (Figs. 17.2 and 17.3). The differences between the solution structure of free IRE-RNA and IRP-tethered IRE-RNA require conformational changes in the RNA and likely of the unliganded protein; while the structure of apo-IRP is not yet known, it is more disordered than in either the RNA or the [4Fe-4S] complexes [65]. In the IRE-RNA/IRP crystal structure [31], a large surface of the IRE-RNA is exposed, free for interactions with other metal ions, RNA, and proteins.

17.4 Metal Ion Sensing to IRE-RNA/IRP

mRNAs encoding proteins of iron traffic and oxidative metabolism are more active when cellular iron increases. Metal ions affect the function of many RNAs such as rRNA; tRNA; ribozymes [17–19, 21, 22, 24, 66]; riboswitches in bacterial mRNAs, where metals contribute both to RNA function and to metal sensing [26, 27]; and possibly hammerhead, mammalian mRNAs [29]. Changes in translation of the iron-responsive mRNAs, dependent on noncoding structures (IRE), are currently attributed entirely to iron effects on the protein repressors, IRP1 and IRP2, that bind IRE-RNAs and that are degraded or modified by increases in cell iron [11, 14]. IRE-RNA binds metal ions at specific sites [30] (Fig. 17.4) [31] as do rRNAs, tRNAs, riboswitches, and ribozymes. Metals alter IRE-mRNA/IRP complexes directly. Khan et al. (2009) [16] had chosen two IRE-RNAs to compare ferritin and mt-aconitase because they varied in structure at the site of metal binding to ferritin IRE-RNA (Fig. 17.4) and because they have quantitatively different responses to the same iron signal in vivo (rats) [67]; the ferritin-IRE evolved before the mt-aconitase IRE [10]. It has been reported that IRP1 binding distinguishes among different IRE structures and that IRE /IRP complexes of the two mRNAs are selectively destabilized by Fe (II), Mn (II) and Mg (II).

The ferritin IRE-RNA/IRP1 complex was more stable than the mt-aconitase IRE-RNA complex in solution (Fig. 17.5), illustrating the effects of phylogenetically conserved differences [10] between

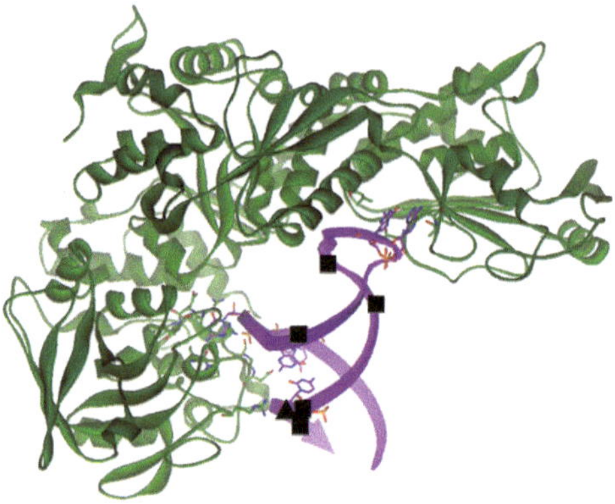

Figure 17.4 IRE-RNA sites influenced by metal binding. The three-dimensional crystal structure of the ferritin-IRE/IRP1 complex modified to show metal sites. -Mg(II) determined by solution NMR; 1.10-phenanthroline cleavage sites; green: protein; purple: RNA. (Figure of the RNA/protein complex taken from Ref. 16.)

the two IRE-RNAs, as well as quantitating, qualitative binding differences observed as the amount of protein required to alter the mobility of IRE-RNA during gel electrophoresis electrophoretic mobility shift assay (EMSA) [62, 64] and can be related to stronger ferritin synthesis response iron in vivo [67]. The K_d value for the ferritin-IRE/IRP1 complex was smaller than for the mt-aconitase IRE/IRP1 complex (Fig. 17.5). The different K_d values indicate that

	Fer H IRE-RNA K_d (nM)	Fer H ΔU6 IRE-RNA, K_d (nM)	Mt acon IRE-RNA K_d (nM)
0.0	14.2 ± 0.3	41.3 ± 0.7	129 ± 3.3
Fe^{2+}(μM)			
5	78.9 ± 4.5		259 ± 17
25	162 ± 6.6		643 ± 43
50	241 ± 8.0		792 ± 59
Mn^{2+}(μM)			
5	57.4 ± 1.1	51.4 ± 1.2	233 ± 8
25	118 ± 2	60.8 ± 1.1	468 ± 19
50	174 ± 4	63.0 ± 1.2	750 ± 32
Mg^{2+}(mM)			
0.5	44 ± 3.0	47.5 ± 1.3	213 ± 9
1.0	61 ± 7.0	54.3 ± 3.5	269 ± 12
2.0	72 ± 6.0	54.0 ± 3.9	307 ± 11
5.0	74 ± 8.0	54.7 ± 5.7	331 ± 14

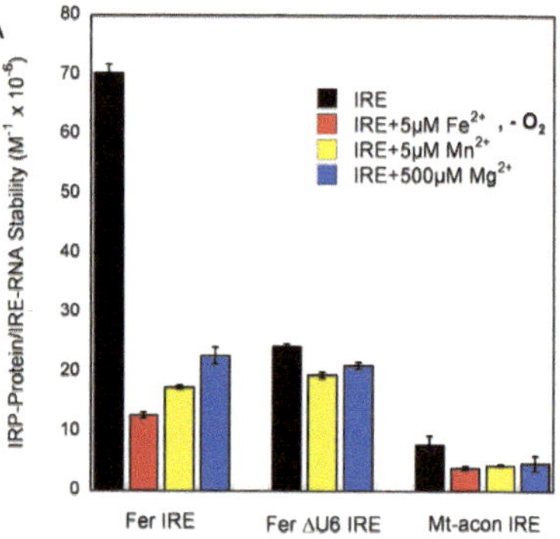

Figure 17.5 Fe (II), Mg (II), and Mn (II) selectively weaken IRE-RNA/IRP1 interactions (fluorescence quenching in solution) (taken from Ref. 16).

a larger fraction of mt-aconitase mRNA will be "free" of IRP, or derepressed, than ferritin mRNA even when iron concentrations are low. Thus, a relatively larger fraction of mt-aconitase mRNA molecules will be translated constitutively, and the relative effect of increased iron on mt-aconitase mRNA will be smaller than on ferritin mRNA, as observed in animal experiments [67]. A structural explanation of the differences in the K_d of the two IRP1–IRE/IRP1 complexes is the U6 bulge, absent in mt-aconitase IRE-RNA [16]. Indeed, the K_d of the ΔU6-ferritin IRE/IRP1 complex is larger by threefold than the K_d for ferritin IRE-mRNA, as is the K_d for mt-aconitase IRE-RNA, and is consistent with qualitative gel electrophoresis data [64]. However, removing the U6 bulge in the ferritin IRE-RNA only partially compensated for difference in IRP binding to mt-aconitase or ferritin IRE-RNA, which indicates that the stability of IRP/IRE-RNA complexes is also influenced by the differences among the phylogenetically conserved helix base pairs in each IRE-RNA.

Metal ions weakened the IRE-RNA/IRP1 interactions for both ferritin and mt-aconitase mRNAs. Mg (II) was studied because of the direct metal interactions observed with the ferritin IRE-RNA by NMR spectroscopy [36], and Mn (II) was studied because it is a good model for air-sensitive Fe (II) and binds to other RNAs [24]. The effects of metals were larger for the ferritin-IRE/RNA, indicating that metal ions recognize different IRE-RNA/IRP interactions. Mg- and Mn-induced destabilization of the RNA/protein complexes was significantly larger for wild-type ferritin and mt-aconitase IRE structures than for the mutant ferritin ΔU6 IRE-RNA [16]. Differences between the ancient ferritin-IRE and the more modern mt-aconitase IRE-RNAs [10] in metal responses could include selective effects of metals.

Iron, the biological regulator, was then examined in the absence of air to stabilize the divalent state. Fe (II) weakened the IRP-RNA/IRP complexes, as did Mg (II) and Mn (II), but the effect on the ferritin IRE-RNA/IRP interaction was larger with Fe (II) than with the other metal ions. Iron increased the K_d for the ferritin IRE-RNA $\sim$tenfold, significantly more than Mn (II), whereas for mt-aconitase the effects of Mn (II) and Fe (II) were comparable. Thus, metal ions appear to amplify further the contributions of variations in the IRE-RNA structure to binding of the IRP1 observed in the absence of metals (Fig. 17.5). Metal induced changes in IRE-RNA/IRP1 interactions measured by fluorescence quenching [16] and gel electrophoresis [64, 68].

Metal ion destabilization of the IRE-RNA/IRP1 complexes competes with the stabilization conferred by the very large number of bonds between the protein and the RNA [31] and the stability of the IRE-RNA fold [61]. RNA/protein destabilization more likely reflects metal ion binding to the RNA or the RNA/protein complex rather than the protein for the following reasons:

- First direct metal binding to selective IRE-RNA sites is known [30, 41].
- Second, the only direct metal binding to IRPs known is [4Fe-4S] cluster binding in IRP1, which confers aconites activity [65, 69, 70], and Cd- or Zn-induced, sites of protein aggregation in crude cell extracts [71]; the molecular signals for iron-induced IRP turnover have not been identified.
- Third, when Fe (II) is added to IRE-RNA, the electrophoretic mobility changes [16]. How metal ions weaken the IRE/IRP complex is not clear from "footprinting"/nuclease protection experiments where the IRE-RNA was completely protected by the protein [59]. However, in a recent crystal structure of the ferritin IRE-RNA/IRP1 complex [31], there are several IRE-RNA sites without protein contacts that could be accessed by solvent or metal ions; the location of known IRE-RNA metal-binding sites [41] is shown superimposed on the crystal structure of the RNA/protein complex. Many molecular conformations are sampled in nuclease protections experiments, while only one conformer, or a few conformers, is sampled in crystals. The different RNA/protein interactions, observed by the two methods, suggest there is more than one IRE-RNA/IRP conformation. Thus, even when IRP is bound to an IRE-RNA, some RNA sites in a fraction of the complexes, possibly kinetically controlled, can be exposed to metal ions that destabilize the IRE-RNA/IRP complex.

Selective iron binding to different IRE-RNAs in IRP complexes as well as other selective metal–RNA interactions [25–27] have well-known parallels in the selectivity of metal–protein interactions and emphasize that three-dimensional structural specificity of folded RNAs and proteins are shared. The lower stability and smaller response to Fe (II) of mt-aconitase IRE-RNA/IRP repressor complexes compared to ferritin IRE-RNA/IRP repressor complexes observed in solution and gel mobility, and the smaller response of mt-aconitase to iron, compared to ferritin in whole-body experiments, illustrate novel features of the IRE-mRNA family. First, phylogenetically conserved variations in the IRE-RNA structure that occur among members of the mRNA family, such as the U6 bulge in ferritin and the AUG initiator codon in mt-aconitase IRE-RNA, contribute to differences in IRP binding in solution and iron responses in vivo. Finally, the weakening of IRE-RNA/IRP interactions by Fe (II) indicates that Fe (II) can has a direct regulatory effect on IRE-RNA/IRP complexes, complementary to iron effects on protein turnover or Fe–S binding, and that IRE-mRNA senses iron ions and possibly other metal ions.

17.5 Metal Ion Sensing to IRE-RNA/eIF4F

IREs in the 5′-UTR of IRE-mRNAs structures bind IRP1 and IRP2, which inhibit translation and regulate iron homeostasis. In current 5′-UTR IRE-RNA regulatory models, the IRP block of mRNA translation (negative regulation) is reversed solely by inactivation of IRPs. Recent studies [72] showed that IRE-RNA plays an active role in both positive and negative regulation. Fe^{2+} increases IRE-mRNA binding of the protein synthesis initiation factor, eIF4F (positive regulation). The large, combined effects of Fe^{2+}-IRE-RNA binding to IRP (negative regulation) and eIF4F (positive regulation) illustrate the impact of IRE riboregulators. An earlier study by Khan et al. (2009) [16] has shown direct IRE-RNA–metal interactions decreased IRP repression.

Iron regulation is controlled at multiple levels in animal cells, including RNA synthesis, mRNA translation, and protein stability. For example, cellular iron regulates ferritin mRNA translation.

Ferritin, required for life, concentrates cellular iron by reactions with oxygen, for cell iron nutrition, and as an antioxidant/iron scavenger after oxidative stress [11, 47, 73]. The mRNA regulatory site is the IRE in the 5′-UTR of the ferritin mRNA, one of a family of mRNAs with highly conserved IRE structures [10, 38]. The IRE-mRNA structures are RNA hairpins, containing a hexaloop and a midhelix bulge C (Fig. 17.1). Loop C/G bases pair to create an AGU triloop [31, 38]. The three main IRP/RNA contact sites, loop A, G, and bulge C (Figs. 17.1 and 17.2) [31], leave a large RNA surface exposed containing metal-binding sites [38]. Protein repressors (IRP1 and IRP2) bind IRE-mRNA structures to inhibit either protein synthesis (5′-UTR IRE) or mRNA degradation (3′-UTR IRE) [14, 47, 49, 74]. Most current IRE-RNA regulatory models focus on iron-dependent changes in concentrations of repressor proteins IRP1 and IRP2, which are degraded by iron-dependent processes [42, 43] or converted to c-aconitase (IRP1) by Fe–S cluster insertion at the RNA site [11–15, 75]. Until recently, the 5′-UTR IRE/IRP regulatory system was described as negative regulation, where IRP binding blocked mRNA translation. Stimulation of protein synthesis was thought to occur solely by IRP inactivation. Khan et al. (2009) [16] observed that Fe^{2+} (anaerobic to prevent oxidation and ferric hydroxide precipitation) bound to IRE-RNA and decreased IRP binding [16]. A recent study showed Fe^{2+} induced positive control by increasing IRE-RNA binding to eIF4F (Fig. 17.6) [72]. All known riboregulators have both negative and positive control components, and some, for example, the flavin mononucleotide (FMN) riboregulator [76], exploit kinetics.

Fe^{2+}-induced release of IRP binding alone is insufficient to account for the increase in translation observed in vitro [45]. A positive control element is necessary to explain the data. To identify IRE-RNA positive control and establish the IRE as a riboregulator, Ma et al. (2012) [72] examined the binding of eIF4F, which includes the cap-binding factor eIF4E and is the rate-limiting step for protein synthesis initiation. Three observations establish IRE-RNA/eIF4F interactions as the positive regulatory branch of IRE-RNA riboregulation. First, IRE mRNA binds eIF-4F with high affinity (Fig. 17.6) in the absence of a cap, explaining older observations of IRE-RNA stimulation of mRNA translation [45]. Second, eIF4F

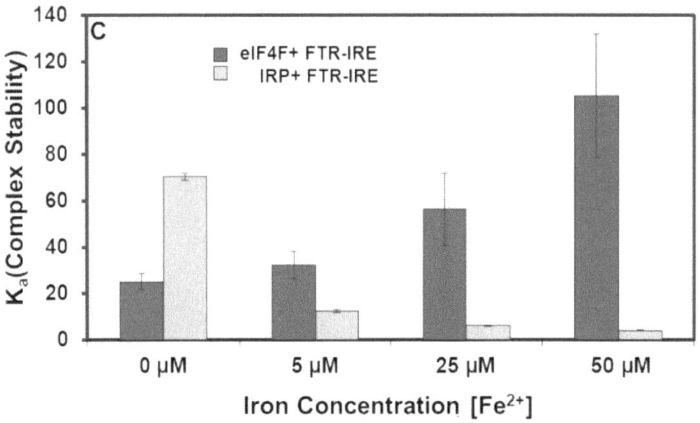

Figure 17.6 Sensing of metal ions for eIF4F binding to IRE-RNA (modification of the figure originally published in Ref. 72).

competes with IRPs for binding to IRE-RNA. Third, Fe^{2+} increases the binding affinity of eIF-4F for IRE-RNA nearly fivefold. Thus, when Fe^{2+} concentrations increase, IRE-RNA/IRP concentrations decrease with IRP release, free IRP concentrations decrease (iron-dependent proteasomal degradation of IRPs [77] and conversion of IRP1 to c-aconitase by Fe–S cluster insertion), and IRE-RNA/eIF4F binding increases. The opposite effects of Fe^{2+} on IRE-mRNA/protein interactions with IRPs and eIF4F lead to loss of negative control (IRP binding, repression) and gain of positive control (eIF4F binding, translation).

The Fe^{2+}-sensitive changes in IRE-RNA/protein interactions for both negative control (IRP binding) and positive control (eIF4F binding) show that the IRE-RNAs are riboregulator (Fig. 17.6) in animal mRNAs. Relating observed effects of [Fe^{2+}] in solution to physiological [Fe^{2+}] effects relies on several assumptions because quantitative data on free cellular iron is limited. In normal mammalian cells, free [Fe^{2+}] is ~0.5–1.0 μM measured by fluorescence analysis [78], while ferritin accumulations occur, measured immunologically, at intracellular [Fe^{2+}] ~5–10 μM. Ma et al. (2012) [72] showed at 5–10 μM [Fe^{2+}] in solution a 6:1 decrease in IRP binding and a 1.5:1 increase in eIF-4F binding. Because eIF4F and IRP bind competitively [72], the IRE/eIF4F complex increases 8:1

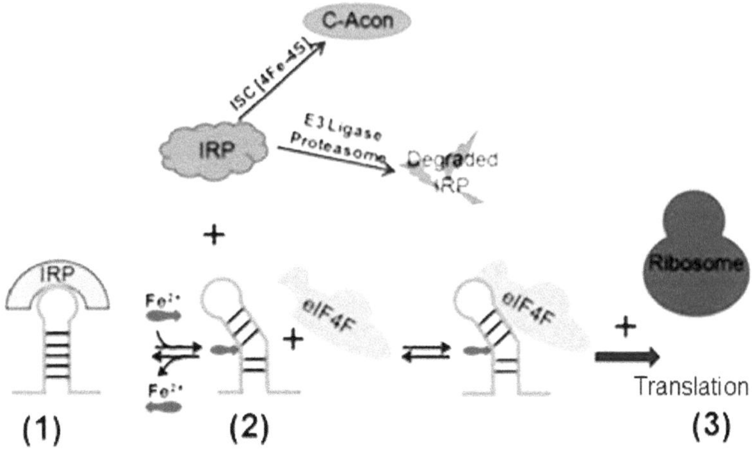

Figure 17.7 A model of Fe^{2+} IRE-RNA riboregulator action. (1) IRE-mRNA repressed by IRP/IRE-RNA binding. (2) Fe^{2+} binding to IRE-RNA causes a conformational change that releases an IRP, enhances eIF4F binding, and increases eIF4F/IRE-RNA; the free IRP is degraded or converted (IRP1) to c-aconitase. (3) eIF4F/IRE-RNA ± cap/eIF-4E recruit ribosomes, and translation (expression) begins. (Modification of the figure originally published in Ref. 72.)

for similar IRP:eIF4F concentrations at 5 μM [Fe^{2+}]. The effect could be 80:1 at 50 μM Fe^{2+}, for example, erythrophagocytic macrophages or hepatocytes in iron overload. A model of IRE-RNA riboregulation (Fig. 17.7) combines Fe^{2+}-regulated IRE-RNA negative control (translational repressor, IRP binding) with IRE-RNA positive control (translation activator, eIF4F binding) and also includes the inactivation of free IRPs described in older models [11–15, 42, 43, 75]. Increasing the cellular Fe^{2+} concentration to bind IRE-RNA lowers RNA/IRP affinity and increases eIF4F binding to facilitate ribosome assembly and translation. Expansion of the IRE-RNA family from the iron metabolic proteins, for example, ferritin, ferroportin, and mt-aconitase, to mRNA encoding a cell cycle protein [79], α–hemoglobin chaperone [80] and amyloid precursor protein [81], suggests that IRE-mRNA translational control and riboregulation impact a number of metabolic processes in animal cells.

References

1. Leibold, E.A., and Munro, H.N. (1988). Cytoplasmic protein binds in vitro to a highly conserved sequence in the 5′ untranslated region of ferritin heavy- and light-subunit mRNAs. *Proc. Natl. Acad. Sci. U S A*, **85**, 2171–2175.
2. Mullner, E.W., Neupert, B., and Kuhn, L.C. (1989). A specific mRNA binding factor regulates the iron-dependent stability of cytoplasmic transferrin receptor mRNA. *Cell*, **58**, 373–382.
3. Henderson, B.R., Seiser, C., and Kuhn, L.C. (1993). Characterization of a second RNA-binding protein in rodents with specificity for iron-responsive elements. *J Biol. Chem.*, **268**, 27327–27334.
4. Guo, B., Yu, Y., and Leibold, E.A. (1994). Iron regulates cytoplasmic levels of a novel iron-responsive element-binding protein without aconitase activity. *J Biol. Chem.*, **269**, 24252–24260.
5. Casey, J.L., et al. (1988). Iron-responsive elements: regulatory RNA sequences that control mRNA levels and translation. *Science*, **240**, 924–928.
6. Gray, N.K., et al. (1996). Translational regulation of mammalian and drosophila citric acid cycle enzymes via iron-responsive elements. *Proc. Natl. Acad. Sci. U S A*, **93**, 4925–4930.
7. Dandekar, T., et al. (1991). Identification of a novel iron-responsive element in murine and human erythroid δ-aminolevulinic acid synthase mRNA. *EMBO J.*, **10**, 1903–1909.
8. Hentze, M.W., and Kuhn, L.C. (1996). Molecular control of vertebrate iron metabolism: mRNA-based circuits operated by iron, nitric oxide, and oxidative stress. *Proc. Natl. Acad. Sci. U S A*, **93**, 8175–8182.
9. Theil, E.C. (1994). IREs: a family of mRNA non-coding sequences. *Biochem. J.*, **304**(Pt 1), 1–11.
10. Piccinelli, P., and Samuelsson, T. (2007). Evolution of the iron-responsive element. *RNA*, **13**(7), 952–966.
11. Wallander, M.L., Leibold, E.A., and Eisenstein, R.S. (2006). Molecular control of vertebrate iron homeostasis by iron regulatory proteins. *Biochim. Biophys. Acta*, **1763**(7), 668–689.
12. Rouault, T.A. (2006). The role of iron regulatory proteins in mammalian iron homeostasis and disease. *Nat. Chem. Biol.*, **2**(8), 406–414.
13. Leipuviene, R., and Theil, E.C. (2007). The family of iron responsive RNA structures regulated by changes in cellular iron and oxygen. *Cell. Mol. Life Sci.*, **64**, 2945–2955.

14. Muckenthaler, M., Galy, B., and Hentze, M.W. (2008). Systemic iron homeostasis and the iron-responsive element/iron-regulatory protein (IRE/IRP) regulatory network. *Annu. Rev. Nutr.*, **28**, 197–213.
15. Goforth, J.B., Anderson, S.A., Nizzi, C.P., and Eisenstein, R.S. (2010). Multiple determinants within iron-reponsive elements dictate iron regulatory protein binding and regulatory hierarchy. *RNA*, **16**, 154–169.
16. Khan, M.A., Walden, W.E., Goss, D.J., and Theil, E.C. (2009). Direct Fe^{2+} sensing by iron-reponsive messenger RNA:repressor complexes weakens binding. *J. Biol. Chem.*, **284**, 30122–30128.
17. Shenvi, C.L., et al. (2005). Accessibility of 18S rRNA in human 40S subunits and 80S ribosomes at physiological magnesium ion concentrations—implications for the study of ribosome dynamic. *RNA*, **11**, 1898–1908.
18. Wang, B., Wilkinson, K.A., and Weeks, K.M. (2008). Complex ligand-induced conformational changes in tRNA(Asp) revealed by single-nucleotide resolution SHAPE chemistry. *Biochemistry*, **47**, 3454–3461.
19. Oshikane, H., et al. (2006). Structural basis of RNA-dependent recruitment of glutamine to the genetic code. *Science*, **312**, 1950–1954.
20. Koculi, E., Thirumalai, D., and Woodson, S.A. (2006). Counterion charge density determines the position and plasticity of RNA folding transition states, *J. Mol. Biol.*, **359**, 446–454.
21. Vicens, Q., et al. (2008). Toward predicting self-splicing and protein-facilitated splicing of group I introns. *RNA*, **14**, 2013–2029.
22. Nelson, J.A., and Uhlenbeck, O.C. (2008). Hammerhead redux: does the new structure fit the old biochemical data. *RNA*, **14**, 605–615.
23. Adilakshmi, T., Bellur, D.L., and Woodson, S.A. (2008). Concurrent nucleation of 16S folding and induced fit in 30S ribosome assembly. *Nature*, **455**, 1268–1272.
24. Vogt, M., et al. (2006). Coordination environment of a site-bound metal ion in the hammerhead ribozyme determined by 15N and 2H ESEEM spectroscopy. *J. Am. Chem. Soc.*, **128**, 16764–16770.
25. Brantl, S. (2007). Metal sensing by RNA in bacteria: exception or rule? *ACS Chem. Biol.*, **2**(10), 656–660.
26. Lipfert, J., et al. (2007). Structural transitions and thermodynamics of a glycine-dependent riboswitch from Vibrio cholerae. *J. Mol. Biol.*, **365**, 1393–1406.
27. Coppins, R.L., Hall, K.B., and Groisman, E.A. (2007). The intricate world of riboswitches. *Curr. Opin. Microbiol.*, **10**, 176–181.

28. Noeske, J., Schwalbe, H., and Wohnert, J. (2007). Metal-ion binding and metal-ion induced folding of the adenine-sensing riboswitch aptamer domain. *Nucleic Acids Res.*, **35**, 5262–5273.
29. Martick, M., Horan, L.H., Noller, H.F., and Scott, W.G. (2008). A discontinuous hammerhead ribozyme embedded in a mammalian messenger RNA. *Nature*, **454**, 899–902.
30. Ke, Y., Sierzputowska-Gracz, H., Gdaniec, Z., and Theil, E.C. (2000). Internal loop/bulge and hairpin loop of the iron-responsive element of ferritin mRNA contribute to maximal iron regulatory protein 2 binding and translational regulation in the iso-iron-responsive element/iso-iron regulatory protein family. *Biochemistry*, **39**(20), 6235–6242.
31. Walden, W.E., et al. (2006). Structure of dual function iron regulatory protein 1 complexed with ferritin IRE-RNA. *Science*, **314**, 1903–1908.
32. Wang, Y.H., et al. (1991). Ferritin mRNA probed, near the iron regulatory region, with protein and chemical (1,10-phenanthroline-Cu) nucleases. A possible role for base-paired flanking regions. *Biol. Met.*, **4**, 56–61.
33. Ciftan, S.A., Theil, E.C., and Thorp, H.H. (1998). Oxidation of guanines in the iron-responsive element RNA: similar structures from chemical modification and recent NMR studies. *Chem. Biol.*, **5**, 679–687.
34. Tibodeau, J.D., et al. (2006). The upregulation of ferritin expression using a small molecule ligand to the native mRNA. *Proc. Natl. Acad. Sci. U S A*, **103**, 253–257.
35. McNaughton, R.L., et al. (2010). Probing in vivo Mn^{2+} speciation and oxidative stress resistance in yeast cells with electron-nuclear double resonance spectroscopy. *Proc. Natl. Acad. Sci. U S A*, **107**, 15335–15339.
36. Gdaniec, Z., Sierzputowska-Gracz, H., and Theil, E.C. (1998). Iron regulatory element and internal loop/bulge structure for ferritin mRNA studied by cobalt(III) hexammine binding, molecular modeling, and NMR spectroscopy. *Biochemistry*, **37**(6), 1505–1512.
37. Chen, O.S., Blemings, K.P., Schalinske, K.L., and Eisenstein, R.S. (1998). Dietary iron intake modulates the activity of iron regulatory proteins and the abundance of ferritin and mitochondrial aconitase in rat liver. *J. Nutr.*, **128**, 525–535.
38. Theil, E.C., and Eisenstein, R.S. (2000). Combinatorial mRNA regulation: iron regulatory proteins and Iso-iron responsive elements (iso-IREs). *J. Biol. Chem.*, **275**, 40659–40662.
39. Addess, K.J., et al. (1997). Structure and dynamics of the iron responsive element RNA: implications for binding of the RNA by iron regulatory binding proteins. *J. Mol. Biol.* **274**, 72–83.

40. Hall, K.B., and Williams, D. (2004). Dynamics of the IRE-RNA hairpin loop probed by 2-aminopurine fluorescene and stochastic dynamics simulations. *RNA*, **10**, 34–47.
41. Wang, Y.-H., Sczekan, S.R., and Theil, E.C. (1990). Structure of the 5′ untranslated regulatory region of ferritin mRNA studied in solution. *Nucleic Acids Res.*, **18**, 4463–4468.
42. Salahudeen, A.A., et al. (2009). An E3 ligase possessing an iron-responsive hemerythrin domain is a regulator of iron homeostasis. *Science*, **326**, 722–726.
43. Vashisht, A.A., et al. (2009). Control of iron homeostasis by an iron-regulated ubiquitin ligase. *Science*, **326**, 718–721.
44. Rouault, T.A. (2006). The role of iron regulatory proteins in mammalian iron homeostasis and disease. *Nat. Chem. Biol.*, **2**(8), 406–414.
45. Dix, D.J., Lin, P.-N., Kimata, Y., and Theil, E.C. (1992). The iron regulatory region of ferritin mRNA is also a positive control element for iron-independent translation. *Biochemistry*, **31**, 2818–2822.
46. Breuer, W., Epsztejn, S., and Cabantchik, Z.I. (1995). Iron acquired from transferrin by K562 cells is delivered into a cytoplasmic pool of chelatable iron(II). *J Biol. Chem.*, **270**(41), 24209–24215.
47. Theil, E.C., and Goss, D.J. (2009). Living with iron (and oxygen): questions and answers about iron homeostasis. *Chem. Rev.*, **109**, 4568–4579.
48. Zahringer, J., Baliga, B.S., and Munro, H.N. (1976). Novel mechanism for translational control in regulation of ferritin synthesis by iron. *Proc. Natl. Acad. Sci. U S A*, **73**, 857–861.
49. Muckenthaler, M., Gray, N.K., and Hentze, M.W. (1998). IRP-1 binding to ferritin mRNA prevents the recruitment of the small ribosomal subunit by the cap-binding complex eIF4F. *Mol. Cell*, **2**(3), 383–388.
50. Johnson-Buck, A.E., McDowell, S.E., and Walter, N.G. (2011). Metal ions: supporting actors in the playbook of small ribozymes. *Met. Ions Life Sci.*, **9**, 175–196.
51. Ferre-D'Amare, A.R., and Winkler, W.C. (2011). The roles of metal ions in regulation by riboswitches. *Met. Ions Life Sci.*, **9**, 141–173.
52. Halder, K., and Hartig, J.S. (2011). RNA quadruplexes. *Met. Ions Life Sci.*, **9**, 125–139.
53. Lammich, S., et al. (2011). Translational repression of the Disintegrin and Metalloprotease ADAM10 by a stable G-quadruplex secondary structure in its 5′-untranslated region. *J. Biol. Chem.* **286**(52), 45063–45072.

54. Merrick, W.C. (2010). Eukaryotic protein synthesis: still a mystery. *J. Biol. Chem.*, **285**(28), 21197–21201.

55. Jackson, R.J., Hellen, C.U., and Pestova, T.V. (2010). The mechanism of eukaryotic translation initiation and principles of its regulation. *Nat. Rev. Mol. Cell Biol.*, **11**(2), 113–127.

56. Lomakin, I.B., Hellen, C.U.T., and Pestova, T.V. (2000). Physical association of eukaryotic initiation factor 4G (eIF4G) with eIF4A strongly enhances binding of eIF4G to the internal ribosomal entry site of encephalomyocarditis virus and is required for internal initiation of translation. *Mol. Cell Biol.*, **20**, 6019–6029.

57. Marintchev, A., et al. (2009). Topology and regulation of the human eIF4A/4G/4H helicase comples in translation initiation. *Cell*, **136**, 447–460.

58. Thorp, H.H., et al. (1996). Cleavage of functionally relevant sites in ferritin mRNA by oxidizing metal complexes. *Inorg. Chem.*, **35**, 2773–2779.

59. Harrell, C.M., et al. (1991). Ferritin mRNA: interactions of iron regulatory element with translational regulator protein P-90 and the effect on base-paired flanking regions. *Proc. Natl. Acad. Sci. U S A*, **88**(10), 4166–4170.

60. Hentze, M.W., et al. (1987). Identification of the iron-responsive element for the translational regulation of the human ferritin mRNA. *Science*, **238**, 1570–1573.

61. McCallum, S.A., and Pardi, A. (2003). Refined solution structure of the iron-responsive element RNA using residual dipolar couplings. *J. Mol. Biol.*, **326**, 1037–1050.

62. Gunshin, H., et al. (2001). Iron-dependent regulation of the divalent metal ion transporter. *FEBS Lett.*, 309–316.

63. Erlitzki, R., Long, J.C., and Theil, E.C. (2002). Multiple, conserved iron responsive elements in the 3′ untranslated region of transferrin receptor mRNA enhance binding of iron regulatory protein 2. *J. Biol. Chem.*, **277**(45), 42579–42587.

64. Ke, Y., Wu, J., Leibold, E.A., Walden, W.E., and Theil, E.C. (1998). Loops and bulge/loops in iron-responsive element isoforms influence iron regulatory protein binding. *J. Biol. Chem.*, **273**(37), 23637–23640.

65. Brazzolotto, X., Timmins, P., Dupont, Y., and Moulis, J.M. (2002). Structural chnges associated with switching activities of human iron regulatory protein1. *J. Biol. Chem.*, **277**, 11995–12000.

66. Rangan, P., Masquida, B., Westhof, E., and Woodson, S.A. (2004). Architecture and folding mechanism of the Azoarcus group I Pre-tRNA. *J. Mol. Biol.*, **339**(1), 41–51.

67. Chen, O.S., Schalinske, K.L., and Eisenstein, R.S. (1997). Dietary iron intake modulates the activity of iron regulatory proteins and the abundance of ferritin and mitochondrial aconitase in rat liver. *J. Nutr.*, **127**(2), 238–248.

68. Butt, J., et al. (1996). Differences in the RNA binding sites of iron regulatory proteins and potential target diversity. *Proc. Natl. Acad. Sci. U S A*, **93**, 4345–4349.

69. Brazzolotto, X., et al. (1999). Human cytoplasmic aconitase (Iron regulatory protein 1) is converted into its [3Fe-4S] form by hydrogen peroxide in vitro but is not activated for iron-responsive element binding. *J. Biol. Chem.*, **274**(31), 21625–21630.

70. Dupuy, J., et al. (2006). Crystal structure of human iron regulatory protein 1 as cytosolic aconitase. *Structure*, **14**(1), 129–139.

71. Martelli, A., and Moulis, J.M. (2004). Zinc and cadmium specifically interfere with RNA-binding activity of human iron regulatory protein 1. *J. Inorg. Biochem.*, **98**(8), 1413–1420.

72. Ma, J., et al. (2012). $Fe2^{+}$ binds iron responsive element-RNA, selectively changing protein-binding affinities and regulating mRNA repression and activation. *Proc. Natl. Acad. Sci. U S A*, **109**(22), 8417–8422.

73. De Domenico, I., McVey Ward, D., Kaplan, J. (2008). Regulation of iron acquisition and storage conseuences for iron-linked disorders. *Nat. Rev. Cell Biol.*, **9**, 72–81.

74. Zahringer, J., Baliga, B.S., and Munro, H.N. (1976). Novel mechanism for translational control in regulation of ferritin synthesis by iron. *Proc. Natl. Acad. Sci. U S A*, **73**, 857–861.

75. Johnson, D.C., Dean, D.R., Smith, A.D., Johnson, M.K. (2005). Structure, function, and formation of biological iron-sulfur clusters. *Annu. Rev. Biochem.*, **74**, 247–281.

76. Shcherbakova, I., and Mitra, S. (2009). *Methods in Enzymology*, Vol. 468, Elsevier, 32–46.

77. Wang, J., et al. (2007). Iron-dependent degradation of Apo-IRP1 by the ubiquitin-proteasome pathway. *Mol. Cell Biol.*, **27**, 2423–2430.

78. Epsztejn, S., et al. (1997). Fluorescence analysis of the labile iron pool of mammalian cells. *Anal. Biochem.*, **248**, 31–40.

79. Sanchez, M., Galy, B., Muckenthaler, M.U., and Hentze, M.W. (2007). Iron-regulatory proteins limit hypoxia-inducible factor-2alpha expression in iron deficiency. *Nat. Struct. Mol. Biol.*, **14**, 420–426.

80. dos Santos, C.O., et al. (2008). An IRE-like stem-loop regulates alpha hemoglobin-stabilizing protein mRNA. *J Biol. Chem.*, **283**, 26956–26964.

81. Cho, H.H., et al. (2010). Selective translational control of the Alzheimer amyloid precursor protein transcript by iron regulatory protein1. *J. Biol. Chem.*, **285**, 31217–31232.

Chapter 18

RNA Information Gene Diseases: Nano-RNA-Based Medical Devices with Corporate Chemotherapy and Gene Therapy

Yoichi Robertus Fujii
Retroviral Genetics Group, Nagoya City University, Nagoya 467-8603, Japan
fatfuji@hotmail.co.jp

> *A feature of the transmissible tumors, which has largely drawn the attention of cancer workers and has modified current theories of cancer origin.*
>
> —Peyton Rous, 1911

RNA sequences would be information. Proteins and approximately 1–1.5% of DNA protein-coding sequences (CDSs) in the human genome correspond to the system. Recently, as part of its information, hundreds of microRNA (miRNA) genes have been identified by high-throughput methods, which were derived from approximately 99% information of human genome sequences. Although aberrant miRNA gene expression causes various cancers, miRNA mimic and anti-miRNA oligonucleotides (AMOs) have been applied for the

RNA Nanotechnology
Edited by Bin Wang

ISBN 978-981-4411-64-6 (Hardcover), 978-981-4411-65-3 (eBook)
www.panstanford.com

therapy of human cancer models as effective agents. It is suggested that cancer may be an RNA gene disease that would be caused by RNA gene information errors on the system frozen. On the contrary, the therapeutic experiment with administration of the viral miRNA mimic against pseudo-HIV-1 target using spuma vector nanoparticles (~100 nm) has been performed and it has been shown that the miRNA mimic was effective to decrease the target virus in vivo, suggesting that the nanoparticles can transmit miRNA information. Before the results, we hypothesized in the RNA wave 2000 model that miRNA would be a mobile genetic element and miRNA could tune both transcription and translation. Recently, nanoparticels (20–100 nm) as chemical substances and nanoRNA (~10 nucleotides [nt] long) information by locked nucleic acid (LNA) have been reported as safer administration; therefore, there is some possibility that the RNA information gene (Rig) could directly be injected intravenously or orally for cancer treatment without vectors. Taking together the results from these papers about miRNA research, we can set criteria for RNA information gene diseases (RigDs) upon RNA waves—"The disease is a system error that can be retuned by correct miRNA information." Thus, on the basis of the criteria, RigDs should be treated with miRNA agents as information technology (IT)-based therapies (ITBTs), such as nanoRNA and AMOs into nanoparticles of corporate chemotherapy and gene therapy as system-based therapy.

18.1 Introduction

Small ribonucleic acids (RNAs) as the RNA gene are biogenerated and directly and indirectly control every biological reaction [1, 2]. MicroRNA (miRNA) is a dominant class of RNA genes, which is noncoding single-stranded RNAs (ssRNAs) of ~22 nts and renders translational repression or activation and/or epigenetically transcriptional control of gene expression as phenotyping, including "somatic genome variations (SGVs)" and homeostatic tuning of cells and adult hosts, such as the broad environment of the metabolic rhythm as well as the small environment of the incisor (Fig. 18.1) [3–7]. In the case of post-transcriptional control, the seed sequences

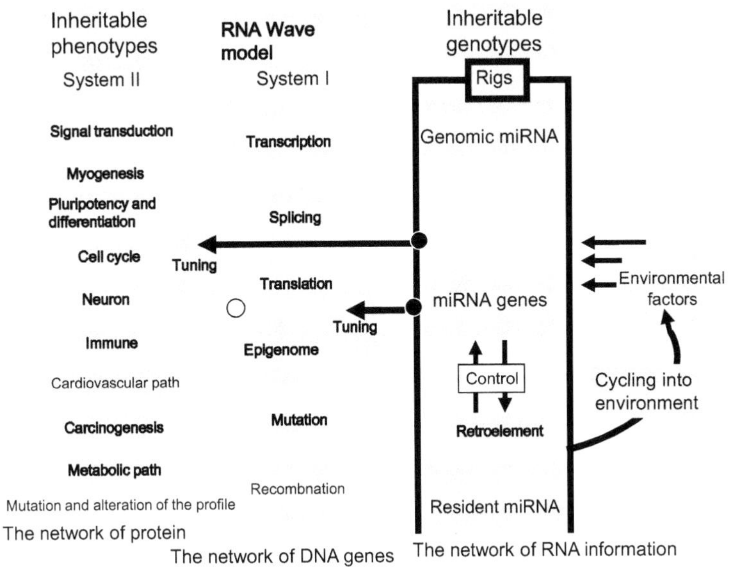

Figure 18.1 Roles of Rigs. miRNAs would be prompt to seize an opportunity of tuning proteincoding gene (DNA genes) expression in the networktype communication but not in a hierarchy of DNA gene and protein systems, such as the central dogma. miRNA is information, not the system; therefore, miRNA is an Rig. Rigs contain the genomic miRNA genes, the resident miRNA genes and the retroelements. Rigs are critical for mediating a biological response to environmental stresses. Further, Rigs may be cycling into the environment from species to species via foods, viruses and dust. The system consisted of a network of the DNA gene as system I and that of protein as system II and these systems are tuned by Rigs as information.

of miRNAs bind with imperfect complementarity to protein-coding messenger RNA (mRNA) 3′-untranslated regions (UTRs), 5′-UTRs, and coding sequences (CDSs) [8, 9]. Although the exact mechanisms for translational and transcriptional repression or activation have still not yet been clarified, dysregulation of miRNAs is implicated in causing diseases, including cancer and infectious diseases [10, 11]. Further, it is well known that plant viroid RNA produces small RNAs [12]. It is not surprising that RNA viruses, including not only retroviruses but also cytoplasmic RNA viruses in mammals, encoded functional miRNAs [13–16]. Thus, treatment of miRNA can specifically intervene in the progression of RNA gene–caused diseases. Further, while the retroelement Alu as an RNA gene in the human genome regulated pri-miRNA transcription, miRNA would regulate about 50% of the retroelement in the human genome and vice versa [17]. Part of the miRNA genes in mammals and plants have been found to be derived from the transposons [18, 19], suggesting that the miRNA genes were born from transposable elements. Therefore, transduction of miRNA using the foamy virus (FV) vector is likely to compromise the RNA gene in RNA genes such as STYLE (Fig. 18.2), which can be the integral usage of gene transduction

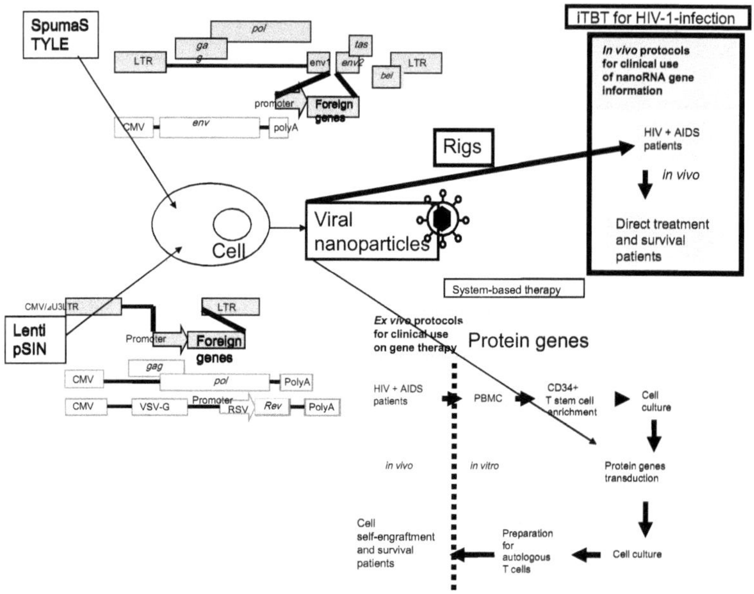

Figure 18.2 ITBT from gene therapy. Diseases are caused by aberrant Rig expression, except for viral infection. Simultaneously, a virus is the native nanoparticle including Rigs, such as RSV because over 100 years of Rous discovery as a transmissible Rig would be introduced in the text. Further, while the Rig has been reported as a quantum language according to quantum computing [7], RigDs could be treated by correct information of Rigs, suggesting that spontaneous information errors by wrong Rig information may induce RigDs. ITBT may be applicable for the treatment of every kind of disease to use gene therapy with lentivector (pSIN) and spumavector (STYLE). *Abbreviation*: RSV, Rous sarcoma virus.

for RNA gene-based therapy. The miRNAs and retrotransposons are information but not a system (see Fig. 18.1).

RNA gene-transducing technology offers a new and effective system for generation of vaccines and therapeutic agents for various diseases. Reports in this field demonstrate that the RNA gene-based vector is at first an efficient tool for delivery of miRNA or short interfering RNA (siRNA) [20–23]. Secondarily the most important issue is the safer approach because of no protein expression as antigens for induced aberrant immunoresponses of hosts such as the pSIN vector (Fig. 18.2). Recently, it has been reported that the level of human hepatitis C virus (HCV) was reduced close to zero by treatment of anti-miRNA gene oligonucleotide (AMO) without a vector system in the liver and in the blood of infected chimpanzees; however, after termination of its treatment, the amounts of HCV are increased again [24]. Further, administration of RNA genes with or without vectors blocked the growth of cancer cells and prevented the onset of tumor formation [25–29]. It is suggested that AMOs and miRNA mimics were effective in short-term administration but viral reserved patients and metastatic tumor bearing patients would require for the long-term treatment with AMOs or miRNA mimics; therefore, the RNA gene-based vector system for gene therapy is also important to control infectious diseases and cancer in the long term according to disease progression.

FV as a nanoparticle is the spumavirus vector used for the RNA gene-based therapeutic experiment in 2004 [29, 30]. The wild-type spumavirus is a nonpathogenic retrovirus in humans or their mammalian hosts; further human-to-human transmission has never been reported, which is very different from human immunodeficiency virus (HIV) and murine leukemia virus (MLV). Therefore, the characters as a lack of transmission and disease by FV phylogenitically resemble for endogenous retroelements in the human genome and have shown that FV has utility as an attractively safe vector system [31–33]. Integration sites of FV into the host chromosome have not located close to mutating and activation cellular genes as compared to those of HIV and MLV [34]. The FV system does not require cell division, and the foreign gene by FV can transfer to stem cells such as hematopoietic and embryo stem cells without activation of their cell cycle ex vivo by use of cytokines [35, 36]. FV can be produced at more than 10^7 units/mL of its titer and free of a helper virus and could be used to transduce more than 9 kb of foreign genes with a wide host range of susceptible cells from different tissues and species [37]. Although these characters of FV are advantageous to use transducing mammalian cells in the case of the second treatment for the second impact (a complication or another second infection) after the first treatment with lentivector (LV) or adeno-associated vector (AAV), it has been difficult to pretarget organs or cells and then induce organ-specific expression of the transduced gene. Recently, it has been reported, to solve the problem, that hematopoietic-specific miR-142-3p inhibited in vivo clotting factor IX (F.IX) expression transduced by LV bearing the miR-142-3p target sequence and eliminated off-target transgene expression of F.IX in hepatocytes, which do not express miR-142-3p [38–40]. This data suggested a concrete possibility that multisusceptible miRNA target sites in FV could be useful for post-targeting and organ-specific RNA gene expression.

For HIV-1 gene therapy on the viral vector, gene transfer under chemotherapeutic agents such as HIV-1 integrase inhibitors (INIs) and/or reverse transcriptase inhibitors (RTIs) is required; therefore, contrary to expectation, efficiency of transduction in the retrovirus vector system would be reduced under chemotherapy because LV is dependent on integration and reversetranscription by the packed HIV-1 integrase (IN) and reverse transcriptase RT. HIV-1 gene therapy on FV would be useful under chemotherapy in infected individuals because of low sensitivity of FV IN against the INI [41, 42]. Further, RT and protease inhibitors have severe side effects. Since environmental factors, including chemotherapies, have been linked to miRNA expression [43, 44], further administration with miRNAs may be involved in being free of these RT and protease side effects after chemotherapy. Therefore, in such a case, direct injection of RNA gene agents to the blood circulation might have an important role for treatment cooperating with chemotherapy. Recently, nanoparticles, such as orally applicable polylactide-*co*-glycolide (PLGA; <500 nm), which is approved by the Food and Drug Administration (FDA) in the United States [45], have been used as a vehicle for pharmaceutical delivery of these RNA genes to treat colorectal tumor in nude mice [46]. Thus, RNA information gene (Rig)-based therapy could be a suitable device, not only for selection of the first choice of agent, but also together with gene therapy and chemotherapy for RNA information gene diseases (RigDs) such as cancer and viral infections.

18.2 RNA Information Genes

18.2.1 *Experimental Identification of Rigs*

miRNAs are small in size and have sequences similar to various small RNAs, low levels of RNA, instability of ssRNA and stage-specific expression; therefore, experimental detection of miRNAs by cloning and sequencing is technically challenging. Initial cloning and sequencing followed by size fractionation were performed to identify miRNA [47]. Then, the polymerase chain reaction (PCR) technique was applied for the cloning of vertebrates and *Caenorhabditis elegans* (*C. elegance*) [48, 49]. The size-fractionated small RNAs were ligated to a 3′ linker and reverse-transcribed. As another method, the 3′-linked RNAs are directly ligated to a 5′ linker [50]. The dC-tailed complementary deoxyribonucleic acid (cDNA) of RNA is amplified by PCR and the isolated cDNAs are cloned and sequenced. Data is analyzed by BLAST-based miRBase (http://microrna.sanger.ac.uk). As a novel miRNAcloning technique, the SAGE-based approach to whole human transcripts confirmed the presence of numerous miRNAs [51]. Although web services for SAGE analysis have been provided, such as SAGEmap [52] and SAGE Genie [53], the expression profiles of known and unknown miRNAs combinations could be analyzed together by web services, such as the miRNA Analysis System (miRAS) (http://e-science.tsinghua.edu.cn/miras). Recently, subtractive hybridization of small RNAs (SHORT) has been reported using transcripts between viralinfected and noninfected cells to identify functional noncoding RNAs (ncRNAs) [54]. This method is based on subtraction and hybridization to avoid the time-consuming and cost-intensive multiple sequencing of currently known and constitutively expressed miRNAs; therefore, the dominant housekeeping transcripts could remove and increase the chance to identify less abundant and unidentified miRNA candidates by priory hybridization. Now, by recent high-throughput pyrosequencing techniques, such as Illumina HiSeq™ and Applied Biosystems SOLID™ platforms we can amplify and we can detect very low levels of miRNAs from the RNA libraries and large amounts of miRNAs in microvesicles from retroelements in the human genome as well as viral miRNAs from HIV-1-infected cells have been identified and sequenced by the pyrosequencing.

For analysis of previously discovered miRNA profiles, microarray technology is suited for high-throughput methods. The 3′ and 5′ adapters are ligated in miRNAs. Then the adapted miRNAs are reverse-transcribed and amplified by PCR with a photolabeled primer [55, 56]. As another method, biotinylated primers are used for the RT step and miRNA levels are detected by streptavidin-bound fluorophores [57]. Altered microarray analysis involves more sensitive methods based on locked nucleic acid (LNA)-modified microarrays [58, 59] (see Fig. 18.2 regarding LNA). With a combination of nanoparticle-amplified surface plasmon resonance imaging (SPRI), the assay had a low level of detection of the concentration of ~fM. The data could be processed with the Siggenes module from Bioconductor (http://www.bioconductor.org/).

More sensitive RT-PCR-based profiling approaches have been reported in real-time RT-PCR [60, 61] and real-time looped RT-PCR [62]. The cDNAs of miRNA are polymerized from the poly(A)-tailed total RNA as templates by an RT primer and RT, and then cDNA is amplified by using the 7500 real-time PCR system. Although real-time PCR requires only ~fg of starting RNA sources, the minor population of miRNA expression profiles might be lost. Thus, a combination of microarray analysis with real-time PCR for detection of the expression level of Rig and its targets has been reported recently [63].

The data from cloning and microarray analysis has been confirmed by northern blotting. Northern blotting has provided approximately a 10-fold and 25-fold increase in sensitivity of detection of miRNA by LNA-modified oligonucleotide probes (Fig. 18.3) [64] and 1-ethyl-3-(3-dimethylaminopropyl) carbodiimide (EDC)-cross-linking [65]. To using these techniques, highly sensitive northern

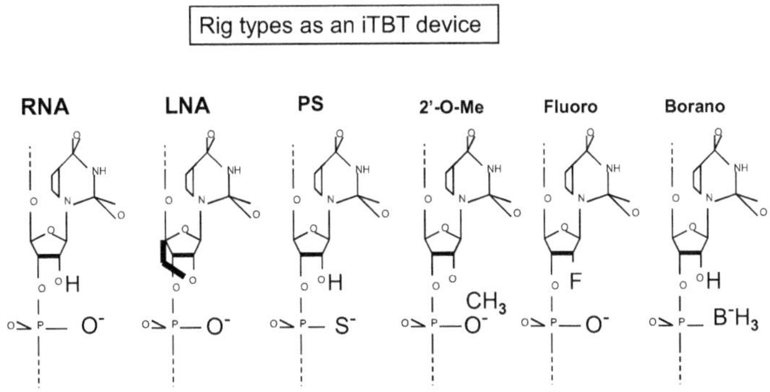

	RNA	LNA	PS	2'-O-Me	Fluoro	Borano
MO and AMO	+	+	?	+	?	+
Toxicity	—	—	+ +	-	N.D.	—
RNase Resistance	- / +	+	+	+	+	+
For array & northern blot	+	+	-	-	-	-
For iTBT	+	+	-	+	-	-

Figure 18.3 Rig types as an ITBT device. Rigs are industrially synthesized with lower costs when compared to the recombinant protein or antibody. ITBT devices are represented in the uridine ribonucleotide (RNA), the locked RNA (LNA), phosphodiester (PO) with phosphorothioate (PS)linkaged RNA, (2′-*O*-Me)modified RNA, 2′-dFUrd and boranophosphate-modified RNA (Borano). All modified RNAs could be resistant to RNase and naive RNA has also been reviewed to be circulating as extracellular Rigs escaped from degradation with RNase activity [268]. Rigs are valued in the potential usefulness of therapeutics, diagnosis, prognostics, prophylactics as well. *Abbreviations*: 2′-*O*-Me; 2′-*O*-methyl; RNase, ribonuclease; PS, phosphorothioate; PO, phosphodiester; 2′-dFUrd, 2′-deoxy-2′-fluorouridine

blotting is needed for the detection of PIWI-interacting RNA (piRNA) [66, 67] and repeat-associated small interfering RNA (rasiRNA) [68] because both do not arise from stem–loop constructs that can be predicted by a computer-based approach. Therefore, although we showed a model of RNA language that miRNA electrospin according to quantum computing would be involved in miRNA tuning [17], it has been reported that a single-nt difference between miRNA was discriminated with a single–quantum dot (QD)-based nanosensor assay [69]. The QD is one of the applications for the measurement of miRNA electrospin; therefore, the miRNA quantum state will be possible to be progressed for identification of each Rig with magnetic resonance imaging (MRI). But the disadvantages of this cuttingedge analysis lie in the fact that the database has not yet been prepared on the web.

18.2.2 *Computer Algorithms for Rig Identification*

The hairpin structure of ncRNA has been predicted by using computing algorithms, such as QRNA [70], ddbRNA [71], MSARI [72], RNAz [73] and RScan [74]. Thousands or millions of conserved hairpin structures as predicted Rigs could have been mapped in the plant, nematode, fly and human genomes. The MiRscan program predicted ~1,500 and ~3,500 hairpins in *C. elegans* and *Homo sapiens* (*H. sapiens*), respectively. To scan the mapped data for putative miRNA hairpins, computational procedures were further developed. The secondary structure of pre-miRNA could be generated by mfold [75] as well as the Vienna package [76] according to RNA base pair matching and thermodynamic calculations. For the *Arabidopsis thaliana* genome, findMiRNA was applied [77]. Thirty new miRNAs were detected and eight were verified. Using MiRscan 35 novel miRNAs were predicted, of which 16 were verified in *C. elegans* [49]. In contrast to the eukaryotic genome, the viral genome is small and viral genomic sequences are not homologous within viral species. Using the same program, 13 miRNAs from the human cytomegalovirus (HCMV) genome were predicted and 5 were verified [78]. In *Drosophila melanogaster* (*D. melanogaster*), 48 new miRNAs were cited by miRseeker and 24 were verified [79]. The phylogenic shadowing approach showed 69 candidates of miRNA in humans, and 16 were verified [80]. By combining PalGrade with a microarray assay, 5,300 miRNA candidates were found in the human genome and 359 positive spots were detected in microarray experiments [81]. HIV-1 miRNA has been investigated with the StemE program [82]. Five miRNA candidates were predicted and one was validated. Gene and miRNA expression profiles are well suited for the assessment of therapies in phase I and II clinical trials; however, the percentage of the verification of these miRNAs is less than 60% at the highest. Furthermore, although there are

10,000 to 35,000 cDNA probes for gene expression, the number of miRNAs currently identified is only in the hundreds. Although in Rig searches and Rigprofiling analyses, recent computing-based prediction is more highly valued progressively, computing search for Rigs may be changed by the language of Rigs under quantum-based RNA waves [7].

18.2.3 *The Version of Rig*

Rigs, which contain at least two categories, are genomic and resident Rigs [1]. The seed sequences of Rigs target mRNA with incomplete pairing, but the siRNA gene has not the seed region at all in plants. The human siRNA gene may also target mRNA with complete pairing [83]. Both are Rigs, but the system of targeting is different. If mi-/siRNA-like RNA genes are evolutionarily present, such as scanning RNA (scnRNA) in *Tetrahynema* [84], there is a question of why single- and multitargeting RNA genes are simultaneously required for regulation of phenotype expression, albeit the seed theory based on Watson–Crick base pairing is partly true and/or something wrong. In the case of mammalian stem cell differentiation, cell lineage–specific miRNAs have been reported to be identified [17]; however, an miRNA is not specific to one target mRNA. An miRNA can target multiple target mRNAs and a mRNA would be targeted by multi-miRNAs. In the Watson–Crick model, a specific DNA gene is specific to one mRNA and a triplet of mRNA is specific to the amino acid, and then the peptide produced has a specific function in gene expression. Rigs may exclude a stereotype scheme from those biological pathways because RNA is information but not the system.

siRNA controls gene expression in plants [85] as well as siRNA and piRNA in humans do [82, 86]. The siRNA genes act in transcriptional gene silencing, and the production of the miRNA gene is required for biogenesis of *trans*-acting siRNA (ta-siRNA). Although spreading of si-/miRNA-directed DNA methylation has been found in plants [87, 88], like siRNAs, piRNAs, and ta-siRNAs, endogenous siRNA (endo-siRNA) and miRNA-offset RNA (moRNA) are produced from transcripts, and these RNAs, including miRNA, could act in transcription of the DNA epigenes in animals. Further, Ago-associated small RNAs recognize genomic sequences and show small RNA-induced heterochromatin formation followed by DNA elimination [89]. In epigenetic control of gene expression by miRNA, siRNA itself can also be methylated in flies by Hen 1 and the genomic miRNA methylation status is also associated with disease progression, such as human cancer [90, 91]. Thus, methylation of Rigs may make variation of information. Although Hen 1 methyltransferase methylates piRNAs in mouse [92, 93] and miRNAs in plants [94], the effects of Hen 1 on Rigs in humans have not been cleared. The possibility remains that human miRNAs may be methylated by Hen 1 homologs [89] or other methyltransferases.

Phenotypical variations by Rigs have been shown in epigenes as well as miRNA–target associations. miRNA is an ssRNA, and the majority of its nt length is about 21 nts or 22 nts. The range of nt sequences in small RNA were 16 nts to 35 nts in mammals and plants, which were thought to be relatively common isoforms of miRNA genes [95]. Further, plus- and minus-sense end-siRNAs were isolated from the transcripts of mammalian transposable elements (TEs) in the genome, endogenous retrovirus L (ERVL-Mm) [96]. The gene variants of miRNAs were also found as single-nucleotide polymorphisms (SNPs) in the human X-linked miRNAs in schizophrenia and autism individuals [97]. The important factor of adenosine-to-inosine (A-to-I) editing by adenosine deamination could make molecular variation of miRNA in Alu elements because most editing sites occur in Alu [98, 99]. The range of nt sequences, the sequences direction, SNPs, and A-to-I editing were found in miRNA target sequences of mRNA as well as in miRNAs and viral RNAs [100]. Further, it has been reported that mRNA targeted by nonsense-mediated mRNA decay (NMD) is stabilized by miRNA-mediated post-translcriptional regulation [101]. This experimental evidence suggests that Rigs and their variations are related with mRNAs and their variations. Alteration of incomplete matching between Rigs and mRNAs should mediate post-transcriptional gene regulation, which could affect expression of phenotypes and diseases [102, 103]. It is commonly assumed that seed sequence–matched miRNA genes could similarly regulate the target mRNA translations; however, only one nt-mismatched miR-310 in the seed in *Drosophila* species has been heterospecifically expressed across the species, and it causes misexpression of genes and reduction of viability [104].

18.2.4 *RNA Gene Targets*

To identify miRNA targets, principles of the computational methods are based on sequence complementarity, thermodynamically free energy of the RNA duplex and statistical inferences. Further, the selective 2′-hydroxyl acetylation analyzed by primer extension (SHAPE) procedure beyond classic seed base pairings was developed on the analysis of HCV:miR-122 interaction (reviewed in Section 18.5) [105]. Now, the miRNA targets are listed based on overall complementarity at the first time and then the seed compensatory has recently been subject to the prediction of miRNA targets; however, SHAPE reveals that the tail sequences of the Rig are also important for virally positive replication to interact on the target sites' sequences.

Computing algorithms for miRNA seed target prediction basically rely on the complementarity of miRNA seed sequences to that of the target sequences and the thermodynamic stability of the miRNA–target RNA duplex. There are many miRNA target prediction programs, such as miRanda, TargetScan, RNAhybrid, DIANA-microT, PicTar, miRU, TargetBoost and RNA22 [106–114]. The validation of computed predictions of miRNA targets is far more challenging compared to that of miRNA [115]; however, it is possible to aim for biologically validated miRNA targets for clinical application of therapeutic agents. Reverse genetic methods are the most common methods for the validation [116–119] because the *let-7* was found as antisense RNA completely pairing the target, which would be an siRNA rather than an miRNA. Further, the finding of *lin-4* sense RNA and *lin-14* 3′-UTR interaction has also been identified in the nematode with reverse genetic methods by using the *luc* reporter [120]. The miRNA target-fused *luc* reporter was coexpressed with the relevant or overexpressed miRNAs in the mammalian cell culture. Although the Dicer enzyme is expressed in peripheral blood monocytic cells (PBMCs), sometimes its expression level is found to be suboptimal. Recently, overexpressed Dicer has been shown to induce miRNA from the TAR element of HIV-1 [121], which implies that in reverse genetic analysis, Dicer and/or Drosha coexpression would be effective to validate miRNA as well as miRNA target prediction in mammalian cells and also validation of mammalian Rigs in plant cells would add material evidence in expression of Rigs [122].

Rig identification methods are various and Rig versions are more multiple, further, targets as well. The diversity would contravene the canons of the Watson–Crick pair in their central dogma. Thus, Rigs are information but not the system. If Rigs imply the system, we, next, would also find the contravention from a complex of various Rigs into human diseases.

18.3 RNA Information Gene Diseases

Cancer, cardiovascular diseases, and infectious diseases are the most common causes of death. miRNA genes could act as not only intracellular genes but also intercellular genes in these mortal diseases [123]. Differences between expression profiles of the intracellular miRNA genes of cancer patients and healthy donors have revealed that the Rigs are deeply responsible for a variety of cancer cellular phenotypes [13]. Further, miRNA genes have been detected, and the amounts of them were measured in plasma, saliva, and urine [124]. The extracellular information by miRNAs is a novel biomarker of the diagnosis in diseases, such as myocarditis, coronary artery disease, cancer, type 2 diabetes, and liver injury [125–129]. Since the miRNA gene is information, a magic bullet of miRNA-based agent for diseases could be applied for therapy.

18.3.1 *Infection and Cancer*

miRNA has been related with both inflammation and cancer through regulation of the immune system [130]. Both immune and signal transduction systems are controlled by miRNAs [131–133]. miR-155 expresses in B cells, T cells, macrophages, and dendritic cells. The typical infectious stimulus, which is implicated in malignancy into gastric organs, has been reported to be identified as *Helicobacter pylori* (*H. pylori*) [134]. Interestingly, the miR-155 gene negatively regulates *H. pylori*–induced inflammation [135, 136]. *BIC*, including miR-155 as an oncomir, has an important role for the oncogenic activation of c-*myc* [137] and suppresses TP53-induced nuclear protein 1 (TP53INP1), which inhibits TP53 activity in pancreatic cancer [138]. Further, *BIC*, in whose which sequences is a proto-oncogene known as retrovirus insertion sites [137], has been reported to be transcriptionally activated by AP1 and nuclear factor kappa B cell (NF-κB) [139, 140], and miR-155 inhibits NF-κB expression with miR-9, miR-146, miR-147, and miR-21 [141]. The miR-146a gene suppresses AP-1 activity and IL-2 production in human T lymphocytes [142]. The miR-9 gene directly targets NF-κB mRNA and regulates human gastric cancer as a tumor suppressor gene [143]. On the contrary, the promoter of the miR-146a gene contains two NF-κB sites, and the Kaposi's sarcoma–associated herpes virus (KSHV) vFLIP K13 protein and the NF-κB up-regulate the miR-146a gene expression [144]. The NF-κB-dependent miR-146a expression targets Fas mRNA and is involved in cytoprotection of rat bone marrow–derived mesenchymal stem cells (MSCs) [145]. KSHV miR-K1 is also sufficient to rescue NF-κB activity and inhibits viral lytic replication [146]; however, the viral miR-K12-1 targets IκBα mRNA to promote NF-κB signaling, which prevents viral activation to continue the state of latent infection of KSHV [147]. This data suggests that it is clear that viral miRNAs via NF-κB activities are dominantly implicated in KSHV-induced tumor, and miR-146a may act as a survival factor for KSHV-infected cells and induce inflammatory and KSHV-associated malignancy pathways in infected individuals.

KSHV-producing protein itself is believed to induce malignancy in host cells [148, 149], and we know that other human tumor-associated viruses, Epstein–Barr virus (EBV), hepatitis B virus (HBV) and HCV, human papillomavirus (HPV), and human T cell lymphotropic virus (HTLV) producing viral oncoproteins are also believed to contribute to human malignancy [150]. But these oncoproteins are not sufficient for oncogenesis, and some unknown factors exist. Alteration of expression amounts of protein is phenotype changing but not a genetic one. Like this, HPV-11 transfected cells changed the profiling of miRNAs compared to nontransfected control ones [151]. As the genetic factor, KSHV producing viral miR-12-11 gene is an ortholog of the seed sequences of miR-155, and the viral miR-12-11 may hijack the target of miR-155, which would induce KSHV-associated lymphomas [152]. The viral miR-K3 can negatively regulate KSHV lytic replication and maintain vital latency [153] and the KSHV miR-K12-10a–suppressed apoptosis by targeting the tumor necrosis factor (TNF)-like weak inducer of the apoptosis receptor (TWEAK) [154]. Further, miR-K12-4-5p enhanced DNA methytransferase (DNMT) 3a and 3b by suppressing the expression of retinoblastoma-like protein 2 (Rblp2) of one of the tumor suppressors [155], suggesting that cellular miRNA as well as KSHV-encoded miRNAs could contribute to carcinogenesis of infected cells; namely, the pathogenic Rigs may disturb housekeeping regulation by host Rigs and induce alteration of profiles of the Rigs; therefore, cancer as well as infection-related cancer should be a RigD [13].

Cervical carcinoma is caused by high-risk HPV types 16, 18, and 31 [156]. Squamous cell carcinoma (SCC) accounts for approximately 80%, and the genome of most cervical SCC cells contains integrated HPV sequences [157]. Although it is well known that HPV E6 and E7 proteins are oncoporteins and contribute to cervical carcinogenesis by inhibiting the tumor suppressor protein TP53 and pRb functions [158, 159], TP53 directly controls the promoter of the tumor suppressor miR-34a gene [160–162], whose expression is down-regulated in most cervical cancer tissues with HPV infection [163]. Further, the HPV E6 protein inhibits miR-34a via destabilization of TP53 [164]. The Rig is an absolute controlling factor of th evirally induced tumor now.

In most of all virus-associated carcinogenic cases, there is the mechanistic possibility that carcinogens are as an intracellular factor, as described above. However, it has recently been reported that exosomal EBV miRNAs may be used in human tumor for intercellular communication [165], and the cellular miRNA genes into microsomes are also actively exported for the cell–cell communication [166]. Further, bovine milk–derived miRNAs in microvesicles have been isolated, and the secreted RNAs from the mammary glands were incorporated into living cells [167]. This data suggests that the communication via miRNAs may work among

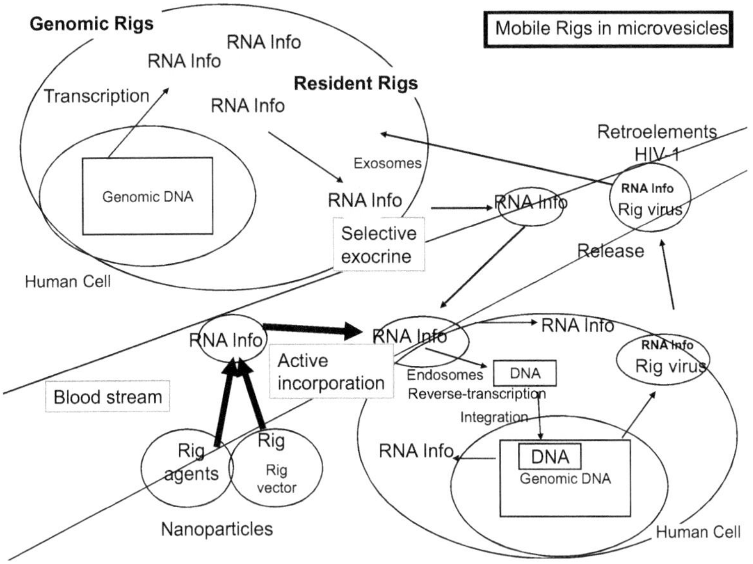

Figure 18.4 Mobile Rigs in microvesicles. The Rig is mobile information (Info) into microvesicles and lipoproteins, and with lipids, the Argonaute (Ago) protein complex. While expression profiles of Rigs are altered by environmental stress factors, such as foods, viruses, chemicals, ultraviolet rays and mobile Rig itself, the alteration of Rigs is transmitted from cell to cell via every body fluids. Therefore, circulating Rigs are biomarker for health and diseases [269]. Further, anti-Rig LNA agents and Rig mimics could be applied for therapy [247, 270]. For Rig agent delivery, nanoRigs in nanoparticles, such as microvesicles and viral particles may be suitable for microtargeting against dysfunction of protein genes.

cells as well as individuals by exosomes', virus particles, and/or foods, RNA devices (Fig. 18.4) [168]. Intriguingly, the extracellular miRNA in blood plasma and cell culture has been reported to be independent of microvesicles and be bound to Ago2 protein [169], of course, although we have to think about tiny exosome-like signatures (20–50 nm) than exosomes [170, 171]. While there is an idea that miRNA-nonspecific Ago protein can be a carrier protein of RNA information, it may be actively and selectively secreted with miRNAs from cells because of the difference between the cells' and the supernatant's miRNA species [168], and the normal human plasma and the patient plasma [172, 173]. Vickers et al. have shown that high-density lipoprotein (HDL) participates in the transport of miRNAs in plasma with stability but not long mRNAs [172]. Since secreted exosome vesicles are derived from endolysosomal compartments called multivesicular bodies (MVBs) and the membrane-associated protein Ago2 can be sorted in a ubiquitin-dependent manner into MVBs by three heteromeric subcomplexs, termed "endosomal sorting complex required for transport" (ESCORT) [174], GW182 associated with miRNA-loaded Ago2 has been reported to be sorted into MVBs for secretion to the extracellular [175]. Therefore, after specific secretion of the miRNA–Ago2 complex from donor cells, the complexes could physico-chemically interact with HDL phospholipids and may be carried to acceptor cells as another delivery pathway. Further, several kinds of RNA genes, including small nuclear RNAs (snRNAs) [7], mRNAs, and endogenous retrovirus RNAs were secreted from cells [176]. These results suggest that the circulating RNA genes in the density particles (d = 1.13–1.19 g/mL) of microvesicles, exosomes (50–100 nm), and retroviral particles (100 nm) and the lower density (1.063–1.21 g/mL) of HDL (8–12 nm) would have an important role for keeping homeostasis in cells and the body condition. Thus, it is possible for us to speculate that disturbance of RNA information by environmental factors, such as viral RNA information, stresses, chemical agents, and microwaves, could change the circulating Rigs and then induce a RigD as cancer involving viral infection–originated tumor, the mechanisms of metastasis occurring as well.

18.3.2 *Endocrinopathy*

Concerning the circulation system, the plasma glucose level has an important role in maintaining homeostasis. Diabetes may be a RigD [177]. Diabetes is two types, type 1 and type 2. The former is caused by a lack of insulin production, and the latter is due to suppression of insulin action and production. The miR-375 gene has been shown to directly suppress glucose-stimulated insulin secretion [178]. Mice miR-375 KO showed hyperglycemia by decreasing pancreatic β-cells [179]. Further, zebrafish miR-375 KO caused deletion of the pancreatic islets [180], suggesting that the function of miR-375 is involved in insulin production and pancreatic islet development [181]; therefore, miR-375 would be related with type 1 and 2 diabetes. On the contrary, miR-124a was specifically expressed in the brain and neurons [182]. Overexpression of miR-124a has been reported to increase insulin secretion in the pancreatic β cells, but under high concentrations of glucose, secretion of insulin was decreased [183]. The brain is also a glucose-dependent tissue; therefore, the miR-124a gene would be implicated in the control of the plasma glucose level. The Goto–Kakizaki (GK) rat is a model of type 2 diabetes. Up-regulated miRNAs, three paralogs of miR-29a, miR-29b, and miR-29c, have been isolated in the GK rats [184]. The miR-29 gene expression repressed insulin-stimulated glucose uptake in 3T3-L1 adipocytes, which is according to insulin resistance. Therefore, the miRNA genes could be involved in insulin action. Although insulin resistance is correlated with human obesity, miR-21 was robustly expressed in human adipose tissue and positively correlated with the body mass index (BMI) [185]. Quite recently, fatty acid regulation of the metabolic syndrome has been reported to be controlled by miR-33a/b [186]. The miR-33a and miR-33b genes are intronically located in the sterol regulatory element-binding protein-2 (SREBP-2) and SREBP-1 genes, respectively. The miR-33a/b genes target fatty acid oxidation enzymes and the insulin receptor substrate 2, which is an essential component of the insulin-signaling pathway in the liver. Fatty acid oxidation as oxidative stress is well known to be related with atherosclerosis, and together, metabolic syndromes, including obesity, diabetes type 2, and atherosclerosis, have traditionally been viewed as lipid storage disorders caused by overnutrition. About increasing adipose tissue mass, the miR-17-92 gene cluster, miR-21, miR-103, miR-143, miR-371, and miR-378/378* were up-regulated as adipogenic markers; in contrast, let-7, miR-27, miR-130, miR-138, miR-369-5p, and miR-448 were down-regulated into adipogenesis [187]. In these miRNA genes, miR-17-92, let-7b, miR-143, and, further, miR-126, miR-150, miR-155, miR-124, and miR-211 were moved in microvesicles between adipocytes and macrophages or circulating macrophages and vascular endothelial cells for communication [188], suggesting that the intercellular communication among tissues via mobile

Rigs may be rationally involved in the systemic inflammation and oxidative stress in the metabolic syndrome.

18.3.3 *Inflammation*

The indication that *bic or* the miR-155 gene combined with two biological phenomena, which are inflammation and cancer, was in a report in 2007 [189]. Aberrant expression of miRNAs induced inflammation and above-mentioned cancer, which is corresponding to our old knowledge as "prolonged exposure to inflammation leading to cancer." The fine-tuning by miRNAs in the immune system is critical for macrophages, granulocytes, NK cells, NK T cells, and dendritic cells upon innate immunity and T cell and B cell functions and differentiation [190]. miR-155 modulates megakaryocytic differentiation of human cord blood $CD34^+$ hematopoietic stem cell progenitors [191]. Most of the virus-associated lymphoid cancers up-regulate miR-155. For example, the infection of $CD4^+$ T cells with HTLV-1 leads to expression of miR-155 [192]. As described above, the infection of B cells with EBV leads to an elevation of miR-155 expression. In mouse macrophages, miR-155 was up-regulated by interferon-gamma (IFN-γ) and polyriboinosinic:polyribocytidylic acid (poly(I:C)) and a variety of Toll-like receptor (TLR) ligands [140, 193]. Thus, miR-155 is involved in the regulation of innate immune response. While miR-155 is predicted to target 5,445, 915, 303, and 199 of transcripts in MICRORNA.ORG, DIANA LAB, Targetscan, and PicTar, respectively, about 20 or more have only been reported to be investigated as the targets mRNA of miR-155 [194]. miR-155 was also up-regulated in a rheumatoid arthritis (RA) patient's circulation and within the RA-inflamed joints [195, 196], miR-16, miR-132, and miR-146a. However, levels of miR-132 expression are increased in PBMCs from RA blood, while they are decreased in RA plasma [195, 197]. RA is a chronic inflammatory autoimmune disorder with progressive systemic inflammatory manifestations in joints. miR-155 is detectable in plasma in RA compared to osteoarthritic donors and healthy donors [197]. Further, miR-155 expression is up-regulated in PBMCs isolated from RA compared to osteoarthritic and healthy donors [198, 199]. This data suggested that plasma levels of miR-155, miR-16, miR-146a, and miR-223 are correlated with disease activity of RA. Although $CD4^+$ cells in synovial fluid, autoantibodies, antigen-presenting macrophages, and RA fibroblast-like synoviocytes (RA-FLS) play an important role in RA pathology, $CD4^+$ T cells in RA recognize donor allogenic class II antigens on donor antigen–presenting cells (APCs) and/or autoantigens presented by self-APCs. T cell receptor (TCR) stimulation induces up-regulation of miR-155. $CD4^+$ T cells with high miR-155 levels are prone to develop into T helper 1 (Th1); on the other hand, low or none of miR-155 levels induce Th2 cells. Th1 cells produce IFN-γ, and interleukin-2 (IL-2) that can have pro-inflammatory effects and promote Treg and cytolytic T lymphocyte (CTL) generation, while Th2 cells produce IL-4, IL-5, IL-10, IL-13, IL-14, and IL-10 that can be immunosuppressive and also facilitate immunoglobulin (Ig) class switching and autoantibody production on autoimmune responses. Recently, miR181a has also been shown to modulate TCR signaling [199] and miR181b-regulated Ig class switch recombination–inactivated B cells [200]. Therefore, autoantibody production is regulated by miRNA genes. miR-155 is required for normal immune function, germinal center response, and generation of Ig class switching, as well as RA by autoantibodies. Taken together, different levels of miRNA such as miR-155 have a different output, such as inflammation or cancer, suggesting that the miRNA gene is different from antibody–antigen, cytokine receptor, and TLR signal transduction systems. The miRNA genes are information rather than innate and adaptive immune-responsing systems; therefore, Rigs are able to bind between the immune system and systemic tumorigenesis.

18.3.4 *Neurodegeneration*

By aberrant Rig expression, the progressive and chronic pathophysiology of RA may be similar to that of neurodegeneration. miRNAs are implicated in neural cell survival and all neurodegenerative diseases [201–203]. Mice on Dicer KO of Purkinje cells were associated with the development of ataxia and cerebellar degeneration [204]. Mice on Dicer ablation of dopaminoceptive neurons displayed ataxia, front and limb clasping, reduced brain size, and smaller neurons [205]. These results suggest that loss of information derived from the matured miRNA genes causes serious defects of the brain system. Alzheimer disease (AD) is a well-known neurodegenerative disease in humans. AD is a progressive disease having characteristics such as aphasia (language disturbance), agnosia (failure to recognize people or objects in the presence of intact sensory function), and apraxia (inability to perform motor acts in the presence of an intact motor system). Synapses and neurons are degenerated by β-amyloid (Aβ) production, which pathologically affects the hippocampus, followed by association cortices and subcortical structures in the brain. The Aβ peptide precursor (APP) protein is processed by the Aβ cleavage enzyme 1 (BACE1) protease and the Aβ is produced by BACE1 and the γ-secretase complex. Approximately 1% of the early onset familial form of AD is due to mutations in the *presenilin* (*PSEN*) 1 and *PSEN2* and *APP* genes, but generally AD is the most common cause of dementia in aged populations. Therefore, dysregulation of the miRNA genes could be aging associated. Using target prediction programs described above, miR-15a, miR-29b, miR-9 and miR-19b could target BACE mRNA; miR-15a, let-7 and miR-106b could bind to APP mRNA; and miR-9 could match PSEN1 mRNA. Recent studies evaluated that up-regulation of the miR-29a/b-1 cluster reduced both BACE1 and Aβ and in AD patients, miR-29a/b-1 was significantly decreased [206, 207]. BACE1 was also targeted by miR-107 [208]. Further, the specific up-regulation of miR-146a, which is an NF-κB-sensitive miRNA, was observed in AD [209]. miR-146a represses the complement factor H (CFH) and CFH protein suppresses the inflammatory response of the brain. Thus, miR-146a would exert proinflammatory effects in AD. The data indicated that AD induction and progression are controlled by Rigs.

Parkinson disease (PD) is also progressive neurogeneration. Cell death of dopaminergic neurons is observed in the substantia nigra of the midbrain. Deletion of Dicer in the mouse substantia nigra leads to similar cell death as in PD [210]. Further, etiology of polyglutamine diseases such as Huntington disease (HD) and spinocerebellar ataxia type 1 (SCA1) are also involved in miRNA tuning [211–214]. From a mega-analysis combining genome-wide schizophrenia association study data with 51,695 individuals, the miR-137 gene and its target genes have been associated with susceptibility to bipolar disorder and schizophrenia [215]. miR-137 is implicated in regulation of neurogenesis and neuromaturation [216, 217], suggesting that dysregulation of miR-137 may induce schizophrenia. Although the above-described RA-related miRNA-132 is also required for normal dendrite maturation in newborn neurons [218], AD, PD, SCA, and schizophrenia are caused by different system errors; even if not so, miRNA-based therapy for neurodegeneration has great potential because the miRNA sequences are just information.

Epigenetic regulation is not "epi-", because both chromatin remodeling and epigenetic modification could be tuned by Rigs. The expression of miR-137 is subjected to epigenetic regulation mediated by MeCP2 [216], which belongs to a family of DNA methyl-CpG-binding domain (MBD) proteins and contributes to the molecular pathogenesis of neuropsychiatric abnormalities, such as Rett syndrome [219]. Prion diseases such as transmissible spongiform encephalopathies are also epigenetic diseases, which comprise a group of neurodegenerative diseases, including Creutzfeldt–Jakob disease (CJD) in humans. Dysregulation of miRNA expression was found in prion-infected mouse brains [220]. miR-342-3p, miR-320, let-7b, miR-328, miR-128, miR-139-5p, and miR-146a were up-regulated by more than 2.5-fold, whereas miR-338-3p and miR-337-3p were down-regulated. Among these, miR-146a is up-regulated in both prion diseases and AD, which was not seen in amyotrophic lateral sclerosis (ALS), schizophrenia, or PD. It is not certain whether miR-146a is related with the etiology and molecular basis of both neurodegenerative disorders; however, epigenes would be controlled by the RNA genes [17]. Although a number of epigenetic drugs, which are in various stages of development, such as 5-azacytidine, decitabine, and *trans*-retinoic acid [221], lack target specificity [222], the fact suggests that the epigenetic phenomenon would specifically be orchestrated by the RNA genes. Thus, Rigs could be rationally used to treat epigenetic diseases as genetic diseases. It is possible that most of all RigDs are retuned by Rigs as system errors.

18.4 Pharmaceutical Agents with Rigs

18.4.1 *Rigs for Information Technology–Based Therapy*

Small RNAs are now industrially synthesized with lower costs than protein or antibody production. The production of miRNAs is, therefore, focused on pharmaceutical properties which can be delivered to a wide range of organs such as the brain. However, the problems of instability in blood and how to specifically target the organ make development of RNA agents challenging. The former may be improved by modification of the backbone and small nts in Fig 18.3. The replacement of PO with PS linkages reduced cleavage by nuclease and increased the half-time of RNA in vivo. PS linkage on one or both strands supported RNAi but extensive PS substitution was toxic to cells (Fig. 18.3) [223]. Substitution with 2′-dFUrd did not affect RNA interference (RNAi) effects of the 2′-dFUrd nt (Fig. 18.3) [223, 224]. RNAi also tolerated LNA [223] and the LNA-modified siRNA caused no IFN response as an off-target effect [225]. LNA is also thought to increase the halflife of miRNA in vivo by enhancing the resistance to RNase. Boranophosphate-modified small RNAs (Borano) were more potent than unmodified siRNAs and showed no toxicity (Fig. 18.3) [226]. But small RNA containing 2′-*O*-Me modification was inactive on RNAi [227], suggesting that 2′-*O*-Me modification could be applied for silencers of miRNAs (Fig. 18.3). Further, the modification of 2′-*O*-Me substitution at position 2 in the seed of the guide strand reduced off-target effects [228]. Antisense oligonucleotides to the miRNA could impair the interaction between an overexpressed miRNA and the target mRNAs [229]. Modification of the RNA backbone into a 2′-*O*-Me group prevents *luc* activity in the reporter assay [178, 230, 231]. Three studies have used 2′-*O*-methoxyethyl (2′-*O*-MOE)-substituted RNA oligonucleotides for validation of predicted miRNA targets [232–234]. Further, the 3′ of miRNA conjugated with cholesterol via a pyrolidine linker was tolerated and increased uptake into the liver by binding with serum albumin [235]. The combination of 2′-*O*-methy ribose and cholesterollinked miRNA was named antagomir [236]. Antagomir can efficiently decrease miRNA levels [237]. LNA has also been reported to be an efficient specific silencer of miRNA expression [236].

In in vivo applications, there are three levels of targeting: tissues, cells and intracellular compartments. Although several approaches have been developed including antibody peptides [238], quite recently, highthroughput methods have been employed. At the level of targeting tissue, tetanus toxin 1 mixed with polyethylenimine (PEI) has been used with neuron cells to develop delivery to the central nervous system (CNS) [239]. In line with this, rabies virus glycoprotein (RVG) mixed with siRNA has been reported to deliver to neuron cells [240]. RVG may provide an approach for delivery of miRNA agents from blood to brain through the blood–brain barrier (BBB). Lactose-conjugated cationic liposome 5 (CL-LA5) was effectively delivered to the liver in vivo [241]. Further, oral administration with siRNA in a thioketal-nanoparticle could target the proinflammatory cytokine TNFα in the mouse intestinal cells and the polymer poly-(1,4)-phenyleneacetone dimethylene thioketal with TNF would selectively degrade in response to reactive oxygen species in the intestinal cells, showing that nanoparticles orally delivered protected mice from experimental colitis induced by dextran sulfate [242]. To further improve target specificity, an effectors–receptors system may be crucial for production of the first preclinical proofs. Further, complications arise due to multitargeting miRNAs and superposing multi-miRNAs against a target; therefore, the above agents should be validated by computer prediction of intracellular miRNA targets identified to date. Moreover, when designing patient-specific therapeutic drugs, the specific development of miRNA itself according to the individual genome can offer valuable information on the presence/absence of infection, other genetic factors and the dietary environment. When local and systemic trafficking of Rig agents to the sites of action is individually established, clinical trials may be scheduled and begun.

18.4.2 *NanoRNA-Based Devices*

The idea of a Rig device in vivo has been shown in the results of repression of viral production by the endogenous miRNAs encoded from proviral HIV-1 *nef*/3′LTR DNA in the human genome infected with HIV-1 [243]. In this experience, a transactivator Tas-dependent FV nanoparticle has been developed for preparation of an in vivo system using miRNA-expressing vector nanoparticles for HIV-1 prevention and therapy [244–246]. That is a spumavirus vector derived from feline origin described previously (see Fig. 18.2). The miR-based therapies have been reviewed as rationales for considering the next generation of pharmaceutical products [21, 23, 247]. Recently, as the seed regions of miRNAs target incomplementary target sites in mRNA by sorting Ago proteins [85], with the LV inserted into the susceptible target site of miRNA, the ability of miRNA to regulate its natural target has been effectively suppressed as a decoy or sponge for the cognate miRNA in the transduced cells [21]. Since chicken anti-influenza miRNAs inhibit influenza virus production, the seed sequences conserved artificial miRNAs in an LV effectively inhibit influenza virus replication in human, dog, mouse, and monkey cells [248]. Together, these results suggest that the seed sequences of RNA are information, and the information could be available for suppression of gene expression, which accidentally means that 8–10 RNA sequences' information would be beyond the origin of species [249].

The Rigs in microvesicles may pass through endothelial cells composing blood vessels. In the brain, cerebral vascular endothelial cells contribute to the BBB in cerebral microvessels (Fig. 18.5). Artificial liposomes and cationic polyethylene glycol (PEG) particles have served as an in vivo delivery system for enhancing the efficacy of various pharmacologically active miRNA genes, as described above. However, liposomes, PEG particles, PEI and viral particles were quickly removed by the reticuloendothelial system (RES) after delivery of them to the blood circulation or the mononuclear macrophage phagocytic system in the liver and spleen. Therefore, viral vectors, such as retrovirus or AAV and PEG liposomes, did not cross the BBB [250, 251]. In experiments with artificial miRNAs (AmiRNAs), hypothalamic arcuate nucleus injection of antineuropeptide Y (NPY) miRNAs in AAV has effectively inhibited severe obesity in adult rats [250]. However, it has recently been reported that recombinant AAV (rAAV) can effectively cross the BBB [252], and the miRNA genes in rAAV9 can intravenously be delivered to the CNS [253]. Further, PEG-conjugated immunoliposomes (PILs) ($\sim$85 nm) bearing the receptor-targeting monoclonal antibodies can cross the BBB [254], and antibodies on the liposomes can attack

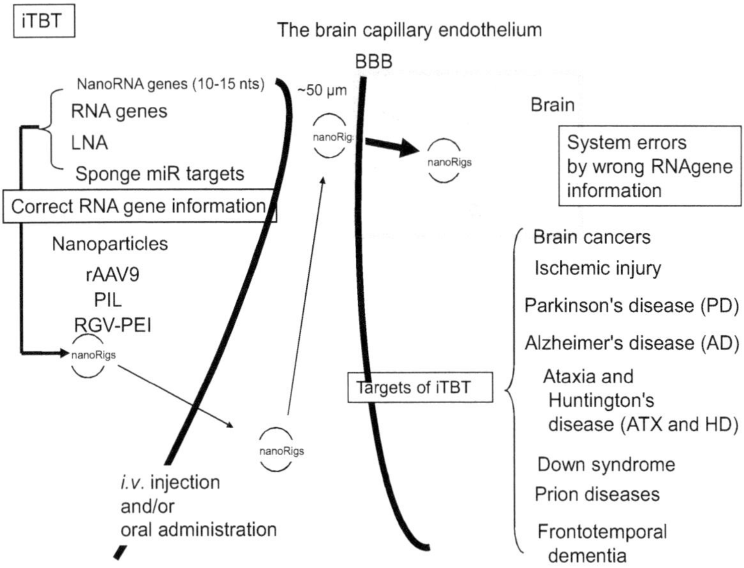

Figure 18.5 The nanodevice with a nanoRig for neurodegeneration. When Rigs' programming languages allow us to express the automated generation of a sequence of elementary operations and to tune the biological phenomenon, correct Rig information could be used for the purpose of ITBT to RigDs, such as brain cancer, ischemic injury, PD, AD, ataxia, HD, Down syndrome, prion diseases, and frontotemporal dementia. Further, viral nanoparticles or exosome-like nanoparticles are possible for delivery of Rigs, RNA gene derivatives, naonRNA genes, LNA, and sponge miR targets, to use through i.v. injection or as oral agents. *Abbreviation*: i.v., intravenous.

the pinpoint of target cells in the brain. Because neurodegeneration appears to be in a brain cell–specific manner, such as entorhinal and neocortical glutamatergic and nucleus basalis cholinergic neurons in AD, nigrostriatal neurons in PD, Purkinje cells in SCA, and spinal motor neurons in ALS, targeting of brain cells is important to deliver the RNA agents. These results showed that modified viral vectors and liposomes can transfer long and double-stranded nts across cerebral vascular endotherial cells. In the case of the carrier of the RNA genes, HDL and exosomes were well known as carriers of apoprotein and phospholipids from the liver cells to the brain cells and vice versa [255–257]. Therefore, the miRNA genes could be transported with HDL and/or exosomes to pass through the

BBB. PEI-based nanoparticles, such as longitudinally unzipping multiwalled carbon nanotubes (MWCNTs) were used for an LNA [258]. To pass through the BBB, the PEI material has been reported to be labeled with RVG, and the RVG-PEI with miR124a has targeted neuronal cells in the mouse brain [259]. This evidence indicates that specific nanoRigs in nanoparticles could exhibit a prolonged circulation time in blood and were incorporated into cells in the brain passing through the BBB. The Rigs in HDLs could pass through endothelial cells composing blood vessels.

Rigs and nanoRigs are similar to the criteria of linguistics. For the first instance, both would be according to double articulation, morphémes and phonémes. Further, Rigs can artificially create, and superposition of a Rig itself would have the generative capacity to operate tuning. Finally, Rig agents in the viral vector could be transmittable from one to another [260]; therefore Rigs may be a language based on quantum theory [7]. What is the meaning of this programming language? Its algorithm would allow expressing the automated generation of a sequence of elementary operations to tune the biological phenomenon [1].

18.5 Programmed Evolution and RigDs

18.5.1 *101 Years of Peyton Rous: Genetic Diseases to Information Ones*

It is well known that the never-ending story about Peyton Rous by Weiss and Vogt contains his discovery of Rous sarcoma virus (RSV) [261]. We can deeply understand how he is a remarkable scientist in the Rockefeller Institute and how important his finding is. He obtained his bachelor of arts degree in 1900 and his doctor of medicine degree in 1905 and only six years later, he published his paper describing RSV, in 1911 [262]. The importance of his finding is twofold. One is, "A malignant tumor could be induced by infection," and the other is "RSV represented a pioneering oncogenic retrovirus for investigating the molecular mechanisms of cancer development." We propose here a third point with the above two, that is RNA information transmission with a viral nanoparticle or by RNA itself because miRNA might have been included in his separable fractions from the infected cells [246, 263, 264]. Although studies in the late 1950s about RNA from tumor cells by Vogt et al. [265, 266] are involved in the logic of the system, such as oncoproteins Src and Myc as parts of cell signaling, we have already introduced that the microRNA gene could tune *src* and *myc* oncogenes [13]. For the cell cycle, cyclin kinase also is under controlling of the RNA genes. Further, integration of oncoretrovirus without oncogenes can induce aberrant expression of the microRNA genes and the overexpression of RNA genes showed the oncogenic potential [17]. Thus, it is suggested that transmittable Rigs would order the protein system encoded by protein genes to transform normal cells by changing the cell cycle regulation, that is one of the programmed evolutions; in turn, RNA language made possible the transmission of the operation into the cells. Therefore, as different species have different RNA languages, "Attempts to so transmit rat, mouse, and dog tumors have never succeeded," Rous described [262]. In his Nobel lecture he said that "cancer is not a genetic disease," [267]; so cancer is a RigD, he would be told if he were here now.

18.6 Conclusions

Given the language of RNA, "evolution" with mutations and recombinations may be at least, in part, operated in its programmed orders by Rigs (see Figs. 18.1 and 18.6). Therefore, diseases are deeply associated with disorder of Rig expression in cells. In turn, abnormal Rigs induce system errors. Since DNA genes in retrovirus vectors could have been applicable for disease therapy as gene therapy, which is still the system to block protein functions or to force-express protein genes. Chemotherapy would also be based on the system to inhibit protein enzyme activity or the metabolic pathway. On the other hand, while the miRNA gene is originally derived from a retrotransposon as a Rig, a Rig can be mobile in every space with nanoparticles according to the RNA wave model. Therefore, spontaneous mutation and recombination with mobile Rigs would induce information errors, and that can cause human diseases. Viral nanoparticles or exosome-like nanoparticles

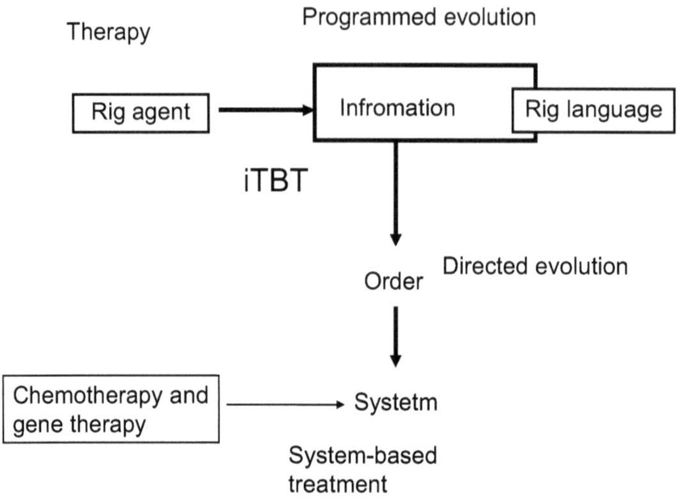

Figure 18.6 Programmed evolution. Diseases are deeply associated with disorders of Rig expression. Since spontaneous mutation and recombination with mobile Rigs would induce information errors and these can cause diseases, Rigs may be applicable for disease therapy. Chemotherapy and DNA gene therapy would be based on the system to inhibit protein enzyme activity or the metabolic pathway, as shown in Fig. 18.2. ITBT with the Rig language might be involved in future progress of the delivery on Rigs with nanotechnology. Its therapeutic application could be contained into the old system-based therapy. This is the new era that Rigs may change all readymade theories of biology and evolution.

are possible for delivery of the Rigs to use through i.v. injection or as oral agents. Taken altogether, inheritable phenotypes may be readily programmed in Rigs written with the Rig algorithm, and the acquired characters in Darwinism could transmit from one to another via Rigs. Therefore, genetic disorders were inherited. Further, the evolution has been programmed by Rigs. It is named programmed evolution. Thus, before Darwinism, they hypothesis was that programmed evolution was directed in its operated orders by the Rig. ITBT might be involved in the future progress of the delivery upon Rigs with nanotechnology for therapeutic application to be contained in the old system-based therapy.

References

1. Fujii, Y.R. (2008). Formulation of new algorithmics for miRNAs. *Open Virol. J.*, **2**, 37–43.
2. Czech, B., and Hannon, G.J. (2010). Small RNA sorting: matchmaking for Argonautes. *Nat. Rev. Genet.*, **12**, 19–31.
3. Astolfi, P.A., Salamini, F., and Sgaramella, V. (2010). Are we genomic mosaics? Variations of the genome of somatic cells can contribute to diversity our phenotypes. *Curr. Genom.*, **11**, 379–386.
4. Jheon, A.H., et al. (2011). Expression of microRNAs in the stem cell niche of the adult mouse incisor. *PLoS ONE*, **6**, e24536.
5. Shende, V.R., Goldrick, M.M., Ramani, S., and Earnest, D. (2011). Expression and rhythmic modulation of circulating microRNAs targeting the clock gene *Bmal1* in mice. *PLoS ONE*, **6**, e22586.
6. Khraiwesh, B., et al. (2009). Transcriptional control of gene expression by microRNAs. *Cell*, **140**, 111–122.
7. Fujii, Y.R. (2012). The RNA gene information: retroelement-microRNA entangling as the RNA quantum code, in *microRNA Protocol*, Ying, S.-Y. (ed.), Humana Press, CA, 47–67.
8. Zhou, X., Duan, X., Qian, J., and Li, F. (2009). Abundant conserved microRNA target sites in the 5′-untranslated region and coding sequence. *Genetica* **137**, 159–164.
9. Schnall-Levin, M., Zhao, Y., Perrimon, N., and Berger, B. (2010). Conserved microRNA targeting in *Drosophila* is as widespread in coding regions as 3′UTRs. *Proc. Natl. Acad. Sci. U S A*, **107**, 15751–15756.
10. Takane, K., and Kanai, A. (2011). Vertibrate virus-encoded microRNAs and their sequence conservation. *Jpn. J. Infect. Dis.*, **64**, 357–366.
11. Farazi, T.A., Spitzer, J.I., Morozov, P., and Tuschl, T. (2010). miRNAs in human cancer. *J. Pathol.*, **223**, 102–115.
12. Navarro, B., et al. (2009). Deep sequencing of viroid-derived small RNAs from grapevine provides new insights on the role of RNA silencing in plant-viroid interaction. *PLoS ONE* **4**, e7686.
13. Fujii, Y.R. (2009). Oncoviruses and pathogenic microRNAs in humans. *Open Virol. J.*, **3**, 37–51.
14. Carpio, L., et al. (2010). microRNA machinary is a integral component of drug-induced transcription inhibition in HIV-1 infection. *J. RNAi Gene Silen.*, **6**, 386–400.

15. Rouha, H., Thuener, C., and Mandl, C.W. (2010). Functional microRNA generated from a cytoplasmic RNA virus. *Nucleic Acids Res.*, **38**, 8328–8337.
16. Varble, A., et al. (2010). Engineered RNA viral synthesis of microRNA. *Proc. Natl. Acad. Sci. U S A*, **107**, 11519–11524.
17. Fujii, Y.R. (2010). RNA genes: retroelements and virally retroposable microRNAs in human embryonic stem cells. *Open Virol. J.*, **4**, 63–75.
18. Li, Y., Li, C., Xia, J., and Jin, Y. (2011). Domestication of transposable elements into microRNA genes in plants. *PLoS ONE*, **6**, e19212.
19. Smalheiser, N.R., and Torvik, V.I. (2005). Mammalian microRNAs derived from genomic repeats. *Trends Genet.*, **21**, 322–326.
20. Fujii, Y.R. (2008). Lost in translation: regulation of HIV-1 by microRNAs and a key enzyme of RNA-directed RNA polymerase, in *MicroRNAs*, Appasani, K. (ed.), Cambridge University Press, New York, 427–442.
21. Brown, B.D., and Naldini, L. (2009). Exploiting and antagonizing microRNA regulation for therapeutic and experimental applications. *Nat. Rev. Genet.*, **10**, 578–585.
22. McDemott, A.M., Heneghan, H.M., Miller, N., and Kerin, M.J. (2011). The therapeutic potential of microRNAs: disease modulators and drug targets. *Pharm. Res.*, **28**, 3016–3029.
23. Liu, Y.P., and Berkhout, B. (2011). miRNA cassetts in viral vectors: problems and solutions. *Biochem. Biophys. Acta*, **1809**, 732–745.
24. Lanford, R.E., et al. (2010). Therapeutic silencing of microRNA-122 in primates with chronic hepatitis C virus infection. *Science*, **327**, 198–201.
25. Kefas, B., et al. (2009). The neuronal microRNA miR-326 acts in a feedback loop with Notch and Has therapeutic potential against brain tumors. *J. Neurosci.*, **29**, 15161–15168.
26. Kumar, M.S., et al. (2008). Suppression of non-small cell lung tumor development by the *let-7* microRNA family. *Proc. Natl. Acad. Sci. U S A*, **105**, 3903–3908.
27. Esquela-Kerscher, A., et al. (2008). The let-7 microRNA reduces tumor growth in mouse models of lung cancer. *Cell Cycle*, **7**, 759–764.
28. Kota, J., et al. (2009). Therapeutic microRNA delivery suppresses tumorigenesis in a murine liver cancer model. *Cell*, **137**, 1005–1017.
29. Henry, J.C., Azevedo-Pouly, A.C.P., and Schmittgen, T.D. (2011). microRNA replacement therapy for cancer. *Pharm. Res.*, **28**, 3030–3042.

30. Fujii, Y.R. (2009). Symphony of AIDS: a microRNA-based therapy, in *Regulation of Gene Expression by Small RNAs*, Gaur, R.K., and Rossi, J.J. (eds.), CRC Press, New York, 333–349.
31. Melering, C.D., and Linial, M.L. (2001). Histrical perspective of foamy virus epidemiology and infection. *Clin. Microbiol. Rev.*, **14**, 165–176.
32. Switzer, W.M., et al. (2004). Frequent simian foamy virus infection in persons occupationally exposed to nonhuman primates. *J. Virol.*, **78**, 2780–2789.
33. Williams, D.A. (2008). Foamy virus vectors come of age. *Mol. Ther.*, **16**, 635.
34. Nowrouzi, A., et al. (2006). Genome-wide mapping of foamy virus vector integrations in a human cell line. *J. Gen. Virol.*, **87**, 1339–1347.
35. Gharwan, H., et al. (2007). Transduction of human embryonic stem cells by foamy virus vectors. *Mol. Ther.*, **15**, 1827–1833.
36. Lo, Y.-T., et al. (2009). The foamy virus genome remains unintegrated in the nucleus of G1/S phase arrested cells and integrase is critical for pre-integration complex transport into the nucleus. *J. Virol.*, **84**, 2832–2842.
37. Bauer, T.R., Jr., et al. (2008). Successful treatment of canine leukocyte adhesion deficiency by foamy virus vectors. *Nat. Med.*, **14**, 93–97.
38. Brown, B.D., et al. (2007). A microRNA-regulated lentiviral vector mediates stable correction of hemophilia B mice. *Blood*, **110**, 4144–4152.
39. Brown, B.D., et al. (2006). Endogenous microRNA regulation suppresses transgene expression in hematopoietic lineages and enables stable gene transfer. *Nat. Med.*, **12**, 585–591.
40. Annoni, A., et al. (2009). In vivo delivery of a microRNA-related transgene induces antigen-specific regulatory T cells and promotes immunologic tolerance. *Gene Ther.*, **114**, 5152–5161.
41. Valkov, E., et al. (2009). Functional and structural characterization of the integrase from the prototype foamy virus. *Nucleic Acids Res.*, **37**, 243–255.
42. Kiem, H.-P., et al. (2010). Foamy combinational anti-HIV vectors with MGMTP140K potently inhibit HIV-1 and SHIV replication and mediate selection in vivo. *Gene Ther.*, **17**, 37–49.
43. Sonkoly, E., and Pivarcsi, A. (2011). MicroRNAs in inflammation and response to injuries induced by environmental pollution. *Mutat. Res.*, **717**, 46–53.

44. Hou, L., Wang, D., and Baccarelli, A. (2011). Environmental chemicals and microRNAs. *Mutat. Res.*, **714**, 105–112.
45. Garinot, M., Fievenz, V., Pourcelle, V., Stoffelbach, F., des Rieux, A., Plapied, L., Theate, I., Freichels, H., Jerome, C., Marchand-Brynaert, J., Schneider, Y.J., and Préat, V. (2007). PEGylated PLGA-based nanoparticles targeting M cells for oral vaccination. *J. Control. Release*, **120**, 195–204.
46. Sureban, S.M., May, R., Mondalek, F.G., Qu, D., Ponnurangam, S., Pantazis, P., Anant, S., Ramanujam, R.P., and Houchen, C.W. (2011). Nanoparticle-based delivery of siDCAMKL-1 increases *microRNA-144* and inhibits colorectal cancer tumor growth via a Notch-1 dependent mechanism. *J. Nanobiotech.*, **9**, 40.
47. Lagos-Quintana, M., Rauhut, R., Lendeckel, W., and Tuschl, T. (2001). Identification of novel genes coding for small expressed RNAs. *Science*, **294**, 853–858.
48. Lim, L.P., Glasner, M.E., Yekta, S., Burge, C.B., and Bartel, D.P. (2003) Vertibrate microRNA genes. *Science*, **299**, 1540.
49. Lim, L.P., Lau, N.C., Weinstein, E.G., Abdelhakim, A., Yekta, S., Rhoades, M.W., Burge, C.B., and Bartel, D.P. (2003). The microRNAs of *Caenorhabditis elegans. gene Dev.*, **17**, 991–1008.
50. Hinas, A., Reimegard, J., Wagner, E.G.H., Nellen, W., Ambros, V.R., and Soderbom, F. (2007). The small RNA repertoire of dictyostelium discoideum and its regulation by components of the RNAi pathway. *Nucleic Acids Res.*, **35**, 6714–6726.
51. Cummins, J.M., He, Y., Leary, R.J., Pagliarini, R., Diaz, L.A., Sjoblom Jr., T., Barad, O., Bentwich, Z., Szafranska, A.E., Labourier, E., Raymond, C.K., Roberts, B.S., Juhl, H., Kinzler, K.W., Vogelstein, B., and Velculescu, V.E. (2006). The colorectal microRNAome. *Proc. Natl. Acad. Sci. U S A*, **103**, 3687–3692.
52. Lash, A.E., Tolstoshev, C.M., Wagner, L., Schuler, G.D., Strausberg, R.L., Riggins, G.J., and Altschul, S.F. (2000). SAGEmap: a public gene expression resource. *Genome Res.*, **10** 1051–1060.
53. Boon, K., Osorio, E.C., Greenhut, S.F., Schaefer, C.F., Shoemaker, J., Polyak, K., Morin, P.J., Buetow, K.H., Strausberg, R.L., De Souza, S.J., and Riggins, G.J. (2002). An anatomy of normal and malignant gene expression. *Proc. Natl. Acad. Sci. U S A*, **99**, 11287–11292.
54. Mrázek, J., Kreutmayer, S.B., Grässer, F.A., Polacek, N., and Hüttenhofer, A. (2007). Substractive hybridization identifies novel differentially expressed ncRNA species in EBV-infected human B cells. *Nucleic Acids Res.*, **35**, e73.

55. Miska, E.A., Alvarez-Saaverdra, E., Townsend, M., Yoshii, A., Sestan, N., Pakic, P., Constantine-Paton, M., and Horvitz, H.R. (2004). Microarray analysis of microRNA expression in the developing mammalian brain. *Genome Biol.*, **5**, R68.

56. Barad, O., Meiri, E., Avniel, A., Aharonov, R., Barzilai, A., Bentwich, I., Einav, U., Gilad, S., Hurban, P., Karov, Y., Lobenhofer, E.K., Sharon, E., Shiboleth, Y.M., Shtutman, M., Bentwich, Z., and Einat, P. (2004). MicroRNA expression detected by oligonucleotide microarrays: system establishment and expression profiling in human tissues. *Genome Res.*, **14**, 2486–2494.

57. Liu, C.G., Calin, G.A., Meloon, B., Gamliel, N., Sevignani, C., Ferracin, M., Dumitru, C.D., Shimizu, M., Zupo, S., Dono, M., Alder, H., Bullrich, F., Negrini, M., and Croce, C.M. (2004). An oligonucleotide microchip for genome-wide microRNA profiling in human and mouse tissues. *Proc. Natl. Acad. Sci. U S A*, **101**, 9740–9744.

58. Castoldi, M., Schmidt, S., Benes, V., Noerholm, M., Kulozik, A.E., Hentze, M.W., and Muckenthaler, M.U. (2006). A sensitive array for microRNA expression profiling (miChip) based on locked nucleic acids (LNA). *RNA*, **12**, 913–920.

59. Fang, S., Lee, H.J., Wark, A.W., and Corn, R.M. (2006) Attomole microarray detection of microRNAs by nanoparticle-amplified SPR imaging measurements of surface polyadenilation reactions. *J. Am. Chem. Soc.*, **128**, 14044–14046.

60. Schmittgen, T.D., Jiang, J., Liu, Q., and Yang, L. (2004). A high-throughput method to monitor the expression of microRNA precursors. *Nucleic Acids Res.*, **32**, e43.

61. Fu, H.J., et al. (2006). A novel method to monitor the expression of microRNAs. *Mol. Biotechnol.*, **32**, 197–204.

62. Gasic, E.V., et al. (2007). Protocol: a highly sensitive RT-PCR method for detection and quantification of microRNAs. *Plant Meth.*, **3**, 12.

63. Liu, T., et al. (2007). Detection of a microRNA signal in an in vivo expression set of mRNAs. *PLoS ONE*, **8**, e804.

64. Valoczi, A., et al. (2004). Sensitive and specific detection of microRNAs by northern blot analysis using LNA-modified oligonucleotide probes. *Nucleic Acids Res.*, **32**, e175.

65. Pall, G.S., et al. (2007). Carbodiimide-mediated cross-linking of RNA to nylon membranes improves the detection of siRNA, miRNA and piRNA by northern blot. *Nucleic Acids Res.*, **35**, e60.

66. Grivna, S.T., Beyret, E., Wang, Z., and Lin, H. (2006). A novel class of small RNAs in mouse spermatogenic cells. *Genes Dev.*, **20**, 1709–1714.
67. Grivna, S.T., Pyhtila, B., and Lin, H. (2006). MIWI associates with translational machinery and PIWI-interacting RNAs (piRNAs) in regulating spermatogenesis. *Proc. Natl. Acad. Sci. U S A*, **103**, 13415–13420.
68. Aravin, A.A., et al. (2003). The small RNA profile during *Drosophila melanogaster* development. *Dev. Cell*, **5**, 337–350.
69. Zhang, Y., and Zhang, C.-Y. (2011). Sensitive detection of microRNA with isothermal amplification and single quantum dot-based nanosensor. *Anal. Chem.*, **84**, 224–231.
70. Rivas, E., and Eddy, S.R. (2001). Noncoding RNA gene detection using comparative sequence analysis. *BMC Bioinf.*, **2**, 8.
71. di Bernardo, D., Down, T., and Hubbard, T. (2003). ddbRNA: detection of conserved secondary structures in multiple alignments. *Bioinformatics*, **19**, 1606–1611.
72. Coventry, A., Kleitman, D.J., and Berger, B. (2004). MSARI: multiple sequence alignments for statistical detection of RNA secondary structure. *Proc. Natl. Acad. Sci. U S A*, **101**, 12102–12107.
73. Washietl, S., Hofacker, I.L., and Stadler, P.F. (2005). Fast and reliable prediction of noncoding RNAs. *Proc. Natl. Acad. Sci. U S A*, **102**, 2454–2459.
74. Xue, C., and Liu, G-.P. (2007). RScan: fast searching structural similarities for structured RNAs in large databases. *BMC Genomics*, **8**, 257.
75. Zuker, M. (2003). Mfold web server for nucleic acid folding and hybridization prediction. *Nucleic Acids Res.*, **31**, 3406–3415.
76. Hofacker, I.L. (2003). Vienna RNA secondary structure server. *Nucleic Acids Res.*, **31**, 3429–3431.
77. Adai, A., et al. (2005). Computational prediction of miRNAs in *Arabidopsis thaliana*. *Genome Res.*, **15**, 78–91.
78. Grey, F., et al. (2005). Identification and characterization of human cytomegalovirus-encoded microRNAs. *J. Virol.*, **79**, 12095–12099.
79. Lai, E.C., Tomancak, P., Williams, R.W., and Rubin, G.M. (2003). Computational identification of Drosophila microRNA genes. *Genome Biol.*, **4**, R42.
80. Berezikov, E., et al. (2005). Phylogenetic shadowing and computational identification of human microRNA genes. *Cell*, **120**, 21–24.

81. Bentwich, I., et al. (2005). Identification of hundreds of conserved and nonconserved human microRNAs. *Nat. Genet.*, **37**, 766–770.

82. Bennasser, Y., Le, S. -Y., Yeung, M.L., and Jeang, K.-T. (2006). MicroRNAs in human immunodeficiency virus-1 infection. *Methods Mol. Biol.*, **342**, 241–253.

83. Meiri, E., et al. (2010). Discovery of microRNAs and other small RNAs in solid tumors. *Nucleic Acids Res.*, **38**, 6234–6246.

84. Kurth, H.M., and Mochizuki, K. (2009). 2′-*O*-methylation stabilizes Piwi-associated small RNAs and ensures DNA elimination in *Tetrahymena*. *RNA*, **15**, 675–685.

85. Czech, B., and Hannon, G.J. (2010). Small RNA sorting: matchmaking for argonates. *Nat. Rev. Genet.*, **12**, 19–31.

86. Esposito, T., Magliocca, S., Formicola, D., and Gianfrancesco, F. (2011). piR_015520 belongs to piwi-associated RNAs regulates expression of the human melatonin receptor 1A gene. *PLoS ONE* 6, e22727.

87. Henderson, I.R., and Jacobsen, S.E. (2008). Tandem repeats upstream of the *Arabidopsis* endogene SDC recruit non-CG DNA methylation and initiate siRNA spreading. *Genes Dev.*, **22**, 1597–1606.

88. Khraiwesh, B., et al. (2010). Transcriptional control of gene expression by microRNAs. *Cell*, **140**, 111–122.

89. Kurth, H.M., and Mochizuki, K. (2009). Non-coding RNA: a bridge between small RNA and DNA. *RNA Biol.*, **6**, 1–3.

90. Ameres, S.L., et al. (2010). Target RNA-directed triming and tailing of small silencing RNAs. *Science*, **328**, 1534–1539.

91. Hildebrandt, M., et al. (2010). Hsa-miR-9 methylation status is associated with cancer development and metastatic in patients with clear cell renal cell carcinoma. *Oncogene*, **29**, 5724–5728.

92. Kirino, Y., and Mourelatos, Z. (2007). Mouse Piwi-interacting RNAs are 2′-O-methylated at their 3′ termini. *Nat. Struct. Mol. Biol.*, **14**, 347–348.

93. Ohara, T., et al. (2007). The 3′ termini of mouse Piwi-interacting RNAs are 2′-O-methylated. *Nat. Struct. Mol. Biol.*, **14**, 349–350.

94. Li, J., et al. (2005). Methylation protects miRNAs and siRNAs from a 3′-end uridylation activity in *Arabidopsis*. *Cur. Biol.*, **15**, 1501–1507.

95. Ebhardt, H.A., Fedynak, A., and Fahlman, R.P. (2010). Naturally occurring variations in sequence length creates microRNA isoforms that differ in argonaute effector complex specificity. *Silence*, **1**, 12.

96. Ohnishi, Y., et al. (2010). Small RNA class transition from siRNA/piRNA to miRNA during pre-implantation mouse development. *Nucleic Acids Res.*, **38**, 5141–5151.

97. Sun, G., et al. (2009). SNPs in human miRNA genes affect biogenesis and function. *RNA*, **15**, 1640–1651.
98. Athanasiadis, A., Rich, A., and Maas, S. (2004). Widespread A-to-I RNA editing of Alu-containing mRNAs in the human transcriptome. *PLoS Biol.*, **2**, e391.
99. Chen, L.L., and Carmichael, G.G. (2008). Gene regulation by SINES and inosines: biological consequences of A-to-I editing of Alu element inverted repeats. *Cell Cycle*, **7**, 3294–3301.
100. Wulff, B.-E., and Nishikura, K. (2010). Substitutional A-to-I RNA editing. WIREs *RNA*, **1**, 90–101.
101. Choe, J., Cho, H., Lee, H.C., and Kim, Y.K. (2010). microRNA/Argonaute 2 regulates nonsense-mediated messenger RNA decay. *EMBO Rep.*, **11**, 380–386.
102. Chuzhanova, N., Cooper, D.N., Férec, C., and Chen, J.-M. (2007). Searching for potential microRNA-binding site mutations amongst known disease-associated 3′ UTR variants. *Genomic Med.*, **1**, 29–33.
103. Kawahara, Y., et al. (2007). Redirection of silencing targets by adenosine-to-inosine editing of miRNAs. *Science*, **315**, 1137–1140.
104. Tang, T., et al. (2010). Adverse interactions between micro-RNAs and target genes from different species. *Proc. Natl. Acad. Sci. U S A*, 107, 12935–12940.
105. Pang, P.S., et al. (2011). Structural map of a microRNA-122: HCV complex. *J. Virol.*, **86**, 1250–1254.
106. Enright, A.J., et al. (2003). MicroRNA targets in *Drosophila*. *Genome Biol.*, **5**, R1.
107. Lewis, B.P., et al. (2003). Prediction of mammalian microRNA targets. *Cell*, **115**, 787–798.
108. Rehmsmeier, M., Steffen, P., Hochsmann, M., and Giegerich, R. (2004). Fast and effective prediction of microRNA/target duplexes. *RNA*, **10**, 1507–1517.
109. Kiriakidou, M., et al. (2004). A combined pomputational-experimental approach predicts human microRNA targets. *Genes Dev.*, **18**, 1165–1178.
110. Lewis, B.P., Burge, C.B., and Bartel, D.P. (2005). Conserved seed pairing, often flanked by adenosines, indicates that thousands of human genes are microRNA targets. *Cell*, **120**, 15–20.
111. Grun, D., et al. (2005). MicroRNA target predictions across seven *Drosophila* species and comparison to mammalian targets. *PLoS Comput. Biol.*, **1**, e13.

112. Zhang, Y. (2005) miRU: automated plant miRNA target prediction server. *Nucleic Acids Res.*, **33**, W701–W704.

113. Saetrom, O., Snove, O., Jr., and Saetrom, P. (2005). Weighted sequence motifs as an improved seeding step in microRNA target prediction algorithms. *RNA*, **11**, 995–1003.

114. Miranda, K.C., et al. (2006). A pattern-based Method for the identification of microRNA binding sites and their corresponding heteroduplexes. *Cell*, **126**, 1203–1217.

115. Salmena, L., et al. (2011). A *ceRNA* hypothesis: the rosetta stone of a hedden RNA language. *Cell*, **146**, 353–358.

116. Lin, S.-L., Chang, D., Wu, D-.Y., and Ying, S.-Y. (2003). A novel RNA splicing-mediated gene silencing mechanism potential for genome evolution. *Biochem. Biophys. Res. Commun.*, **310**, 754–760.

117. Chang, S., et al. (2004). MicroRNAs act sequentially and asymmetrically to control chemosensory laterality in the nematode. *Nature*, **430**, 785–789.

118. Esau, C., et al. (2004). MicroRNA-143 regulates adipocyte differentiation. *J. Biol. Chem.*, **279**, 52361–52365.

119. Mansfield, J.H., et al. (2004). MicroRNA-responsive "sensor" transgenes uncover Hox-like and other developmentally regulated patterns of vertebrate microRNA expression. *Nat. Genet.*, **36**, 1079–1083.

120. Wightman, B., Ha, I., and Ruvkun, G. (1993). Posttranscriptional regulation of the heterochronic gene *lin-14* by *lin-4* mediates temporal pattern formation in *C. elegans*. *Cell*, **75**, 855–862.

121. Klase, Z., et al. (2007). HIV-1 TAR element Is processed by Dicer to yield a viral micro-RNA involoved in chromatin remodeling of the viral LTR. *BMC Mol. Biol.*, **8**, 63.

122. Lombardi, R., et al. (2009). High-level HIV-1 Nef transient expression in *Nicotiana benthamiana* using the P19 gene silencing suppressor protein of artichoke mottled crinckle virus. *BMC Biotech.*, **9**, 96.

123. Wang, K., et al. (2010). Export of microRNAs and microRNA-protective protein by mammalian cells. *Nucleic Acids Res.*, **38**, 7248–7259.

124. Hanson, E.K., Lubenow, H., and Ballantyne, J. (2009). Identification of forensically relevant body fluids using a panel of differentially expressed microRNAs. *Anal. Biochem.*, **387**, 303–314.

125. Hanke, M., et al. (2009). A robust methodology to study urine microRNA as tumor marker: microRNA-126 and microRNA-182 are related to urinary bladder cancer. *Urol. Oncol.*, **28**, 655–661.

126. Fichtlscherer, S., et al. (2010). Circulating microRNAs in patients with coronary artery disease. *Circ. Res.*, **107**, 677–684.
127. Ai, J., et al. (2009). Circulating microRNA-1 as a potential novel biomarker for acute myocardial infraction. *Biochem. Biophys. Res. Commun.*, **391**, 73–77.
128. Zampetaki, A., et al. (2010). Plasma microRNA profiling reveals loss of endothelial miR-126 and other microRNAs in type 2 diabetes. *Circ. Res.*, **107**, 810–817.
129. Wang, K., et al. (2009). Circulating microRNAs, potential biomarkers for drug-induced liver injury. *Proc. Natl. Acad. Sci. U S A*, **106**, 4402–4407.
130. Schetter, A.J., Heegaard, N.H.H., and Harris, C.C. (2009). Inflammation and cancer: interweaving microRNA, free radical, cytokine and p53 pathways. *Carcinogenesis*, **31**, 37–49.
131. Jeker, L.T., and Bluestone, J.A. (2010). Small RNA regulators of T cell-mediated autoimmunity. *J. Clin. Immunol.*, **30**, 347–357.
132. Inui, M., Martello, G., and Piccolo, S. (2010). MicroRNA control of signal transduction. *Nat. Rev. Mol. Cell Biol.*, **11**, 252–263.
133. O'Connell, R.M., Rao, D.S., Chaudhuri, A.A., and Baltimore, D. (2010). Physiological and pathological roles for microRNAs in the immune system. *Nat. Rev. Immunol.*, **10**, 111–122.
134. Blaser, M.J., and Berg, D.E. (2001). *Helicobacter pylori* genetic discovery and risk of human disease. *J. Clin. Invest.*, **107**, 767–773.
135. Xiao, B., et al. (2009). Induction of microRNA-155 during *Helicobacter pulori* infection and its negative regulatory role in the inflammatory response. *J. Infect. Dis.*, **200**, 916–925.
136. Tang, B., et al. (2010). Identification of MyD88 as a novel target of miR-155, involved in negative regulation of Helicobacter pylori-induced inflammation. *FEBS Lett.*, **584**, 1481–1486.
137. Tam, W., Hughes, S.H., Hayward, W.S., and Besmer, P. (2002). Avian bic, a isolated from common retroviral site in avian leukosis virus-induced lymphomas that encoding RNA, cooperates with c-myc in lymphomagenesis. *J. Virol.*, **76**, 4275–4286.
138. Gironella, M., et al. (2007). Tumor protein 53-i9nduced nuclear protein 1 expression is repressed by miR-155, and its restoration inhibits pancreatic tumor development. *Proc. Natl. Acad. Sci. U S A*, **104**, 16170–16175.

139. Yin, Q., et al. (2008). B-cell receptor activation induces BIC/miR-155 expression through a conserved AP-1 element. *J. Biol. Chem.*, **283**, 2654–2662.

140. O'Connell, R.M., et al. (2007). MicroRNA-155 is induced during the macrophage inflammatory response. *Proc. Natl. Acad. Sci. U S A*, **104**, 1604–1609.

141. Lu, F., et al. (2008). Epstein-Barr virus-induced miR-155 attenuates NK-κB signaling and stabilizes latent virus persistence. *J. Virol.*, **82**, 10436–10443.

142. Curtale, G., et al. (2010). An emreging player in the adaptive immune response: microRNA-146a is a modulator of IL-2 expression and activation-induced cell death in T lymphocytes. *Blood*, **115**, 265–273.

143. Wan, H.-Y., et al. (2010). Regulation of the transcription factor NF-B1 by microRNA-9 in human gastric adenocarcinoma. *Mol. Cancer*, **9**, 16.

144. Punj, V., et al. (2010). Kaposi's sarcoma-associated herpesvirus-encoded viral FLICE inhibitory protein (vFLIP) K13 suppresses CXCR4 expression by upregulating miR-146a. *Oncogene*, **29**, 1835–1844.

145. Suzuki, Y., Kim, H.W., Ashraf, M., and Haider, H.K. (2010). Diazoxide potentiates mesenchymal stem cell survival via NF-κB-dependent miR-146a expression by targeting Fas. *Am. J. Physiol. Heart Circ. Physiol.*, **299**, H1077–H1082.

146. Lei, X., et al. (2010). Regulation of NF-κB inhibitor Iκ-Bα and viral replication by a KSHV microRNA. *Nat. Cell Biol.*, **12**, 193–199.

147. Lei, X., et al. (2010). MicroRNAs control herpesviral domancy. *Cell Cycle*, **9**, 1225–1226.

148. Wen, K.W., and Damania, B. (2010). Kaposi sarcoma-associated herpesvirus (KSHV): molecular biology and oncogenesis. *Cancer Lett.*, **289**, 140–150.

149. Mesri, E.A., Cesarman, E., and Boshoff, C. (2010) Kaposi's sarcoma and its associated herpesvirus. *Nat. Rev. Cancer*, **10**, 707–719.

150. Martin, D., and Gutkind, J.S. (2010). Human tumor-associated viruses and new insights into the molecular mechnisms of cancer. *Oncogene*, **27**, S31–S42.

151. Dreher, A., et al. (2010). Differential expression of cellular microRNAs in HPV-11 transfected cells. An analysis by three different array platforms and qRT-PCR. *Biochem. Biophys. Res. Commun.*, **412**, 20–25.

152. Bai, Z.-Q., Lei, X.-F., Wang, L.-D., and Gao, S.-J. (2008). Identification and function of microRNAs encoded by herpesviruses. *Virol. Sin.* **23**, 459–472.

153. Lu, C.-C., et al. (2010). MicroRNAs encoded by Kaposi's sarcoma-associated herpesvirus regulate viral life cycle. *EMBO Rep.*, **11**, 784–790.

154. Abend, J.R., Uldrick, T., and Ziegelbauer, J.M. (2010). Regulation of TWEAKR expression by KSHV microRNA prevents TWEAK-induced apoptosis and inflammatory cytokine expression. *J. Virol.*, **84**, 12139–12151.

155. Lu, F., et al. (2010). Epigenetic regulation of Kaposi's sarcoma-associated herpesvirus latency by virus-encoded microRNAs that target Rta and the cellular Rbl2-DNMT pathway. *J. Virol.*, **84**, 2697–2706.

156. Zur Hausen, H. (2002). Papillomaviruses and cancer: from basic studies to clinical application. *Nat. Rev. Cancer*, **2**, 342–350.

157. Melsheimer, P., et al. (2004). DNA aneuploidy and integration of human papillomavirus type 16 e6/e7 oncogenes in intraepithelial neoplasia and invasive squamous cell carcinoma of the cervix uteri. *Clin. Cancer Res.*, **10**, 3059–3063.

158. Scheffner, M., Werness, B.A., Huibregtse, K., and Harlow, E. (1990). The E6 oncoprotein encoded by human papillomavirus types 16 and 18 promotes the degradation of p53. *Cell*, **63**, 1129–1136.

159. Dyson, N., Howley, P.M., Munger, K., and Harlow, E. (1989). The human papillomavirus-16 E7 oncoprotein is able to bind to the retinoblastoma gene product. *Science*, **243**, 934–937.

160. Chang, T.C., et al. (2007). Transactivation of miR-34a by p53 broadly influences gene expression and promotes apoptosis. *Mol. Cell*, **26**, 745–752.

161. He, L., He, X., Lowe, S.W., and Hannon, G.J. (2007). microRNAs join the p53 network-another piece in the tumour-suppression puzzle. *Nat. Rev. Cancer*, **7**, 819–822.

162. Raver-Shapira, N., et al. (2007). Transcriptional activation of miR-34a contributes to p53-mediated apoptosis. *Mol. Cell*, **26**, 731–743.

163. Wang, X., et al. (2008). Aberrant expression of oncogenic and tumor-suppressive microRNAs in cervical cancer is required for cancer cell growth. *PLoS ONE*, **3**, e2557.

164. Wang, X., et al. (2009). Oncogenic HPV infection interrupts the expression of tumor-suppressive miR-34a through viral oncoprotein E6. *RNA*, **15**, 637–647.

165. Meckes, D.G., Jr., et al. (2010). Human tumor virus utilizes exosomes for intercellular communication. *Proc. Natl. Acad. Sci. U S A*, **107**, 20370–20375.

166. Wang, K., et al. (2010). Export of microRNAs and microRNA-protective protein by mammalian cells. *Nucleic Acids Res.*, **38**, 7248–7259.

167. Hata, T., et al. (2010). Isolation of bovine milk-derived microvesicles carring mRNAa and microRNAs. *Biochem. Biophys. Res. Commun.*, **396**, 528–533.

168. Fujii, Y.R., and Saksena, N.K. (2008). Viral infection-related microRNAs in viral and host genomic evolution, in *RNA and the Regulation of Gene Expression*, Morris, K. V. (ed.), Horizon Scientific Press, London, 91–107.

169. Turchinovich, A., Weiz, L., Langheinz, A., and Burwinkel, B. (2011). Characterization of extracellular circulating microRNA. *Nuclic Acids Res.*, **39**, 7223–7233.

170. Karlsson, M., et al. (2001). "Tolerosomes" are produced by intestinal epitherial cells. *Eur. J. Immunol.*, **31**, 2892–2900.

171. Hawari, F.I., et al. (2004). Release of full-length 55-kDa TNF receptro 1 in exosome like vesicles: a mechanism for generation of soluble cytokine receptors. *Proc. Natl. Acad. Sci. U S A*, **101**, 1297–1302.

172. Vickers, K.C., et al. (2011). MicroRNAs are transported in plasma and delivered to recipient cells by high-density lipoproteins. *Nat. Cell Biol.*, **13**, 423–435.

173. Kosaka, N., Iguchi, H., and Ochiya, T. (2010). Circulating microRNA in body fluid: a new potential biomarker for cancer diagnosis and prognosis. *Cancer Sci.*, **101**, 2087–2092.

174. Chaput , N., and Théry, C. (2010). Exosomes: immune properties and potentila clinical implementations. *Semin. Immunopathol.*, **33**, 419–440.

175. Gibbings, D.J., Ciaudo, C., Erhardt, M., and Voinnet, O. (2009). Multivesicular bodies associate with components of miRNA effector complexes and modulate miRNA activity. *Nature Cell Biol.*, **11**, 1143–1151.

176. Balaj, L., et al. (2011). Tumour microvesicles contain retrotransposon elements and amplified oncogene sequences. *Nat. Commun.*, **2**, 180.

177. Tang, X., Tang, G., and Özcan, S. (2008). Role of microRNAs in diabetes. *Biochem. Biophys. Acta*, **1779**, 697–701.

178. Poy, M.N., et al. (2004). A pancreatic islet-specific microRNA regulates insulin secretion. *Nature*, **432**, 226–230.

179. Poy, M.N., Spranger, M., and Stoffel, M. (2007). microRNAs and the regulation of glucose and lipid metabolism. *Diabetes Obes. Metab.*, **9**, 67–73.

180. Kloosterman, W.P., et al. (2007). Targeting inhibition of miRNA maturation with morpholinos reveals a role for miR-375 in pancreatic islet development. *PLoS Biol.*, **5**, e203.

181. Poy, M.N., et al. (2009). *miR-375* maintains normal pancreatic α- and β-cell mass. *Proc. Natl. Acad. Sci. U S A*, **106**, 5813–5818.

182. Krichevsky, A.M., Sonntag, K.C., Isacson, O., and Kosik, K.S. (2006)., Specific microRNAs modulate embryonic stem cell-derived neurogenesis. *Stem Cell* **24**, 857–864.

183. Lovis, P., Gattesco, S., and Regazzi, R. (2008). Regulation of the expression of components of the exocytotic machinery of insulin-secreting cells by microRNAs. *Biol. Chem.*, **389**, 305–312.

184. He, A., et al. (2007). Overexpression of micro ribonucleic acid 29, highly up-regulated in diabetic rats, leads to insulin resistance in 3T3-L1 adipocytes. *Mol. Endocrinol.*, **21**, 2785–2794.

185. Keller, P., et al. (2011). Gene-chip studies of adipogenesis-regulated microRNAs in mouse primary adipocytes and human obesity. *BMC End. Disorders*, **11**, 7.

186. Dávalos, A., et al. (2011). miR-33a/b contribute to the regulation of fatty acid metabolism and insulin signaling. *Proc. Natl. Acad. Sci. U S A*, **108**, 9232–9237.

187. Hulsmans, M., de Keyzer, D., and Holvoet, P. (2011). MicroRNAs regulating oxidative stress and inflammation in relation to obesity and atherosclerosis. *FASEB J.* **25**, 15–17.

188. Ogawa, R., et al. (2010). Adipocyte-derived microvasicles contain RNA that is transported into macrophages and might be secreted into blood circulation. *Biochem. Biophys. Res. Commun.*, **398**, 723–729.

189. Rodriguez, A., et al. (2007). Requirement of *bic*/microRNA-155 for normal immune function. *Science*, **316**, 608–611.

190. Ha, T.-Y. (2011). The role of microRNAs in regulatory T cells and in the immune response. *Immune Net.*, **1**, 11–41.

191. Romania, P., et al. (2008). MicroRNA-155 modulates megakaryopoiesis at progenitor and precursor level by targeting ETS-1 and Meis 1 transcription factors. *British J. Hemato.*, **143**, 570–580.

192. Pichler, K., Schneider, G., and Grassmann, R. (2008). MicroRNA miR-146a and further oncogenesis-related cellular microRNAs are dysregulated in HTLV-1 transformed T lymphocytes. *Retrovirology*, **5**, 1742–1745.

193. Tili, E., et al. (2007). Modulation of miR-155 and miR-125b levels following lipopolysaccharide/TNF-alpha stimulation and their possible roles in regulating the response to endotoxin shock. *J. Immunol.*, **179**, 5082–5089.
194. Tili, E., and Croce, C.M. (2009). *miR-155*: on the crosstalk between inflammation and cancer. *Inter. Rev. Immunol.*, **28**, 264–284.
195. Pauley, K.M., et al. (2008). Upregulated miR-146a expression in peripheral blood mononuclear cells from rheumatoid arthritis patients. *Arthritis Res. Ther.*, **10**, R101.
196. Nakasa, T., et al. (2008). Expression of microRNA-146 in rheumatoid arthritis synovial tissue. *Arthritis Rheum.*, **58**, 1284–1292.
197. Murata, K., et al. (2010). Plasma and synovial fluid microRNAs as potential biomerkers of rheumatoid arthritis and osteoarhtritis. *Arthritis Res. Ther.*, **12**, R86.
198. Niimoto, T., et al. (2010). MicroRNA-146a expresses in interleukin-17 producing T cells in rheumatoid arthritis patients. *BMC Musculoskelet. Disord.*, **11**, 209.
199. Li, Q.J., et al. (2007). miR-181a is an intrinsic modulator of T cell sensitivity and selection. *Cell*, **129**, 147–161.
200. de Yebens, V.G., et al. (2008). miR-181b negatively regulates activation-induced cytidine deaminase in B cells. *J. Exp. Med.*, **205**, 2199–2206.
201. Eacker, S.M., Dawson, T.M., and Dawson, V.L. (2009). Understanding microRNAs in neurodegeneration. *Nat. Rev. Neurosci.*, **10**, 837–841.
202. Christensen, M., and Schratt, G.M. (2009). microRNA involvement in developmental and functional aspects of the nervous system and in neurological diseases. *Neurosci. Lett.*, **466**, 55–62.
203. Yelamanchili, S.V., and Fox, H.S. (2010). Defining larger roles for "tiny" RNA molecules: role of miRNAs in neurodegeneration research. *J. Neuroimmune. Pharmacol.*, **5**, 63–69.
204. Schaefer, A., et al. (2007). Cerebellar neurodegeneration in the absence of microRNAs. *J. Exp. Med.*, **204**, 1553–1558.
205. Cuellar, T.L., et al. (2008). Dicer loss in striatal neurons prodused behavioral and neuroanatomical phenotypes in the absence of neurodegeneration. *Proc. Natl. Acad. Sci. U S A*, **105**, 5614–5619.
206. Hébert, S.S., et al. (2008). Loss of microRNA cluster miR-29a/b-1 in sporadic Alzheimer's disease correlates with increased BACE1/β-secretase expression. *Proc. Natl. Acad. Sci. U S A*, **17**, 6415–6420.

207. Cogswell, J.P., et al. (2008). Identification of miRNA changes in Alzheimer's disease brain and CSF yields putative biomarkers and insights into disease pathways. *J. Alzheimers Dis.*, **14**, 27–41.
208. Wang, W.X., et al. (2008). The expression of microRNA miR-107 decreases early in Alzheimer's disease and may accelerate disease progression through regulation of betasite amyloid precusor protein-cleaving enzyme 1. *J. Neurosci.*, **28**, 1213–1223.
209. Lukiw, W.J., Zhao, Y., and Cui, J.G. (2008). An NF-kappaB-sensitive microRNA-146a-mediated inflammatory circuit in Alzheimer disease and stressed human brain cells. *J. Biol. Chem.*, **283**, 31315–31322.
210. Kim, J., et al. (2007). A microRNA feedback circuit in midbrain dopamine neurons. *Science*, **317**, 1220–1224.
211. Lee, S.-T., et al. (2011). Altered microRNA regulation in Huntington's disease models. *Exp. Neurol.*, **227**, 172–179.
212. Packer, A.N., et al. (2008). The bifunctional microRNA miR-9/miR-9* regulates REST and CoREST and is downregulated in Huntington's disease. *J. Neurosci.*, **28**, 14341–14346.
213. Johnson, R., et al. (2008). A microRNA-based gene dysregulation pathway in Huntington's disease. *Neurobiol. Dis.*, **29**, 438–445.
214. Lee, Y., et al. (2008). miR-19, miR-101 and miR-130 co-regulate ATXN1 levels to potentially modulate SCA1 pathogenesis. *Nat. Neurosci.*, **11**, 1137–1139.
215. Ripke, S., et al. (2011). Genome-wide association study identifies five new schizophrenia loci. *Nat. Genet.*, **43**, 969–976.
216. Szulwach, K.E., et al. (2010). Cross talk between microRNA and epigenetic regulation in adult neurogenesis. *J. Cell Biol.*, **189**, 127–141.
217. Smrt, R.D., et al. (2010). MicroRNA miR-137 regulates neuronal maturation by targeting ubiquitin ligase mind bomb-1 *Stem Cells*, **28**, 1060–1070.
218. Magill, S.T., et al. (2010). microRNA-132 regulates dendritic growth and arborization of new born neurons in the adult hippocampus. *Proc. Natl. Acad. Sci. U S A*, **23**, 20382–20387.
219. Chahrour, M., and Zoghbi, H.Y. (2007). The stroy of Rett syndrome: from clinic to neurobiology. *Neuron*, **56**, 422–437.
220. Saba, R., et al. (2008). A miRNA signature of prion induced neurodegeneration. *PLoS ONE*, **3**, e3652.
221. Costa, F.F. (2010). Epigenomics in cancer management. *Cancer Manag. Res.*, **2**, 255–265.

222. Ptak, C., and Petronis, A. (2008). Epigenetics and complex disease: from etiology to new therapeutics. *Annu. Rev. Pharmacol. Toxicol.*, **48**, 257–276.

223. Parrish, S., et al. (2001). Functional anatomy of a dsRNA trigger: differential requirement for the two trigger strands in RNA interference. *Mol. Cell*, **6**, 1077–1087.

224. Braasch, D.A., et al. (2003). RNA interference in mammalian cells by chemically-modified RNA. *Biochemistry*, **42**, 7967–7975.

225. Dahlgren, C., Wahlestedt, C., and Thonberg, H. (2006). No induction of anti-viral responses in human cell lines HeLa and MICF-7 when transfecting with siRNA or siLNA. *Biochem. Biophys. Res. Commun.*, **341**, 1211–1217.

226. Hall, A.H.S., et al. (2006). High Potency Silencing by Single-Stranded Boanophosphate siRNA. *Nucleic Acids Res.*, **9**, 2773–2781.

227. Elbashir, S.M., et al. (2001). Functional Anatomy of siRNAs for Mediating Efficient RNAi in *Drosophila melanogaster* Embryo Lysate. *EMBO J.*, **20**, 6877–6888.

228. Jackson, A.L., et al. (2006). Position-specific chemical modification of siRNAs reduces "off-target" transcript silencing. *RNA*, **12**, 1–9.

229. Weiler, J., Hunziker, J., and Hall, J. ka do. (2005). Anti-miRNA oligonucleotides (AMOs): ammunition to target miRNAs implicated in human disease? *Gene Ther.*, **13**, 496–502.

230. Chen, J.F., et al. (2006). The role of microRNA-1 and microRNA-133 in skeletal muscle proliferation and differentiation. *Nat. Genet.*, **38**, 228–233.

231. Schratt, G.M., et al. (2006). A brain-specific microRNA regulates dendritic spine development. *Nature*, **439**, 283–289.

232. Esau, C., et al. (2004). MicroRNA-143 regulates adipocyte differentiation. *J. Biol. Chem.*, **279**, 52361–52365.

233. Esau, C., et al. (2006). mir-122 regulation of lipid metabolism revealed by in vivo antisense targeting. *Cell Metab.*, **3**, 87–98.

234. Davis, S., Lollo, B., Freier, S., and Esau, C. (2006). Improved targeting of miRNA with antisense oligonucleotides. *Nculeic Acids Res.*, **34**, 2294–2304.

235. Soutschek, J., et al. (2004). Therapeutic silencing of an endogenous gene by systemic administration of modified siRNAs. *Nature*, **432**, 173–178.

236. Krutzfeld, J., et al. (2005). Silencing of microRNAs in vivo with "antagomirs." *Nature*, **438**, 685–689.

237. Krutzfeld, J., et al. (2007). Specificity, duplex degradation and subcellular localization of antagomirs. *Nucleic Acids Res.*, **35**, 2885–2892.
238. Song, E., et al. (2005). Antibody mediated in vivo delivery of small interfering RNAs via cell-surface receptors. *Nat. Biotech.*, **23**, 709–717.
239. Park, I.K., et al. (2007). Neuron-specific delivery of nucleic acids mediated by Tet1-modified poly(ethylenimine). *J. Gen. Med.*, **9**, 691–702.
240. Kumar, M.S., et al. (2008). Suppression of non-small cell lung tumor development by the let-7 microRNA family. *Proc. Natl. Acad Sci. U S A*, **105**, 3903–3908.
241. Watanabe, T., et al. (2007). Liver target delivery of small interfering RNA to the HCV gene by lactosylated cationic liposome. *J. Hepatol.*, **47**, 744–750.
242. Wilson, D.S., et al. (2010). Orally delivered thioketal-nanoparticles loaded with TNFα-siRNA target inflammation and inhibit gene expression in the intestines. *Nat. Mater.*, **9**, 923–928.
243. Yamamoto, T., et al. (2002). Double-stranded *nef* RNA interferes with human immunodeficiency virus type 1 replication. *Microbiol. Immunol.*, **46**, 809–817.
244. Omoto, S., et al. (2004). HIV-1 nef suppression by virally encoded microRNA. *Retrovirology*, **1**, 44.
245. Schopman, N.C.T., et al. (2011). Deep sequencing of virus-infected cells reveals HIV-encoded small RNAs. *Nucleic Acids Res.*, **40**, 414–427.
246. Klase, Z., Houzet, L., and Jeang, K.-T. (2011). Replication competent HIV-1 viruses that express intragenomic microRNA reveal discrete RNA-interference mechanisms that affect viral replication. *Cell BioSci.*, **1**, 38.
247. Bader, A.G. (2012). miR-34 - a microRNA replacement therapy is headed to the clinic. *Front. Genet.*, **3**, 120.
248. Chen, S.C.-Y., Stern, P., Guo, Z., and Chen, J. (2011). Expression of multiple artificial microRNAs from a chicken miRNA 126-based lentiviral vector. *PLoS ONE*, **6**, e22437.
249. Darwin, C. (1859). *On the Origin of Species by Means of Natural Selection, or the Preservation of Favoured Races in the Struggle for Life*, John Murray Press.
250. Sousa-Ferreira, L., et al. (2011). Moderate long-term modulation of neuropeptide Y in hypothalamic arcuate nucleus induces energy balance alteratons in adult rats. *PLoS ONE*, **6**, e22333.

251. Papahadjopoulos, D., et al. (1991). Sterically stabilized liposomes: improvements in pharmacokinetics and antitumor therapeutic efficacy. *Proc. Natl. Acad. Sci. U S A*, **88**, 11460–11464.

252. Gray, S.J., et al. (2010). Directed evolution of a novel adeno-associated virus (AAV) vector that crosses the seizure-compromised blood-brain barrier (BBB). *Mol. Ther.*, **18**, 570–578.

253. Xie, J., et al. (2011). MicroRNA-regulated, systemically delivered rAAV9: a step closer to CNS-restricted transgene expression. *Mol. Ther.*, **19**, 526–535.

254. Zhang, Y., et al. (2004). Intravenous RNA interference gene therapy targeting the human epidermal growth factor receptro prolongs survival in intracranial brain cancer. *Clin. Cancer Res.*, **10**, 3667–3677.

255. Lund, E.G., Guileyardo, J.M., and Russel, D.W. (1999). cDNA cloning of chelesterol 24-hydroxylase, a mediator of choresterol homeostasis in the brain. *Proc. Natl. Acad. Sci. U S A*, **96**, 7238–7243.

256. Bjorkhem, I., et al. (1998). Cholesterol homeostasis in human brain: turnover of 24S-hydroxycholesterol and evidence for a cerebral origin of most of this oxysterol in the circulation. *J. Lipid Res.*, **39**, 1594–1600.

257. Graner, M.W., et al. (2009). Proteomic and immunologic analysis of brain tumor exosomes. *FASEB J.*, **23**, 1541–1557.

258. Dong, H., et al. (2011). The use of polyethyleneimine-grafted graphene nanoribbon for cellular delivery of locked nucleic acid modified molecular beacon for recognition of microRNA. *Biomaterials*, **32**, 3875–3882.

259. Hwang do, W., et al. (2011). A brain-targeted rabies virus glycoprotein-disulfide linked PEI nanocarrier for delivery of neurogenic microRNA. *Biomaterials* **32**, 4968–4975.

260. Pan, Q., et al. (2011). Hepatic cell-to-cell transmission of small silencing RNA can extend the therapeutic reach of RNA interference (RNAi). *Gut*, **61**, 1330–1339.

261. Weiss, R.A., and Vogt, P.K. (2011). 100 years of Rouse sarcoma virus. *J. Exp. Med.*, **208**, 2351–2355.

262. Rous, P. (1911). A sarcoma of the fowl transmissible by an agent separable from the tumor cells. *J. Exp. Med.*, **13**, 397–411.

263. Kosaka, N., et al. (2011). Competitive interactions of cancer cells and normal cells via secretory microRNAs. *J. Biol. Chem.*, **287**, 1397–1405.

264. Schopman, N.C., et al. (2012). Selective packaging of cellular miRNAs in HIV-1 particles. *Virus Res.*, **169**, 438–447.

265. Duesberg, P.H., and Vogt, P.K. (1970). Differences between the ribonucleic acids is transforming and nontransforming avian tumor viruses. *Proc. Natl. Acad. Sci. U S A*, **67**, 1673–1680.

266. Martin, G.S. (1970). Rous sarcoma virus: a function required for the maintenance of the transformed state. *Nature*, **227**, 1021–1023.

267. Rous, P. (1967). The challenge to man of the neoplastic cell. *Science*, **157**, 24–28.

268. Kosaka, N., Iguchi, H., and Ochiya, T. (2010). Circulating microRNA in body fluid: a new potential biomarker for cancer diagnosis and prognosis. *Cancer Sci.*, **101**, 2087–2092.

269. Velu, V.K., Ramesh, R., and Srinvasan, A.R. (2012). Circulating micro-RNAs as biomarkers in health and disease. *J. Clin. Diagn. Res.*, **6**, 1791–1795.

270. Fujii, Y.R. (2013). RNA wave for the HIV therapy: foods, stem cells and the RNA information gene. *World J. AIDS*, **3**, 131–146.

Index